Authenticity of Foods
of Animal Origin

Food Biology Series

Authenticity of Foods of Animal Origin

Ioannis S. Arvanitoyannis
School of Agricultural Sciences
University of Thessaly
Thessaly, Hellas, Greece

CRC Press
Taylor & Francis Group
Boca Raton London New York

CRC Press is an imprint of the
Taylor & Francis Group, an **informa** business

A SCIENCE PUBLISHERS BOOK

Dedication

This book is dedicated to
My wife Nikol Katsiki for her continuous support
and our three children
Jason
Artemis-Eleni for bearing with me
Nefeli-Kallisti

Ioannis

Preface to the Series

Food is the essential source of nutrients (such as carbohydrates, proteins, fats, vitamins, and minerals) for all living organisms to sustain life. A large part of daily human efforts is concentrated on food production, processing, packaging and marketing, product development, preservation, storage, and ensuring food safety and quality. It is obvious therefore, our food supply chain can contain microorganisms that interact with the food, thereby interfering in the ecology of food substrates. The microbe-food interaction can be mostly beneficial (as in the case of many fermented foods such as cheese, butter, sausage, etc.) or in some cases, it is detrimental (spoilage of food, mycotoxin, etc.). The *Food Biology* series aims at bringing all these aspects of microbe-food interactions in form of topical volumes, covering food microbiology, food mycology, biochemistry, microbial ecology, food biotechnology and bio-processing, new food product developments with microbial interventions, food nutrification with nutraceuticals, food authenticity, food origin traceability, and food science and technology. Special emphasis is laid on new molecular techniques relevant to food biology research or to monitoring and assessing food safety and quality, multiple hurdle food preservation techniques, as well as new interventions in biotechnological applications in food processing and development.

The series is broadly broken up into food fermentation, food safety and hygiene, food authenticity and traceability, microbial interventions in food bio-processing and food additive development, sensory science, molecular diagnostic methods in detecting food borne pathogens and food policy, etc. Leading international authorities with background in academia, research, industry and government have been drawn into the series either as authors or as editors. The series will be a useful reference resource base in food microbiology, biochemistry, biotechnology, food science and technology for researchers, teachers, students and food science and technology practitioners.

Ramesh C. Ray
Series Editor

Preface

In the last 20 to 30 years, the number of food poisoning incidents has increased considerably and this had disastrous effects both on consumers and the food industry. Financially, it has resulted in liable, with huge sums of money being paid as compensation. These incidents have triggered the consumers' and governments' awareness in food crises. In the EU for instance, an organization called European Food Safety Authority (EFSA) was established to deal with food and feed crises within the frame of Rapid Alert System for Feeds and Foods (RASFF). This aims at informing, in time, all the directly involved partners (food companies, laboratories, government officers, etc.) and, also monitor the measures/actions undertaken to minimize both the extent and the intensity of the resultant defect.

The best policy to solve a crisis is to recall of the affected products. However, an effective and rapid recall assumes the proper functioning of traceability. The latter is a prerequisite of current legislation (Regulation 178/2002). Traceability can be based either on European Article Number (EAN) -8, -9, -12, -13 and -128 or Radio Frequency Identification (RF-ID). Although RF-ID is a more expensive process than EAN, it is preferred in the case of animals and fish. Once the animal has been examined, all the information stored in the RF-ID can be passed onto EAN-128.

This book addresses five major issues: (i) food traceability, (ii) food authenticity techniques and methods to detect potential adulteration, (iii) application of authenticity techniques and methods to foods of animal origin, (iv) legislation relating to traceability and authenticity in EU, USA, Canada, Japan and Australia-New Zealand, and (v) Trends and suggestions for further research in food traceability and authenticity.

This book will be useful to academicians, industrialists, students of food science and technology, veterinarians and technicians specialized in food technology.

Ioannis S. Arvanitoyannis (Dr., Ph.D.)
Full Professor of Food Technology, Quality and Safety
University of Thessaly, Volos, Hellas, Greece

Contents

List of Abbreviations

2-ACBs	2-alkylcyclobutanones
2,3,7,8 –TCDD	2,3,7,8-Tetrachlorodibenzo-p-dioxin
2D	Two-dimensional
AAS	Atomic Absorption Spectrophotometry
ABTS	2,20-azinobis (3-ethylbenzothiazolin)-6-sulfphonate
ACP	Antigen-coated plate
AFGP	2-acetylfuran-3-glucopyranoside
AFLP	Amplified fragment length polymorphism
ANN	Artificial neural network
AP-PCR	Arbitrarily primed Polymerase chain reaction
APGC-MS/MS	Atmospheric Pressure Gas Chromatography in conjunction with tandem MS
APLSR	ANOVA partial least squares regression
BBN	Bayesian belief networks
BDCYX	1,4-bisdesoxycyadox
BIA	Biosensor immunoassay
BNCRC	Brazilian National Residue and Contaminants Control Plan
BPNN	Back-propagation neural networks
BSA	Blood Serum Albumin
BSE	Bovine spongiform encephalopathy
CW	Continuous Wave
C12-LAS	Dodecylbenzenesulfonate
CA	Cluster Analysis
CAP	Chloramphenicol
CAR/PDMS	Carboxenpolydimethylsiloxane
CDA	Canonical discriminant analysis
CE	Capillary Electrophoresis
CF	Chicken forward
CF-IRMS	Continuous flow Isotope Ratio Mass Spectrometry
cis-2-dDeCB	cis-2-(dodec-5'-enyl)-cyclobutanones
CL	Chemiluminescent
CMP	Caseinomacropeptide
CNS	Central nervous system

CO	Country of origin
COI	Cytochrome c oxidase I
COIII	Cytochrome oxidase subunit III
CoxI	Cytochrome oxidase subunit I
CR	Chicken reverse
CSC	Certificate of specific character
CSCB	Same type of bacon
CYX	Cyadox
DAS	Double antibody sandwich
DCB	Dodecylcyclobutanone
DCD	Dicyandiamide
DeA	Decanoic acid
DEHP	Di(2-ethylhexyl) phthalate
DES	Diethylstilbesterol
DES-MCPE-BSA	Diethylstilbesterol-mono-caroxyl-propyl-ethyl-bovine-serum-albumin
DESIR	Dry extract spectroscopy by infrared reflection
DHS	Dynamic headspace
DI-IRMS	Dual-inlet Isotope Ratio Mass Spectrometry
DL-PCBs	Dioxin-Like PolyChlorinated Biphenyls
DMZ	Dimetridazole
DMZOH	Hydroxydimetridazole
DNC	Dinitrocarbanilide
DOE	Design of experiment
DPLS	Discriminant Partial Least Square
DPPH	Diphenyl-2-picrylhydrazyl
DSC	Differential Scanning Calorimetry
DVP/CAR/PDMS	Divinylbenzene/carboxen/polydimethylsiloxane
EA–IRMS	Elemental Analyzer–Isotope Ratio Mass Spectrometry
EC	Electrochemical detection
EDCS	Endocrine Disrupting Chemicals
EGDM	Ethylene glycol dimethyl ether
EI	Electron impact
EI-MS	Electron ionization mass analyser
ELISA	Enzyme Linked Immunosorbent Assay
ENRO	Enrofloxacin
EPA	Environmental Protection Agency
EPS	Expanded polystyrene foams
ERP	Entreprise Resource Planning
ESI–IT	Electrospray ionization–ion trap
FDA	Factorial Discriminant Analysis
FINS	Forensically informative nucleotide sequencing
FLP-GC-MS/MS	Fast Low Pressure-GC with tandem MS
Flu	Flumequine
FR	Feeding regime

FRAP	Ferric reducing antioxidant power
FSIS	Food Safety and Inspection Service
FT-IR	Fourier Transform Infrared
FT-MIR spectroscopy	Fourier Transform-Mid-Infrared Spectroscopy
FT-NIR	Fourier Transform Near-infrared Spectroscopy
FT-Raman	Fourier-Transform Raman Spectroscopy
GA-MLR	Genetic algorithms-multiple linear regression
GC	Gas Chromatography
GC-HRMS	Gas Chromatography-High Resolution Mass Spectrometry
GC-MS	Gas Chromatography-Mass Spectrometry
GC-MS/MS	Gas Chromatography-tandem Mass Spectrometry
GCC	Gross chemical composition
GCxGC	Two-dimensional gas chromatography
GCxGC-TOF/MS	Gas Chromatography coupled to Time of Flight Mass Spectrometry
GFAP	Glial fibrillary acidic protein
GLC	Gas-liquid chromatography
GMO	Genetically modified organisms
HATR	Horizontal attenuated total reflectance
HBCD or HBCDDs	HexaBromoCycloDodecane
HCA	Hierarchical cluster analysis
HEWL	Hen-egg-white lysozyme
HFCS	High Fructose Corn Syrup
HILIC	Hydrophilic interaction liquid chromatography
HMF	Hydroxymethylfurfural
HPAED-PAD	High performance anion-exchange chromatography and pulsed amperometric detection
HPIC	High Performance Ion Chromatography
HPLC	High Performance Liquid Chromatography
HPLC-EC	High-performance liquid chromatography method with electrochemical detection
HPLC-FLDA	High-performance liquid chromatography -fluorescence detection
HR-GC/MS	High Resolution Gas Chromatography-Mass Spectrometry
HR-MAS	High-resolution magic angle spinning
HRP	Horseradish peroxidase
HTT	High Throughput Transmission
HyHEL	Hybridomaantihen-egg-white lysozyme
ICA	Independent Component Analysis
ICP-AES	Inductively Coupled Plasma Atomic Emission Spectroscopy
ICP-AES	Inductively coupled plasma atomic emission spectrometer

ICP-HRMS	Inductively coupled plasma high resolution mass spectrometry
ICP-MS	Inductively coupled plasma mass spectrometry
ICP-SFMS	Inductively coupled plasma double focusing sector field mass spectrometry
ICP–OES	Inductively coupled plasma optical emission spectroscopy
ID-LC-MS/MS	Isotope dilution-liquid chromatography-tandem mass spectrometry
IEF	Isoelectric Focusing
IGAC	Industry-Government Advisory Committee
IHC	International Honey Commission
INDEX	Inside-needle dynamic extraction
IPZ	Ipronidazole
IR	Infrared
IRMS	Isotope Ratio Mass Spectrometry
ISCIRA	Internal standard isotope ratio analysis
ISSR	Inter-simple sequence repeat
ITC	Isothermal titration Calorimetry
LC	Liquid Chromatography
LC-MS	Liquid Chromatography–Mass Spectrometry
LC-QTOF–MS	LC–hybrid quadrupole time-of-flight mass spectrometry
LC–MS/MS	Liquid Chromatography–tandem Mass Spectrometry
LDA	Linear discriminant analysis
LF NMR	Low field Nuclear Magnetic Resonance
LOQ	Limits of quantification
LP-GC/MS-MS	Low pressure vacuum outlet gas chromatography tandem mass spectrometry
LP-RAPD	Long-primer random amplified polymorphic DNA
LR-GC/MS	Low Resolution Gas Chromatography-Mass Spectrometry
LRI	Linear retention indices
LS-SVM	Least squares support vector machine
MA	Modified atmosphere
MAE	Microwave-Assisted Extraction
MALDI–TOF	Matrix-assisted laser desorption/ionization time-of-flight
MAP	Modified atmosphere packaging
MBM	Meat and bone meal
MBP	Myelin basic protein
MC-ICP–MS	Multi-collector inductively coupled plasma mass spectrometry

MD-DA	Mahalanobis-distance discriminant analysis
MGB	Minor groove binding
MIR	Mid Infrared
MLP	Multilayer Perceptron
MLP	Multilayer perceptron
MLR	Multiple linear regression
MOFSET	Metal oxide semi-conductor field-effect transistors
MPI	Ministry of Primary Industries
MPLS	Multi-way partial least-squares regression
MRI	Magnetic resonance imaging
MRLs	Maximum residue limits
MS	Mass spectroscopy
MVA	Multivariate analysis
NADH	Nicotinamide adenine dinucleotide
NAFTS	National Agriculture and Food Traceability System
NDL-PCB	Non-dioxin-like polychlorinated bisphenyls
NF	Neurofilament
NFR	Novel flame retardants
NIR	Near Infrared
NIRR	Near infrared reflectance reflection
NIRT	Near infrared reflectance transmission
NIT	Near infrared transmittance
NMR	Nuclear Magnetic Resonance
NP-HPLC	Normal Phase High Performance Liquid Chromatography
NPSD	Nitrogen purge and steam distillation
NSE	Neuron-specific enolase
OCDD/F	Octachlorinateddibenzodioxin/furans
OHP	Organohalogen Pollutants
PAD	Photodiode array detection
PAD	Pulsed amperometric detectors
PAHs	Polycyclic Aromatic Hydrocarbons
PBBs	Polybrominated biphenyls
PBDEs	PolyBrominatedDiphenyl Ethers
PCA	Principal Component Analysis
PCBs	Polychlorinated Biphenyls
PCBs/PBDEs	Polychlorinated Biphenyls/ PolyBrominatedDiphenyl Ethers
PCDD/DFs	Polychlorinated dibenzo-p-dioxins and dibenzofurans
PCDD/Fs	Polychlorodibenzo-p-dioxins and polychlorinated dibenzofurans
PCDDs	PolyChlorinatedDibenzoDioxins
PCDFs	PolyChlorinatedDibenzoFurans

PCR	Polymerase Chain Reaction
PCR-RFLP	Polymerase chain reaction-restriction fragment length polymorphism
PDA	Photodiode array detectors
PDMS	Poly(dimethylsiloxane)
PDO	Protected Designation of Origin
PFAAs	perfluorinated alkyl acids
PGI	Protected geographical indication
PLE	Pressurised Liquid Extraction
PLOT	Porous layer open tubular
PLS	Partial Least Square
PLS-DA	Partial least-squares discriminant analysis
POPs	Persistent organic pollutants
PTs	Proficiency Testing
QC-PCR	Quantitative competitive polymerase chain reaction
QCA	Quinoxaline-2-carboxylic acid
RAPD	Randomly amplified polymorphic DNA
RE	Restriction endonucleases
RF	Random Forest
RfD	Reference dose
RFLP	Restriction fragment length polymorphism
RIO	Radial Immunodiffusion
RMSECV	Root Mean Square Error of Cross-Validation
RNZ/RON	Ronidazole
RP-HPLC	Reversed phase-High Performance Liquid Chromatography
RSD	Relative Standard Deviation
RT-NASBA	Real-time nucleic acid sequence-based amplification
RT-PCR	Real time-polymerase chain reaction
SCAR	Sequence characterized amplified region
SCIRA	Stable Carbon Isotope Ratio Analysis
SDE	Simultaneous distillation extraction
SDS-PAGE	Sodium dodecylsulphate polyacrylamide gel electrophoresis
SE	Solvent extraction
SEC	Size-exclusion Chromatography
SECV	Standard error of cross-validation
SEP	Standard error of validation
SEP	Size Exclusion Purification
SIMCA	Soft Independent Modeling of Class Analogy
SINE	Short interspersed nuclear element
SNDV	Standard normal variance and detrend
SPDE	Solid phase dynamic extraction
SPE	Solid Phase Extraction
SPE-GC-MS	Solid Phase Extraction-Gas Chromatography-Mass Spectroscopy

SPME	Solid phase micro extraction
SPME	Solid phase matrix extraction
SPME-GC/MS	Solid Phase Microextraction-Gas Chromatography-Mass Spectrometry
SSCP	Single-stranded conformational polymorphism
SUPRAS	Supramolecular solvent
SVM	Support vector machines
SW-NIR	Short-wavelength near-infrared
SYBR Green I	N',N'-dimethyl-N-[4-[(E)-(3-methyl-1,3-benzothiazol-2-ylidene)methyl]-1-phenylquinolin-1-ium-2-yl]-N-propylpropane-1,3-diamine
T&T	Tracking and Tracing
TBA	2-thiobarbituric acid
TBARS	2-thiobarbituric acid reactive substances
TCDF	tetra chlorinated dibenzofuran
TDS	Total Diet Study
TEQ	Toxic equivalency factor
TF	Turkey forward
TFA	Trifluoracetic acid
TMA	Trimethylamine
TMA-N	Trimethylamine nitrogen
TOFMS	Time-of-flight mass spectrometry
TR	Turkey reverse
TR-FIA	Time-resolved fluoroimmunoassay
TSB	Trypticase soy broth
TVB-N	Total volatile basic nitrogen
UPLG-MS/MS	Ultra performance liquid chromatography-tandem mass spectrometer
Urea-IEF	Urea-isoelectric focusing
urea-PAGE	urea-Polyacrylamide Gel Electrophoresis
USDA-ARS	US Department of Agriculture, Agricultural Research Service
UV-Vis	Ultraviolet-visible
VIS	Visible infrared spectrospopy
VIS/NIR	Visible/near infrared spectroscopy
VOCs	Volatile organic compounds
VP	Vacuum package
WBSF	Warner-Braztler shear force
WHC	Water holding capacity
WOF	Warmed-over flavour
XPS	Extruded polystyrene foams
ZDV	Zero dead volume
µATR	Micro Attenuated Total Reflection

PART A

Introduction to Authenticity and Traceability of Foods

1

Methods and Techniques for Detecting Food Authenticity

*Ioannis S. Arvanitoyannis** and *Konstantinos V. Kotsanopoulos*

1.1 Introduction to Food Authentication

The word "authentic" is commonly defined as something reliable, or genuine. Therefore, where foods of animal origin are concerned, the exact identification of the species is of paramount importance, from a food authenticity point of view, and should be conducted by relying on factors not heavily altered during food processing. The identification of meat species to ensure the authentication of meat products is only an example of how useful the analytical techniques can be (Lüthy, 1999). The assurance of food authenticity and the detection of adulteration are critical issues in the food industry that attract increasing attention. As regards meat and meat products, the basic authenticity concerns involve the substitution of high value raw materials with lower value materials such as cheaper pieces of meat, mechanically recovered meat, offal, blood, water, eggs, gluten or various other protein sources of animal or vegetable origin. Furthermore, the latter can also give rise to various food safety implications since addition of these products can lead to allergic reactions in certain individuals. Another aspect is the differentiation of frozen-and-thawed meat from fresh meat. Legislation has been established to impose severe economic consequences aiming at restricting these types of phenomena. However, each case is assessed on an individual basis. In the UK, the 1984 Meat and Meat Products legislation specifically prohibits the adulteration of meat with meat derived from other species. Some countries proscribe the consumption of certain types of meat, such as pork, for religious reasons. As a result, analytical techniques have been developed to identify different meat species in raw, cooked and processed products (Al-Jowder et al., 1997).

School of Agricultural Sciences, Department of Agriculture, Ichthyology and Aquatic Environment, University of Thessaly, Fytoko St., 38446 Nea Ionia Magnesias, Volos, Hellas, Greece.
* Corresponding author

The establishment of traceability systems became compulsory under the General Food Law in January 2005. Nevertheless, the compliance of a company with this law is not necessarily translated into added value or reduced costs in the food chain. It is only a prerequisite for obtaining a license to produce. Thus, legal obligations alone cannot offer adequate motivation to food companies to invest in tracking and tracing (T&T) systems. A commercial and strategic approach is required to highlight the benefits of adopting these systems. The determination of the goals is the starting point for a quality based T&T system. Conventional T&T systems usually act as tools for effectively using collected data about the location and identification of products, to enable recall management. On the other hand, a quality based T&T system will also collect information about relevant qualitative factors such as temperature and relative humidity throughout the supply chain. A combination of T&T and quality data is continuously used (active approach) for controlling and managing the flow of products through the food chain, usually improving the performance of the system. Due to the fact that the quality of perishable food products is highly variable by nature, the objective and precise registration of its value and forecast of its future development still remains very difficult to define. Moreover, apart from the technical aspects, ethical aspects could also affect any decisions taken (Scheer 2006).

Throughout history, attempts have been made to control the quality and safety of the food supply. The massive urbanization trend of the twentieth century was accompanied by major improvements in this area. Food laws developed simultaneously with urbanization, since in urban areas most people do not produce their own food. However, the fact that several stages are involved in the supply chain of foods (growers, processors, wholesalers, distributors) increases the potential of mishandling, fraudulent practices, and misrepresentation. Advances in science and technology significantly increased the need for the establishment of regulations in the food supply system. The extensive use of chemicals in food production and processing has dramatically increased crop yields and improved the shelf life and quality of products, while new products have also been developed. However, it has also created problems related to chemicals' overuse or misuse, thus triggering the establishment of strict government controls (Puckett, 2004).

Hoorfar et al. (2011) reported that food chain integrity covers all stages of the food chain from producers to consumers. It relied on the assurance of microbial and chemical food safety, authenticity of origin, fraud detection and quality. It is however expected that consumer demands will be expanded to include not only safety, healthiness and taste, but also sustainability, minimization of carbon footprint, animal welfare, freedom of child labour, etc. Therefore, research is demanded to convert these subjective perceptions into commercial and operational criteria to produce foods of high integrity.

Fraud has proved to be an increasingly frequent phenomenon, mainly strengthened by the opening of international markets and global competition. The main reason of carrying out these illegal activities is extra profit. At present, the use of accurate control methods to ensure the compliance of food materials is imperative and the only way to effectively eliminate these risks of falsification. Markets such as the beekeeping market, are not excluded from this trend and control organisations continuously face an increasing number of cases of non-compliance. Honey fraud incidents, for example, usually involve either the non-compliance of an origin name resulting from mixing (voluntary or not) of honeys from different varieties, or a non-compliance, which results by deliberately adding adulterating syrup. Direct addition of syrups to the honey can be carried out after harvesting or feeding bees to improve yield. The increasing frequency of frauds, currently being rather often encountered, can finally have major economic consequences for the honest sector (Cotte et al., 2003).

In the food and pharmaceutical production chain, high-price products are mixed with lower quality products for several reasons. These practices cannot be easily detected mainly due to the very low amounts of high degraded DNA (due to processing) analyzed and thus new methods need to be developed and optimized to assist in such investigations (Schubbert et al., 2008). According to Woolfe and Primrose (2004), "The fraudulent misdescription of food contents on product labels is a widespread problem, particularly with high added-value products commanding a premium price. Proving conclusively that fraud has occurred requires the detection and quantification of food constituents. These are often biochemically similar to the materials they replace, making their detection and identification extremely difficult. Despite the fact that food matrices are extremely complex and variable, a variety of the molecular markers used to physically map genomes have now been successfully adapted for detection of food substitution. These successes include the speciation of meats, fish and fruit in processed food products, the identification of the geographical origin of olive oil, the detection of dilution of Basmati rice with non-Basmati varieties and the quantitative detection of neuronal tissue and offal in processed meat".

The UK Food Standards Agency's authenticity programme has funded numerous projects that aim at developing technologies that can be used for the effective determination of the authenticity of foods. Such analyses are not easily performed, considering the complexity of foods. The determination of appropriate markers for the detection or measurement of food authenticity is a difficult task. However, results from the food authenticity programme supported by the UK FSA clearly indicate that this problem can be resolved. Stable isotope analysis, for example, in combination with trace element measurement could be very effective for determining the geographical origin of foods, while in some cases, it can also provide information about the production methods. DNA analysis can provide conclusive data with regards to whether a particular species or variety is present or absent in a product and in some cases quantitative results can be given. Proteomics has been relatively recently used in the food industry but it is adequately specific and precise to perform authenticity measurements. Metabolomics and lectin chips have also shown promise but more research on these methods is still needed (Primrose et al., 2010). Moreover, due to the continuously increasing load of information given on labels with regards to product composition and quality, the development and standardisation of analytical methods to either confirm the information given on the label or to detect fraudulent activities is demanded (Martinez et al., 2003).

Recent advances in research in relation to the use of nucleic acids, proteins, and small molecules have been accompanied by advances in technology and instrumentation. These techniques are continuously improved to become more user-friendly, while instrumentation is now available at lower cost. As a result, more laboratories can now perform detailed analyses of food composition. Through this trend the need to validate and standardise analytical methods, certify laboratory performance through both time and location, and establish reliable standards and certified reference materials, has been highlighted. It is obvious that all these areas need to be improved although analytical techniques, such as profiling, will further develop and improve independently of the need to use these methods for the analysis of, e.g., genetically modified and genetically engineered foods (National Academy of Sciences, 2004).

With regards to heavy metals, food manufacturers and processors are not only interested in confirming that their products do not contain toxic metals, or essential trace elements, above certain limits or meet various legal requirements or codes of practices, but they also support the development of methods that will be used to confirm that the products do not contain metals that may deteriorate the quality of the products. As a result, surveillance methods for quantifying and evaluating metals intake through consumption of foods are also required (Reilly, 2002).

Furthermore, the increasing interest of Muslims in the meat products they consume has rendered the proper product description highly important due to the need to give consumers complete and accurate information to enable them to make informed choices and to ensure fair trade, particularly in the ever growing halal food sector. On a global scale, Muslim consumers are concerned about several factors concerning meat and meat products such as substituted pork, undeclared blood plasma, use of prohibited ingredients, pork intestine casings and non-halal approved methods of slaughter. Specific and accurate analytical techniques have been developed to assist in different issues, but it is commonly accepted that the most suitable technique for any particular sample is usually determined by the nature of the sample itself (Nakyinsige et al., 2012).

1.2 Common Methods and Techniques Used for Food Authenticity and Quality Assurance

1.2.1 Infrared Spectroscopy

Since the 19th century, many regions of the electromagnetic spectrum have been examined and analytical instrumentation based on the energies of these regions has been provided. Many of these methods, and particularly those that are non-destructive, are currently applied, or could be applied, for analyzing food characteristics and components (Scotter, 1997). One of the most important techniques used nowadays is Infrared Spectroscopy (IR). IR spectra are acquired on an instrument, known as IR spectrometer. IR is applied both to collect information about the structural characteristics of a compound and as an analytical technique for the assessment of the purity of a substance (http://orgchem.colorado.edu/Spectroscopy/irtutor/IRtheory.pdf).

IR spectroscopy covers the electromagnetic spectrum between the red end of the visible range and the microwave region. Standard spectrometers can be used for scanning the mid-IR (MIR) region between 4000 and 400 cm^{-1} (2.5–25 μm). The near-IR extends from 12500–4000 cm^{-1} (0.8–2.5 μm) and the region from 400–50 cm^{-1} (25–200 μm) is commonly known as the far-IR. IR radiation leads to transitions between vibrational energy levels as well as between rotational energy levels. The manifestation of transitions between rotational energy levels is carried out in the gas phase, but spectra in the liquid or solid state show broad bands, which are characterised by un-resolved peaks, arising from the individual rotational transitions. Functional groups within molecules can be identified by examining the frequency of absorption due to transitions between vibrational energy levels. The vibrations of molecules can also be effectively examined using Raman scattering (Gordon and Macrae, 1987). The absorption of IR radiation by organic molecules is followed by its conversion into energy of molecular vibration. In IR spectroscopy, an organic molecule is exposed to IR radiation and at the time that the radiant energy becomes equal to the energy of a specific molecular vibration, absorption takes place (http://orgchem.colorado.edu/Spectroscopy/irtutor/IRtheory.pdf). The frequency of the vibration in conjunction with the energy indicates the bond strength. The examination of the vibrational frequencies can thus provide information about the structure of the compound of interest and this is the basis for the application of vibrational spectroscopy (http://www.metrohmsiam.com/processtalk/PT_04/PT04_MEP_Monograph_NIRS_81085026EN.pdf).

An IR spectrophotometer is an instrument that is used to pass IR light through an organic molecule leading to the production of a plot of the amount of light transmitted, on the vertical axis against the wavelength of IR radiation on the horizontal axis. In IR spectra, the absorption peaks point downward since the vertical axis is the percentage transmittance of the radiation through the sample examined. The absorption of radiation reduces the percentage transmittance value. Due to the interaction of all bonds of an organic molecule with IR radiation, IR spectra give a significant amount of structural data. Currently, there are four types of photometers that can be used to measure IR absorption:

- Dispersive grating spectrophotometers that are used to provide qualitative information
- Non-dispersive photometers that are used to quantitatively determine organic species in the atmosphere
- Reflectance photometers that are used for analysing solids and
- Fourier transform IR (FT-IR) instruments used to both qualitatively and quantitatively analyse the samples.

The instrumentation used to measure IR absorption depends on the availability of a source of continuous IR radiation and a sensitive IR transducer, or detector. IR sources are made available through an inert solid that is electrically heated to a temperature between 1500 and 2200 K. Emission of IR radiation from the heated material will then occur (http://hiq.linde-gas.com/en/analytical_methods/Infrared_spectroscopy/infrared_spectroscopy.html).

1.2.2 Mid-infrared (MIR) Spectroscopy

MIR spectroscopy is a linear spectroscopy method widely applied for control analysis in the pharmaceutical and chemical industries. Similarly to non-linear spectroscopy, MIR spectroscopy targets molecular vibrations. Most gaseous chemical substances give fundamental vibrational absorption bands in the MIR spectral region ($\approx$ 2–25 µm), and the absorption degree of light by these fundamental bands can be used to identify them. MIR spectroscopy can be most effectively used for identifying materials, to monitor the process through increase or decrease of a functional absorption band or for determining a specific substance in a mixture (http://www.geniaphotonics.

com/business-markets/industrial/i-mid-ir-spectroscopy/). The MIR region is of high importance for the analytical chemistry. It is defined as the region of wavelengths between 3×10^{-4} and 3×10^{-3} nm. Since chemists usually work with numbers easy to be written and read, IR spectra are usually reported in μm, although the unit, ν (nu bar or wavenumber), is also widely used. A wavenumber is described as the inverse of the wavelength in cm: $ν = 1/λ$, where ν is in units of cm^{-1}, λ is in units of cm and E $= hcv$. In wavenumbers, the MIR range is 4000–400 cm^{-1}. An increase in the energy will lead to wavenumber increase (http://orgchem.colorado.edu/Spectroscopy/irtutor/IRtheory.pdf).

Generally, the fundamental vibrational bands exhibited in this region are characterized by stronger line strengths in comparison to the overtone and combination bands commonly exhibited in the visible and MIR regions. Furthermore, spectra are not as congested, and thus the target substance can be detected even when analyzing a large number of molecules. The 3 μm region is highly important due to the strong vibrational bands related with the C-H, N-H, and O-H stretches (http://ritchie.chem.ox.ac.uk/research_files/midIR.htm).IR spectral measurements have been used to cover a vast variety of applications, such as the analysis of the composition of gas and liquid mixtures, or trace components (mainly to define gas purity). The latest FDA directives indicate that monitoring of pharmaceutical and chemical processes should be carried out on a stage by stage basis, and it has therefore become obvious that the detection technique currently in use is not as sensitive and quick as required to effectively perform the monitoring. The increasing use of IR spectroscopy as a detection technique has revealed that the MIR region is the spectral region where the production of the majority of the fundamental structural information is carried out. While near-Infrared (NIR) has been extensively applied to cover different industrial process needs, MIR spectroscopy can provide more useful analytical information which can more easily be used in Process Analytical Technology applications (http://www.geniaphotonics.com/business-markets/industrial/i-mid-ir-spectroscopy/).

1.2.3 Near-infrared (NIR) Spectroscopy

NIR technology is nowadays very important for quality control processes and monitoring of raw materials and finished products. This technology can be used to quickly, safely and reliably analyze raw ingredients and finished products and it is therefore a very useful analytical tool that enables control over the manufacturing process and assists in the optimization of the use of materials and ingredients, aiming at reducing off-specification products and thus reprocessing and disposal costs. NIR is a secondary technique. Its calibration is based on the analysis of a set of known samples. The analysis time required is approximately 50 seconds per sample and can thus be effectively used to give real time results that can have a direct impact on the manufacturing process. NIR spectroscopy is applied both to quantitatively (e.g., determining concentrations) and qualitatively (e.g., identifying materials, intermediates, etc.) analyse products as well as process control. NIR spectrophotometers are used by food, agricultural, dairy, pharmaceutical, chemical, petrochemical, feed, and other industries to confirm that raw materials and finished products meet their specification requirements (http://www.unityscientific.com.au/page.asp?id=28).

NIR spectroscopy has found many applications in different industrial sectors as well as in research, and it has been widely accepted and used to assess the quality of foods and beverages. The technology is relatively cheap, rapid, non-destructive and can be used for taking simultaneous measurements of several components. The main NIR applications in the food and beverage industry are on proximate quality assessment, and research. There is also an increasing interest in the use of NIR spectroscopy for microbiological research and analysis. However, further optimisations need to be made to enable NIR to be used for microbiological monitoring ecological and physiological analyses. Research focused on the use of the technology for these microbiological monitoring functions is demanded since it could lead to the development of a valuable and cost-effective technique that could be used in addition to proximate systems currently used in the food industry (http://www.foodsafetycentre.com.au/docs/Applications%20of%20Near%20Infrared%20Spectroscopic%20Analysis%20in%20the%20Food%20Industry%20and%20Research.pdf).

The use of NIR eliminates the need for dilution or the requirement of short optical path lengths or dispersion in non-absorbing matrices that are required when traditional spectroscopic analyses [e.g., Ultraviolet-visible (UV/VIs) or MIR spectroscopies] are employed. The collection of NIR spectra can be achieved in either transmittance or reflectance mode. Generally, transmittance (log(1/T)) measurements are carried out on translucent samples. Diffuse reflectance (log(1/R)) measurements are carried out on opaque or light scattering matrices, which may include slurries, suspensions, pastes or solids. For NIR spectra collected in either measurement mode, the identification of unique spectral features related to individual chemical substances within a matrix, is not always easy. To enhance spectral features and compensate for baseline offsets, the spectra information is usually mathematically pre-treated. Frequently, calculation of the second-derivative of the absorbance data, with respect to wavelength, is carried out. In the second-derivative data, conversion of absorbance maxima to minima with positive side-lobes is performed. The apparent spectral bandwidth is also sharply reduced, increasing the resolution potential of overlapping peaks. Elimination of most baseline differences between spectra is achieved (http://www.metrohmsiam.com/processtalk/PT_04/PT04_MEP_Monograph_NIRS_81085026EN.pdf).

According to Pojić et al. (2012): "The application of near-infrared spectroscopic technique for the quantitative analysis of food products and commodities is nowadays widely accepted. However, 160 years passed from the discovery of near-infrared part of the spectrum to its first analytical application which is related to the work of Karl Norris who firstly demonstrated the potential of the NIR spectroscopy in quantitative analysis particularly for prediction of moisture and protein content in wheat. The intense development of this technique during the last 50 years has been challenged by the development of powerful computers, software and chemometric tools, since the NIRS data processing is quite demanding task".

Among the most important advantages of NIR spectroscopy include the minimal sample preparation needed, which can be carried out *in situ* in many cases, short analysis time, sample preservation, and much deeper sample penetration in comparison to far or MIR radiation. Moreover, NIR analysis allows measurement of many constituents at the same time. NIR analysis is also not as expensive as other technologies used for the same purposes. Current uses of NIR spectroscopy in the food and beverage industries are mainly for the purposes of quality assessment, providing both qualitative and quantitative measurements. NIR spectroscopic qualitative assessment can give measurements with regards to the consistency (or presence) of both raw materials and final product. Quantitative NIR assessment can be employed for measuring the level of specific ingredients within complex mixtures *in situ*, thus minimising the need for sample preparation, limiting the relevant costs and saving considerable time (Nilsson, 2014).

1.2.4 Fourier Transform Near-infrared (FT-NIR) Spectroscopy

FT-NIR spectroscopy is widely used nowadays for quality control applications in all industrial sectors, such as pharmaceutical, food, agricultural and chemical industries. Its main advantage is the shorter time required for the analyses in comparison to the wet chemical and chromatographic techniques. FT-NIR is a non-destructive technique, which requires no sample preparation or use of hazardous chemicals. All these factors render the method quick and reliable for both quantitative and qualitative analysis. FT-NIR can be used to rapidly identify materials while it can also be successfully employed to carry out accurate multi-component quantitative analysis (http://www. bruker.com/products/infrared-near-infrared-and-raman-spectroscopy/ft-nir.html). FT-NIR spectrometers are therefore used for chemical identification and analysis. Their function is based on the characteristics of the functional groups of every molecule, each of which can produce a unique absorption or transmission spectrum that can be used for their identification. Interferometers are used to split the light source and the individual beams are then recombined, generating an interferogram, which is subjected to FT to produce a spectrum. The instruments can also be used for measuring the entire spectrum simultaneously, both rapidly and sensitively with low noise levels, while they also feature internal calibration on every scan. FT-NIR spectra are less intense than FT-IR spectra, but they require no sample preparation, saving a lot of time (http:// www.azom.com/materials-equipment.aspx?cat=125).

The use of FT-NIR spectroscopy presents numerous benefits over traditional dispersive techniques, which become clearly obvious when the analysis of NIR spectra of chemicals, polymers, and pharmaceuticals is required. FT-NIR spectrometers are mechanically simpler than traditional dispersive instruments due to the fact that they have no moving parts apart from the mirror of the interferometer. The breakdown possibility is therefore significantly reduced. The majority of dispersive instruments involve movement of gratings and filters, generating a spectrum. The simpler mechanical structure is translated to a more reliable, robust scanning mechanism and thus a more reliable analyzer. Dispersive NIR instruments rely on a prism or grating for the separation of the NIR frequencies. Under optimum conditions and using the best instruments, only frequencies of 50 cm^{-1} could separate apart. However, the majority of pharmaceutical, chemical, and polymer samples have spectral information that resolves at 8 cm^{-1}. Therefore, significant spectral data for these types of samples cannot be analysed using dispersive instruments and thus a slit mechanism is used to increase the resolution. Since the slit lowers the amount of beam that is measured, a significant amount of energy is lost, and the measurement of samples at higher resolutions becomes extremely difficult. Because the stroke length of the moving mirror reflects on the resolution on an FT-NIR system, there is no degradation of optical throughput as is caused by the slits in dispersive instruments. The lack of performance degradation in FT-NIR systems, allows for quick and easy measurements of high resolution spectra (https://static.thermoscientific.com/images/D14181~.pdf).

1.2.5 Raman Spectroscopy

During scattering of light from a molecule or crystal, most photons are elastically scattered. The scattered photons have the same energy (frequency), and thus the same wavelength, as the incident photons. However, a low number of photons (approximately 1 in 10^7) is scattered at optical frequencies that differ from and are usually lower than the frequency of the incident photons. The process that results in this inelastic scatter is termed the Raman Effect. Raman scattering can take place with a change in vibrational, rotational or electronic energy of a molecule. When elastic scattering occurs, the process is called Rayleigh scattering. In the case of non-elastic scattering, the process is called Raman scattering (http://instructor.physics.lsa.umich.edu/advlabs/ Raman_Spectroscopy/Raman_spect.pdf).

Raman spectroscopy is a spectroscopic technique, relying on the inelastic scattering of monochromatic light, usually from a laser source. Inelastic scattering takes place when the frequency of photons in monochromatic light changes by interacting with a sample. The absorption of the photons of the laser light by the sample is followed by re-emission. The frequency of the reemitted photons is lower or higher compared to the original monochromatic frequency, a phenomenon called the Raman Effect. This shift provides information about vibrational, rotational and other low frequency transitions in molecules. Raman spectroscopy can be effectively applied in the analysis of solid, liquid and gaseous samples (http://content.piacton.com/Uploads/Princeton/ Documents/Library/UpdatedLibrary/Raman_Spectroscopy_Basics.pdf).

The scattering of the photons by a molecule is related to the molecule's polarisability. The polarisability, α is the ability of an applied electric field, E, to induce a dipole moment, μ_{in}, in an atom or molecule: $\mu_{in} = \alpha E$ (http://www.spectroscopyonline. com/spectroscopy/data/articlestandard/spectroscopy/442001/836/article.pdf).

The Nobel prize was given in 1928 to V.C. Raman for the discovery of the Raman Effect. Upon illumination of a sample using monochromatic light, such as the light of a laser, a spectrum is generated that consists of a strong line (the exciting line) of the same frequency as the incident illumination and with weaker lines on either side shifted from the strong line by frequencies that range from a few to about 3500 cm^{-1}. The lines of frequencies lower than the exciting lines are called Stokes lines, while all the others are named anti-Stokes lines. Raman spectroscopy is a technique of high importance, which is used to rapidly identify molecules and minerals. Other uses of Raman spectroscopy include the study of molecular structures (http://instructor.physics. lsa.umich.edu/advlabs/Raman_Spectroscopy/Raman_spect.pdf).

According to Li-Chan (1996): "The diversity of applications and high content of molecular structure information provided, combined with recent advances in instrumentation, have rekindled interest in this technique in many diverse disciplines, including food science. Suitable analytes cover the entire range of food constituents, including the macro-components (proteins, lipids, carbohydrates and water) as well as minor components such as carotenoid pigments or synthetic dyes, and even microorganisms or packaging materials in contact with foods. Raman spectroscopy may be used as a tool for quality control, for compositional identification or for the detection of adulteration, as well as for basic research in the elucidation of structural or conformational changes that occur during processing of foods".

In the study of Herrero (2008), a comparison of Raman spectroscopy data with data obtained using traditional methodologies was made. The traditional techniques included protein solubility, apparent viscosity, water holding capacity, instrumental texture methods, dimethylamine content, peroxide values, and fatty acid composition, which are all widely used for the determination of the quality of fish and meat muscle, processed and stored under various conditions. It was proved that Raman spectroscopy data were in agreement with the results obtained using traditional quality methods and could therefore be applied for evaluating muscle food quality. Furthermore, Raman spectroscopy provided useful information about the changes in the structures of proteins, water and lipids of meat and fish, while it could also be used to identify meat products. The most important advantages of the method are the facts that it is a direct and non-invasive technique and only very small portions of sample are required for the analysis.

FT-NIR Raman spectroscopy has been proved to be a very useful spectroscopic method for the molecular structural analysis of chemical substances. FT Raman spectroscopy has been suggested for use in the pharmaceutical sciences (Tudor et al., 1990) but it has also been used for the analysis of foods. In the study of Ozaki et al. (1992), the 1064-nm excited FT-NIR spectra have been measured *in situ* for different foods aiming at investigating the potential of FT-NIR spectroscopy in food analysis. It was shown that FT-NIR spectroscopy can be successfully used for the detection of substances present at trace levels in foodstuffs, for the estimation of the degree of unsaturation of fatty acids in foods, for investigating the structure of food components, and for monitoring quality changes in foods. Carotenoids in foods give rise to two intense bands near 1530 and 1160 cm^{-1} via the pre-resonance Raman effect in the NIR FT-Raman spectra, and therefore, the technique can be used for their non-destructive detection. Foods containing very high amounts of lipids such as oils, tallow, and butter give bands near 1658 and 1443 cm^{-1} due to C=C stretching modes of *cis* unsaturated fatty acid parts and CH_2 scissoring modes of saturated fatty acid parts, respectively. It was also proved that the different types of lipid-containing foods show a linear relation between the iodine value and the intensity ratio of two bands at 1658 and 1443 cm^{-1} (I_{1658}/I_{1443}). This ratio could therefore be employed to practically estimate the unsaturation degree of various fatty foods. Raman spectra of raw and boiled egg white were compared and it was shown that the amide I band shifts from 1666 to 1677 cm^{-1} and the intensity of the amide III band at 1275 cm^{-1} decreases upon boiling. These results are an indication of the structural changes most α-helixes undergo to become unordered structures.

1.2.6 Fluorescence and Ultraviolet-visible (UV-Vis) Spectroscopy

Fluorescence spectroscopy is based on the measurement of the intensity of photons emitted after the absorption of photons by a sample. Fluorescent molecules are mainly aromatic. Fluorescence is a technique of high importance for various analytical science fields, due to its high sensitivity and selectivity. It has been successfully used to examine real-time structure and dynamics both in solutions and using microscopes, mainly for bio-molecular systems (http://www2.warwick.ac.uk/services/rss/business/analyticalguide/fluorescence/).

A hot body emitting radiation solely due to its high temperature exhibits "incandescence". All other types of light emission are defined as luminescence. In the case of luminescence, energy is lost from the system, and therefore, the supply of some form of energy from an external source is necessary to continue the emission. Thus the emission of radio-luminescence from a luminous clock face is performed by high energy particles from the radioactive material in the phosphor, while the electroluminescence of a gas discharge lamp is provided through the passage of an electric current through an ionized gas. Similar phenomena include chemiluminescence, which is supplied by the energy of a chemical reaction, and bioluminescence, which is chemiluminescence supplied by the energy of reactions that occur within living organisms. When the external source of energy derives from the absorption of infrared, visible or ultraviolet light, the emitted light is called photoluminescence and this is the procedure carried out in fluorimetric analyses. To establish a proper measurement unit for the processes of absorption and emission of light, it is assumed that radiant energy can only be carried out in definite units, or quanta. The energy, E, carried by any one quantum is proportional to its frequency of oscillation, that is $E = h\nu = hc/\lambda$ ergs, where ν is the frequency, λ the related wavelength and h = Planck's constant $(6.624 \times 10^{-27}$ ergs/second) (http://physweb.bgu.ac.il/~bogomole/Books/An%20Introduction%20to%20Fluorescence%20Spectroscopy.pdf).

To effectively perform fluorescence experiments, good attention to experimental details and good knowledge of the instrumentation is required. Various potential artifacts, however, can distort the data. The detection of light can be carried out with high sensitivity and thus the gain or amplification of instruments can usually be increased to provide signals that can be easily observed, even if the sample is nearly nonfluorescent. These signals observed at high amplification may not derive from the fluorophore of interest. The interference can be caused by background fluorescence derived from solvents, light leaks in the instruments, emission from the optical components, stray light that passes through the optics, light scattered by turbid solutions, Rayleigh and/or Raman scatter, etc. (Lakowicz, 2006). A typical Fluorescence Spectrometer (Spectrofluorometer) is composed of a sample holder, an incident photon source (usually a xenon lamp), monochromators employed to select specific incident wavelengths, focusing optics, photon-collecting detector (single- or multi-channel), a control software unit and an emission monochromator or cut-off filters. The detector is typically set at 90 degrees to the light source. Due to the fact that fluorescence is intrinsically highly sensible, care must be taken to record a true fluorescence signal of the analyte examined. A fluorescence emission spectrum is recorded when the excitation wavelength of light remains constant and scanning of the emission beam is carried out as a function of wavelength. On the other hand, an excitation spectrum is recorded when the emission light remains constant and the excitation light is scanned as a function of wavelength. The shape of the excitation spectrum is usually quite similar to the shape of the absorbance spectrum. Most materials do not have natural fluorescence. However, information, particularly in fluorescence microscopy, can be obtained by staining non-fluorophores with an active label (http://www2.warwick. ac.uk/services/rss/business/analyticalguide/fluorescence/).

According to Adhikary (2010): "Steady-state and time-resolved fluorescence methods are commonly used to characterize emissive properties of fluorophores. Time-resolved fluorescence measurements are generally more informative about the molecular environment of the fluorophore than steady-state fluorescence measurements because the competing or perturbing kinetic processes such as collisional quenching, solvent relaxation, energy transfer, and rotational reorientation, which affect the fluorescence, occur on the timescale of the flurophore's lifetime (10^{-9} s). Thus, time-resolved fluorescence spectroscopy can be used to quantify these processes and gain insight into the chemical surroundings of the fluorophore".

Ultraviolet or visible light is absorbed by various molecules. The absorbance of a solution increases by increasing the attenuation of the beam. The absorbance is directly proportional to the path length, b, and the concentration, c, of the absorbing species. Beer's Law states that $A = ebc$, where e is a constant of proportionality, defined as "absorbtivity" (http://teaching.shu.ac.uk/hwb/chemistry/tutorials/molspec/uvvisab1. htm). The wavelength of the radiation absorbed depends on the type of molecule. An absorption spectrum is characterized by different absorption bands that correspond to structural groups within the molecule. For example, the absorption that is observed in the UV region for the carbonyl group in acetone is of the same wavelength as the absorption from the carbonyl group in diethyl ketone. UV-Vis spectroscopy (uv = 200–400 nm, visible = 400–800 nm) is based on electronic excitations between energy levels that correspond to the molecular orbitals of the systems. Specifically, transitions that involve π orbitals and lone pairs (n = non-bonding) are of high importance and thus UV-Vis spectroscopy is commonly used for the identification of conjugated systems characterized by stronger absorptions (http://www.chem.ucalgary.ca/courses/351/ Carey5th/Ch13/ch13-uvvis.html).

UV/Vis spectroscopy is usually used to quantitatively determine solutions of transition metal ions highly conjugated organic compounds, and biological macromolecules. Colouring of solutions of transition metal ions (i.e., via absorption of visible light) can be achieved due to excitations of d electrons within the metal atoms from one electronic state to another. The colour of metal ion solutions highly depends on the presence of other species (e.g., other anions or ligands). For example, the colour of a solution of copper sulfate is light blue, but the addition of ammonia enhances the intensity of the colour, changing the wavelength of maximum absorption (λ_{max}). Organic compounds, and particularly those with a high degree of conjugation, also absorb light in the UV or visible regions of the electromagnetic spectrum. The solvents most widely used for these analyses include water for water-soluble compounds, or ethanol for organic-soluble compounds. Although organic solvents are usually characterized by high UV absorption, not all solvents can be used in UV spectroscopy. Ethanol, for example, has a very weak absorption at most wavelengths. Factors such as solvent polarity and pH can affect drastically the absorption spectrum of an organic substance. Tyrosine, for example, increases in absorption maxima and molar extinction coefficient by increasing the pH from 6 to 13 or by decreasing the solvent polarity. While charge transfer complexes also cause colour changes, the colours can often too intense to be used for quantitative measurements (http://www.princeton.edu/~achaney/tmve/wiki100k/docs/Ultraviolet-visible_spectroscopy.html).

The functioning of a UV-Vis spectroscoper is quite simple. Separation of a beam of light from a visible and/or UV light source (coloured red) into its component wavelengths is carried out by a prism or diffraction grating and each monochromatic beam separates into two equally intense beams by a half-mirrored device. One beam, the sample beam (coloured magenta), is directed through a small transparent container (cuvette) that contains a solution of the compound of interest in a transparent solvent. The other beam, the reference (coloured blue), is directed through an identical cuvette that contains only solvent. Measurements of the intensities of these light beams are taken using electronic detectors and a comparison is made. The intensity of the reference beam, which should have been lightly or not absorbed, is defined as I_0, while the intensity of the sample beam is defined as I. Automatic scanning of all component wavelengths is rapidly performed by the spectrometer as previously described. The scanning usually includes the UV region from 200 to 400 nm, while the Vis region is from 400 to 800 nm. When no absorption by the same is observed, I becomes equal to I_0. However, if the sample compound absorbs light, then I is less than I_0, and this difference may be plotted on a graph versus wavelength. Absorption may be presented as transmittance ($T = I/I_0$) or absorbance ($A = \log I_0/I$). When no radiation has been absorbed, $T = 1.0$ and $A = 0$. The majority of spectrometers display absorbance on the vertical axis, with a usually observed range of 0 (100% transmittance) to 2 (1% transmittance). The wavelength of maximum absorbance is λ_{max}. Compounds with strong absorbance need to be diluted before the analysis, so that sufficient light energy is received by the detector, and only totally transparent solvents can be used for this purpose. Solvents frequently used include water, ethanol, hexane and cyclohexane, while solvents that have double or triple bonds, or heavy atoms (such as S, Br and I) are not usually preferred. Since the absorbance of a sample will be proportionally related to its molar concentration in the sample cuvette, a corrected absorption value, defined as molar absorptivity is used during the comparison of the spectra of different compounds. Specifically, Molar Absorptivity, $\varepsilon = A/cl$, where $A =$ absorbance, $c =$ sample concentration in moles/litre and $l =$ length of light path through the cuvette in cm (http://www2.chemistry.msu.edu/faculty/reusch/VirtTxtJml/Spectrpy/UV-Vis/uvspec.htm#uv1).

1.2.7 NMR Techniques

Nuclear Magnetic Resonance (NMR) spectroscopy is a technique used widely in the chemical analysis for both quality control and research purposes, and specifically for the determination of the content and purity of a sample as well as its molecular structure. For example, NMR can be used for the quantitative analysis of mixtures that contain known compounds. In the case of unknown compounds, NMR can either be used as a tool for confirming a match against spectral libraries or can give information about the basic structure. Based on the basic structure, NMR can be used for the determination of a molecular conformation in solution or for examining the physical characteristics (e.g., conformational exchange, phase changes, solubility, and diffusion) at a molecular level. Nowadays, numerous NMR techniques can be used to meet different needs for analysis (http://chem.ch.huji.ac.il/nmr/whatisnmr/whatisnmr.html). According to Gordon and Macrae (1987): "NMR spectroscopy is a technique that has developed rapidly from early experiments in late 1940s until the present day, when it plays a key role in structural analysis and identification of organic molecules. Although NMR is less sensitive than some techniques, such as UV or fluorescence spectrophotometry, it can be applied much more widely and provides specific structural information that is often adequate for the unambiguous identification of organic molecules. NMR depends on the absorption of radiation by the nuclei of certain isotopes. The technique requires a strong magnetic field to remove the degeneracy of the nuclear energy levels and a radio frequency field to induce transitions between the energy levels. Isotopes with suitable nuclear magnetic properties include ^1H, ^{13}C, ^{15}N and ^{31}P."

Nuclear Magnetic Spectroscopy is based on the fact that when a population of magnetic nuclei is exposed to an external magnetic field, the nuclei become aligned in a certain and defined number of orientations. In the case of ^1H there are two orientations. In the first orientation, there is an alignment of protons with the external magnetic field, while in the other the nuclei are aligned against the field. The alignment with the field is defined as the "alpha" orientation and the alignment against the field is defined as "beta" orientation. It should be mentioned that before the nuclei are exposed to the magnetic field they have random orientation. Since the alpha orientation is the preferred one, more nuclei are aligned with the field than against the field. The NMR spectroscopy is based on the use of energy in the form of electromagnetic radiation for pumping the excess alpha oriented nuclei into the beta state. After the removal of the energy, the nuclei relax back to the alpha state. These changes of the magnetic field that are related to the relaxation process are called resonance. Detection of the resonance can be achieved and these data are converted into the peaks of the NMR spectrum (http://www.brynmawr.edu/chemistry/Chem/mnerzsto/The_Basics_Nuclear_Magnetic_Resonance%20_Spectroscopy_2.htm).

NMR instrumentation is categorized into two generic groups: continuous wave (C.W.) and Fourier transform (FT). The earliest experiments were carried out with C.W. instruments, and in 1970 the first FT instruments were used and finally prevailed over C.W. NMR spectrometers follow similar principles with optical spectrometers. The sample is retained under a strong magnetic field, and the frequency of the source is slowly scanned (the source frequency is often held constant, and the field is scanned). As regards FT NMR spectrometers, it is important to mention that the energy changes involved in NMR spectroscopy are very small, and this lowers the sensitivity of the method. The sensitivity can increase through recording many spectra, and adding them together. Since noise is random, it adds as the square root of the number of spectra involved. This means that if one hundred spectra of a substance were summed, the noise would increase by a factor of ten, but the signal would increase in magnitude by a factor of one hundred, thus significantly increasing the sensitivity. However, if a continuous wave instrument was employed for this purpose, the time required for the collection of the spectra would be very long (based on the fact that one scan takes two to eight minutes). In FT-NMR, simultaneous irradiation of all frequencies in a spectrum is carried out with a radio frequency pulse. Following the pulse, the nuclei return to thermal equilibrium. Record of a time domain emission signal is taken by the instrument as the nuclei relax. A frequency domain spectrum is obtained by Fourier transformation (http://teaching.shu.ac.uk/hwb/chemistry/tutorials/molspec/nmr3.htm).

NMR has been used in the food industry to support various applications that require characterization of the energetic status of cells in order to monitor the fermentation of yoghurts (using phosphorus 31 LR NMR), examination of cell cultures in the mashing of beer (using proton NMR), and thermal processing of various types of rice (proton NMR). Although NMR is not characterized by high sensitivity in these process applications, it can be used to examine bulk properties. It can be applied for the examination of processes and textures, while it does not require the use of any marker compounds or any physical incursion apart from a magnetic field. NMR data can be used to investigate the route of processes such as enzymatic and conventional chemical reactions as well as for quality and process control processes during production (http://www.foodprocessing-technology.com/projects/nmr-technology/).

1.2.8 Stable Isotope Ratio Mass Spectrometry (IRMS)

Isotope Ratio Mass Spectrometry (IRMS) is a spectrometric technique used to obtain information about the geographic, chemical and biological origins of compounds. The effectiveness of the technique in determining the origin of organic substances derives from the relative isotopic abundances of the elements, which comprise the material. Due to the fact that the isotope ratios of elements such as carbon, hydrogen, oxygen, sulphur and nitrogen can locally increase or decrease through various kinetic and thermodynamic factors, measurement of the isotope ratios can be used for differentiating between samples which otherwise share identical chemical compositions. Numerous sample introduction methods can now be used for commercial isotope ratio mass spectrometers. Combustion is usually used for bulk isotopic analysis, while gas and liquid chromatography are mainly employed in the case of real-time isotopic analysis of specific compounds within a sample (Muccio and Jackson, 2009).

The establishment of an isotopic "profile" or "signature" of a compound is carried out through measurement of the ratios of the stable isotopes of various elements such as $^2H/^1H$, $^{13}C/^{12}C$, $^{15}N/^{14}N$ and $^{18}O/^{16}O$. The isotopic levels of these elements were established during the formation of the earth and, on a global scale, they have remained unchanged. Small variations in the isotopic composition of materials can occur during biological, chemical and physical processes. IRMS can be used for the measurement of the relative abundance of isotopes in different materials. The variations in the natural abundance of stable isotopes can be expressed using delta (δ) notation as shown in the following equations.

Ratio (R) = abundance of the heavy isotope/abundance of the light isotope

$$\delta = [(R_{samp}/R_{Sts})-1]$$

R_{samp}: Ratio of sample

R_{Sts}: Ratio of the international standard

δ-values are usually multiplied by 1000 in order to be expresssed in parts per thousand (‰ or per mil) or by 1000,000, to be reported in parts per million (ppm) (Carter and Barwick, 2011).

According to Sulzman (2007): "A mass spectrometer is an instrument that separates charged atoms or molecules on the basis of their mass-to-charge-ratio, m/z. There are two basic types of IRMS, dual-inlet (DI-IRMS) and continuous flow (CF-IRMS). In general, precision is higher when samples are analyzed with a dual-inlet system. A continuous flow system allows a researcher to introduce multiple component samples (e.g., atmospheric air, soil, leaves) and obtain isotopic information for individual elements or compounds within the mixture."

With DI-IRMS, the preparation of the samples for analysis is carried out off-line. This off-line preparation involves the use of special instrumentation such as vacuum lines, compression pumps, concentrators, reaction furnaces, and micro-distillation equipment. This technique is quite time-consuming, and large samples are usually required, while contamination and isotopic fractionation can be observed at all steps. The CF-IRMS sample introduction technique consists of a helium carrier gas that carries the analyte gas into the ion source of the IRMS. This technique is employed for connecting an IRMS to several automated sample preparation devices. While dual inlet gives usually the most accurate results for stable isotope ratio measurements, continuous flow mass spectrometry offers the advantage of on-line sample preparation, while it is simpler, faster, cost effective, requires the use of smaller samples and offers the possibility of interfacing with other preparation techniques, such as elemental analysis, gas chromatography (GC), and liquid chromatography (LC) (Luykx and Van Ruth, 2008). A representative example of the use of this technique is given in the study of Schmidt et al. (2005), in which it was suggested that the analysis of natural stable isotope compositions of carbon, nitrogen and sulphur could be used for verifying the geographical origin and feeding history of beef cattle.

1.2.9 Gas Chromatography

GC is a separation analytical technique that is used for the analysis of volatile substances in the gas phase. In GC, the components of a sample are dissolved in a solvent and vapourization occurs to separate the analytes through distribution of the sample between two phases: a stationary phase and a mobile phase. The mobile phase is commonly a chemically inert gas with the ability of carrying the molecules of the analyte through the heated column. GC is a type of chromatography in which there is no interaction between the mobile phase and the analyte. The stationary phase can be a solid adsorbent, termed gas-solid chromatography (GSC), or a liquid on an inert support, termed gas-liquid chromatography (GLC) (http://chemwiki.ucdavis.edu/Analytical_Chemistry/Instrumental_Analysis/Chromatography/Gas_Chromatography). All the gases used for GC analysis have to be very pure and special care must be taken for the removal of trace levels of moisture, oxygen and hydrocarbons from the gas before its use. Oxygen is considered one of the most important contaminants that need to be removed, due to the fact that it causes column degradation and problems with the more sophisticated detectors such as electron capture and mass spectrometer detectors. The removal of the contaminants is carried out by passing the gas through cartridges of absorbent. These are commonly 'self indicating' to allow the analyst to check whether the absorbent bed is exhausted. The supply of the gases is carried out in cylinders at high pressures and similarly to High

Performance Liquid Chromatography (HPLC), there is no need for using pumps for the facilitation of the movement of the mobile phase through the instrument. Precise attenuation and regulation of the cylinder pressures is required to ensure a constant pressure at the front end of the column. Mass flow controllers are used to ensure that the pre-set flow-rate of gas remains constant and not affected by the column oven temperature (Stuart and Prichard, 2003).

The technique was developed by Martin and Synge (1941), who suggested that gas-liquid partition chromatograms should be used for analytical purposes. Through their work using liquid-liquid partition chromatography, they realized that the mobile phase could also be a vapour. Very refined separations of volatile substances in a column in which a permanent gas flows over gel impregnated with non-volatile solvent would be much faster thus increasing the efficiency of the columns and reducing the separation times required (http://forensicscienceeducation.org/wp-content/uploads/2013/02/Theory_and_Instrumentation_Of_GC_Introduction.pdf). The development of GC is frequently related with advances in the development of measuring techniques used to monitor substances after their separation. Although the separation process has been thoroughly examined, less attention has been paid to the accuracy degree of the generated results and the stability of the measuring devices. These measuring devices are known as "detectors" and form an independent part of the gas chromatograph. Their use depends on the physical and chemical characteristics of the eluted substances or compounds that are formed by their reaction inside the detector (Ševčík, 1976). A GC and specifically GLC analysis is based on the vapourization of a sample and injection onto the head of the chromatographic column. The sample is transferred through the column by the flow of inert, gaseous mobile phase. A liquid stationary phase is contained in the column. This liquid phase is adsorbed onto the surface of an inert solid. It is highly important that the carrier gas is chemically inert. Some of the most widely used gases are nitrogen, helium, argon and carbon dioxide. The carrier gas used usually depends on the type of detector. The carrier gas system also contains a molecular sieve for the removal of moisture and other impurities.

To optimize the efficiency of the column, the sample should be quite small, and its introduction onto the column should be carried out rapidly since slow injections of large samples can lead to band broadening and loss of resolution. The most commonly used injection method involves the use of a micro-syringe for the injection of the sample through a rubber septum into a flash vapouriser port at the head of the column. The temperature of the sample port is commonly about 50°C higher than the boiling point of the least volatile component of the sample. In the case of packed columns, the sample size can range from a few tenths of a microlitre up to 20 microlitres. Capillary columns, however, need much less sample, typically around 10^{-3} mL. For capillary GC, split/splitless injection is used (http://teaching.shu.ac.uk/hwb/chemistry/tutorials/chrom/gaschrm.htm). According to Lehotay and Hajšlová (2002): "GC is used widely in applications involving food analysis. Typical applications pertain to the quantitative and/or qualitative analysis of food composition, natural products, food additives, flavour and aroma components, a variety of transformation products, and contaminants, such as pesticides, fumigants, environmental pollutants, natural toxins, veterinary drugs, and packaging materials."

1.2.10 High-performance Liquid Chromatography (HPLC)

Although HPLC is not as sensitive as GC, it can be used for more applications. Transfer of the analyte is performed through the stationary phase of a column (either packed or capillary) with the mobile phase, and separation of the components of the mixture is carried out on the column. When polar columns are used (e.g., silica based), and the mobile phases are nonpolar, the HPLC method is considered to be normal phase (NP-HPLC). When the column is non-polar and the mobile phase polar, HPLC is considered to be reversed phase (RP-HPLC). The majority of the applications today are based on the use of RP-HPLC using a stationary phase that is a C_8 or C_{18} bonded phase. Pumping rates can vary from nL to mL min^{-1}, with column dimensions of 250 × 4.6 mm i.d., 3 to 10 μm particle diameters, and pressures operating up to ca. 3000 psi. As regards the equipment used, there is a reservoir that holds the solvents or mobile phases and a high-pressure pump or other delivery system that controls the flow rate and pressure. The detector is commonly a multiwavelength UV (sensitivity around 0.1–1 ng) in tandem with multiwavelength fluorescence detection (sensitivity 0.001–0.01 ng). In the case of applications, such as the analysis of sugars, pulsed amperometric detectors (PAD) are employed. Their sensitivity ranges from 0.01 to 1 ng. When the analysis of plant pigments is required, photodiode array detectors (PDA) can be used for a very preliminary identification of peaks based on the fact that carotenoids do not show fluorescence, while chlorophylls show fluorescence and absorbance (Bianchi and Canuel, 2011).

The time required for a substance to travel through the column to the detector is called "retention time". Measurement of this time is carried out from the time of the injection of the sample to the point at which the display presents a maximum peak height for that compound. Different compounds are characterized by different retention times. Each compound will have a different retention time, which depends on: (1) the pressure used (since the pressure affects the flow rate of the solvent), (2) the nature of the stationary phase (the material used and the particle size), (3) the composition of the solvent, and (4) the temperature of the column. It is therefore obvious that all conditions have to be carefully controlled whenever retention times are used as a means of identifying compounds (http://www.chemguide.co.uk/analysis/chromatography/hplc.html). Stainless steel is commonly used to make columns, which are between 50 and 300 mm long and have an internal diameter of between 2 and 5 mm. The particle size of the stationary phases used for these columns is usually 3–10 μm. Columns that have internal diameters of less than 2 mm are known as microbore columns. Ideally the temperature of the mobile phase and the column should be constant during the analysis. The majority of the analyses are carried out at ambient temperature but columns may be heated using, for example, a water-bath, a heating block or a column oven in order to increase the efficiency.

Full efficiency of an analytical column cannot be achieved due to the design limitations imposed by the pumps, injectors and detectors. The connections between injector/column, column/detector and/or detector/detector can reduce the efficiency of the system. Any fittings should be of the "zero dead volume" (ZDV) type. It is believed that minimum lengths of capillary tubing with a maximum internal diameter of 0.25 mm is the best choice for these fittings in order to minimize band broadening.

Collection of signals from the detector can be performed on chart recorders or electronic integrators. These instruments can usually store limited amount of data. Modern data stations use computers and have a large storage capacity for collecting, processing and storing data that can potentially be reprocessed. Integration of peak areas and the setting of threshold levels can usually be easily achieved in an assay since the peak of the substance subjected to the analysis should be free of interference. However, when impurities are tested, the selection of the peak area integrator parameters is of high importance, especially when baseline separations are not always attainable. If baseline separations cannot be achieved, valley-to-valley integration can be used for peaks of similar size, while tangent skimming can be used for peaks of different sizes. Although HPLC allows limits to be established for specific impurities and for the sum of impurities, there must be a threshold below which peaks should not be integrated. This "disregard level" or "reporting threshold" is related with the area of the peak in the chromatogram of the prescribed reference solution and is usually equal to 0.1% or 0.05% of the substance analyzed (http://apps.who.int/phint/en/p/docf/).

1.2.11 Polymerase Chain Reaction

According to Brooks (2011): "PCR methods, which are more sensitive and detect the DNA molecules of these allergens, can be used in raw and cooked products and are not affected by the heating process because DNA typically remains intact after being exposed to the cooking temperatures of most foods. PCR methods are also not subject to the typical interferences that inhibit Enzyme-linked immunosorbent assay (ELISA)-based methods because the DNA is purified away from these inhibitors before analysis begins. PCR, however, cannot be used on all products. Oils and other products, such as milk or egg whites, cannot be tested by PCR because they do not contain DNA. These products must instead be tested using ELISA-based methods for detection."

PCR was developed by Kary Mullis in the 1980s. This method is based on the use of DNA polymerase for synthesizing new strands of DNA complementary to the offered template strand. Since DNA polymerase can be used to add a nucleotide only onto a preexisting 3'-OH group, a primer is required to which the first nucleotide can be added. This requirement makes it possible to delineate a particular region of template sequence, which needs to be amplified. At the end of the PCR reaction, billions of copies of a particular sequence are produced. These copies are called amplicons (http://www.ncbi.nlm.nih.gov/projects/genome/probe/doc/TechPCR.shtml). The method is carried out through synthesis by polymerase of a complementary sequence of bases to any single strand of DNA providing it has a double stranded starting point. This however offers the advantage of the selection of a particular gene that will be amplified by polymerase in a mixed DNA sample through the addition of small fragments of DNA, complimentary to the gene of interest. These DNA fragments are known as primers because they prime the DNA sample ready for the polymerase to bind and initiate the copying the gene of interest. During PCR, temperature alterations are used for controlling the activity of the enzyme polymerase and the binding of primers. The reaction initiates by raising the temperature to 95°C. Under these temperature conditions, all double stranded DNA is converted into single strands. By lowering the temperature to ~50°C, subsequent binding of the primers to the gene of interest occurs. The copying of the DNA strand can then begin. The optimal temperature for the polymerase to operate is 72°C and, therefore, at this point the temperature is usually raised to 72°C to accelerate the enzyme activity. This procedure doubles the copies of the gene of interest. By repeating these consequent temperature changes, the number of copies is continuously doubled. Following amplification of the gene of interest into millions of copies, the amplified DNA can be run out on an agarose gel and visualized after staining it with a dye. The term chain reaction has been given to this process since only a small fragment of DNA needs to be identified and acts as the template for the production of the primers that initiate the reaction. Polymerases, the enzymes used to make DNA copies, string together individual DNA building blocks and produce long molecular strands (http://amrita.vlab.co.in/?sub=3&brch=186&si m=321&cnt=1). In practice, the PCR technique can be far more complicated and it is frequently very difficult to achieve the desired result. For example, non-specific binding of primers to multiple places on the DNA strand, leading to a mixture of amplified product, is a common problem. Furthermore, several DNA impurities can terminate the reaction (http://www.highveld.com/pcr/pcr-basics.html). Since the discovery of PCR, DNA polymerases other than the original Taq have been found. Some of these are characterized by better "proof-reading" ability or higher stability at high temperatures, thus rendering the PCR technique more specific and reducing errors from insertion of incorrect dNTP. A few variations of the PCR technique have been developed for specific applications and are currently employed by molecular genetic laboratories (http://biotech.about.com/od/pcr/a/PCRtheory_2.htm).

Real-Time-PCR is a technique based on PCR and is applied for amplifying and simultaneously quantifying a targeted DNA molecule. PCR is used to exponentially amplify short DNA sequences (commonly 100 to 600 bases) within a longer double stranded DNA molecule. A pair of primers of about 20 nucleotides is necessary for running the PCR. These primers are complementary to a defined sequence of two strands of DNA. Melting of the double stranded DNA is carried out by raising the temperature to > 90°C and by subsequently lowering the temperature to allow the primers to bind to the denatured DNA. Extension of the primers is carried out by a thermo-stable DNA polymerase (Taq polymerase) and thus a copy of the sequence of interest is formed. The same primers can then be re-used to logarithmically amplify the DNA sequence. In Real-Time-PCR fluorescent molecules are attached to the amplified cDNA. Real Time PCR instruments can measure this fluorescence and quantification of the amplified DNA is carried out as it accumulates in the reaction in real time after each reaction cycle. Increase of copied DNA amount increases the fluorescence and these changes can be displayed on a graph (http://www.kdanalytical.com/instruments/technology/pcr.aspx).

To conclude with, PCR is a relatively new technique that since its development in 1983 has been used in several different applications including food analysis. It has been extensively used in the field of food microbiology, although it has recently found applications in other areas, such as food hygiene and food toxicology (http://cdn.intechopen.com/pdfs-wm/37268.pdf).

1.2.12 Enzyme-linked Immunosorbent Assay (ELISA)

ELISA is a rapid immunochemical test. It is based on the use of an enzyme (a protein that catalyzes a biochemical reaction) and an antibody or antigen (immunologic molecules). ELISA tests are used for the detection of substances with antigenic properties, such as proteins. Some of these substances may include hormones, bacterial antigens and antibodies (http://www.medterms.com/script/main/art.asp?articlekey=9100). The technique was designed and developed by Peter Perlmann and Eva Engvall at Stockholm University in Sweden (http://www.clinchem.org/content/51/12/2415.full).

ELISAs can be carried out with a number of modifications to the principal method. The most important stage, immobilization of the antigen of interest, can be performed directly via adsorption to the assay plate or indirectly via a capture antibody that has been attached to the plate. Subsequent detection of the antigen is carried out either directly (labeled primary antibody) or indirectly (labeled secondary antibody). The most effective ELISA assay format is the sandwich assay. This type of ELISA termed as a "sandwich ELISA" because the analyte of interest is bound between two primary antibodies—the capture antibody and the detection antibody. This sandwich format is used because it significantly increases the sensitivity of the assay. An ELISA can also be carried out as a competitive assay. This is usually the case when the antigen is small and has only one epitope, or antibody binding site. One variation of this method is based on labeling purified antigen instead of the antibody. Unlabeled antigen from samples and the labeled antigen compete for binding to the capture antibody. A decrease in signal from purified antigen will indicate that antigen is present in samples when compared with assay wells with labeled antigen alone. Fluorescent tags and other alternatives to enzyme based detection can be employed for plate-based assays. Although they do not involve reporter-enzymes, these techniques are usually referred to as ELISAs. Similarly, wherever detectable probes and specific protein binding interactions can be used in a plate-based method, these assays are commonly called ELISAs although they do not involve antibodies (http://www.piercenet.com/method/overview-elisa).

According to Asensio et al. (2008): "People suffering from food allergies are dependent on accurate food labeling, as an avoidance diet is the only effective countermeasure. Even a small amount of allergenic protein can trigger severe reactions in highly sensitized patients. Therefore, sensitive and reliable tests such as ELISA are needed to detect potential cross-contamination. ELISA is well suited for the determination of food authenticity because it is sensitive and specific; fast and cheap; easy to perform and the investment in equipment is much less than other techniques. Although ELISA limitations for genetically modified food detection have been shown, this methodology could allow, together with other analytical methods such as DNA-based methods, consumers protection against fraudulent practices in the food industry."

Since 2005, it is required that manufacturers use allergen declarations with regards to eight major allergens: Milk, Eggs, Fish, Shellfish, Peanuts, Wheat, Soybeans and Tree Nuts. These allergens cause more than 90% of the documented food allergen-related incidents (http://www.elisa-antibody.com/ELISA-applications/food-industry). It is estimated by FDA that approximately 2% of adults and 5% of children in the population have an allergy to one or more of these types of foods. Although the majority of allergies reveal only minor symptoms, some allergens can lead to severe reactions. Millions of people are affected every year by food-related allergies, while more than 30,000 people require emergency hospitalization, and about 150 die. The symptomatology of an allergic reaction can include flushed skin or rashes (hives), swelling of the tongue, mouth, throat or face, tingling or itchy sensations in the mouth, severe coughing or difficulty in breathing, and/or loss of consciousness. One or more of these symptoms presented within a few minutes to a few hours after eating could indicate the occurrence of a food-related allergic reaction (Brooks, 2011). In the food industry, ELISA and PCR are most commonly used for detecting allergens. ELISA methods can be applied for the detection of the allergen protein molecule. Specifically, during an ELISA assay, antibodies bind to an allergen and an enzyme-linked conjugate is used to create a measurable colorimetric change. There are cases, however, that ELISA methods are not effective. Interference of some matrices (e.g., chocolate) with the ELISA method has been observed. Other matrices can cross-react. Also, the method is also not the most suitable for cooked or heated products since the denaturation of protein molecules render the allergen not detectable, while it can still cause allergic reactions to sensitive individuals (http://www.elisa-antibody.com/ELISA-applications/food-industry).

1.2.13 Capillary Zone Electrophoresis

Capillary electrophoresis (CE) is an analytical technique that is used for the separation of ions based on their electrophoretic mobility using an applied voltage. The electrophoretic mobility depends on the charge of the molecule, the viscosity, and the atom's radius. The rate of particle movement is directly proportional to the applied electric field, and specifically, the greater the field strength, the faster the mobility. Since the charge of the molecule is one of the parameters that determine mobility, neutral species are not affected. Between two ions of the same size, the one with greater charge will move faster. For ions of the same charge, the smaller particle has less friction and thus faster migration rate. CE is mainly used because it can lead to faster results and gives high-resolution separation (http://chemwiki.ucdavis.edu/Analytical_Chemistry/Instrumental_Analysis/Capillary_Electrophoresis). The technique is based on the use of a high voltage power supply (0 to 30 kV), a fused silica (SiO_2) capillary, two buffer reservoirs, two electrodes, and an on-column detector. The sample is injected by temporarily replacing one of the buffer reservoirs with a sample vial. Introduction of a specific amount of sample is achieved by controlling either the injection voltage or the injection pressure. The unprecedented resolution of CE is the result of the extremely high efficiency of the technique. Modelling of the separation efficiency of CE and other high-resolution techniques such as chromatography and field-flow fractionation is performed by the van Deemter equation, which gives the relation between the plate height, H and the velocity, v_x, of the carrier gas or liquid along the separation axis, x.

$$H = A + (B/v_x) + Cv_x.$$

In this equation, A, B and C are constants. By lowering the H value, the separation increases (Xu, 1996). CE and related capillary techniques offer several advantages that are particularly useful in the field of biomedicine. Specifically, they offer high separation power, separations with high numbers of theoretical plates, and good selectivity under certain conditions. They are also versatile and can be used for the separation of the most diverse types of compounds. However, it should be mentioned that different equipment has to be used to separate each compound. Furthermore, capillary electromigration methods require very small amounts of samples, a fact which is considered as one of the biggest advantages of the method, especially as regards its applications in clinical biochemistry (e.g., microanalyses on biopsies) (Deyl et al., 1994).

1.2.14 Differential Scanning Calorimetry

Differential scanning calorimetry (DSC) is used for monitoring heat effects related with phase transitions and chemical reactions as a function of temperature and to provide information about the physical characteristics of a compound. In DSC, the heat flow difference between a sample and a reference at the same temperature is recorded as a function of temperature. The reference is an inert material such as alumina, or even an empty aluminum pan. The temperature of both the sample and reference is constantly increased. Since the DSC is at constant pressure, heat flow equals to enthalpy changes: $(dq/dt)_p = dH/dt$. In this equation, dH/dt is the heat flow in mcal/sec. The heat flow difference between the sample and the reference is: $\Delta dH/dt = (dH/dt)$ sample $- (dH/dt)$ reference, and can be either positive or negative. The majority of phase transitions are endothermic processes in which heat is absorbed and thus the heat flow to the sample is higher than that to the reference, leading to positive $\Delta dH/dt$. Other endothermic processes include helix-coil transitions in DNA, protein denaturation, dehydrations, reduction reactions, and some decomposition reactions. In the case of exothermic processes (e.g., crystallization some cross-linking processes, oxidation reactions and some decomposition reactions), the heat flow to the reference is higher than the heat flow to the sample and the dH/dt is negative (http://particle.dk/methods-analytical-laboratory/dsc-differential-scanning-calorimetry/dsc-theory/). Molecular recognition is of paramount importance for all biological phenomena. It can be either intermolecular (e.g., ligand binding to a macromolecule) or intramolecular (e.g., protein folding). Therefore, understanding the connection between the structure of proteins and the energetics of their stability and binding with other (bio)molecules is very important for the fields of biochemistry and biotechnology. A factor commonly used for the characterization of the stability of a system (e.g., the folded and unfolded state of the protein) is the equilibrium constant (K) or the free energy (DG°), which is defined as the sum of enthalpy (DH°) and entropy (DS°). These parameters depend on the temperature through the heat capacity change (DCp). The thermodynamic parameters DH° and DCp can be obtained through spectroscopic experiments, using the van't Hoff method. Alternatively, calorimetry can be used for their direct measurement. Along with isothermal titration calorimetry (ITC), DSC is a powerful method, less described than ITC, used for the direct measurement of the thermodynamic parameters that characterizes biomolecules (Bruylants et al., 2005).

1.2.15 Electronic Noses

At present significant efforts are made towards the development of new techniques for the assessment of food quality. Quality control of product fragrance and flavour is commonly made by comparing sensory, instrumental, analytical, and if required microbiological data, with standards and specifications. Chromatography and spectroscopy as well as combined technique [such as Gas-Chromatography Mass Spectrometry (GC-MS)] and sensory analysis are commonly employed for these purposes. The "aroma" of a foodstuff is a complex mixture that consists of hundreds of different chemical volatile species, known as "chemical patterns". Separation of these chemical patterns can be carried out using analytical techniques, and identification and quantification of individual components can then be achieved. However, these techniques are complex, expensive and time consuming. On the other hand, sensory analysis can be used to evaluate a food product based on the impact of its odours/flavours on human senses. Again, it can be very expensive to maintain a trained sensory panel, and there are always limitations as to the number of replicate samples, that can be examined due to olfactory adaptation to odours (Capone et al., 2005). According to Peris and Escuder-Gilabert (2009): "The complexity of most food aromas make them difficult to be characterized with conventional flavour analysis techniques such as gas chromatography or GColfactometry. Nevertheless, sensory analysis by a panel of experts is a costly process since it requires trained people who can work for only relatively short periods of time; additional problems such as the subjectivity of human response to odours and the variability between individuals are also to be considered. Hence, the need of an instrument such as the electronic nose (EN), whose strengths include high sensitivity and correlation with data from human sensory panels for several specific applications in food control. Because they are easy to build, cost-effective and as they provide a short time of analysis, ENs are becoming more and more popular as objective automated non-destructive techniques to characterize food flavours. However, there is much research still to be done especially with regard to sensors technology, data processing, interpretation of results and validation studies."

The two basic components of an EN are (1) the sensing system and (2) the automated pattern recognition system. The role of the sensing system can be an array of a lot of different sensing elements or a single device or a combination of both. When volatile organic compounds (VOCs) come in contact with the sensor array, a signature or pattern is produced, which is characteristic of the vapour. By presenting many different chemicals to the sensor array, a database of signatures can be generated. Data analysis and pattern recognition in particular, are also essential parts of any sensor array system (Berna, 2010).

ENs use several types of electronic gas sensors that are partially specific and have a pattern recognition technique that can recognize simple and complex odours (Zohora et al., 2013). ENs have recently found many applications in food quality control and traceability processes, including microbial contamination diagnosis. It has been suggested that ENs could be used to screen microbial contamination of food through the analysis of the pattern of volatile compounds formed by microbial metabolism. The fingerprint variation can be a result of the appearance of new chemical substances (primary or secondary metabolites) or the result of changes in the relative amounts of the original volatile compounds. Food-contamination can be detected using standard microbial plate count methods but these techniques are both time-consuming and expensive. Also, improper sampling of the food product can easily lead to misleading results since the culture-based methods rely on the site of sampling. ENs can both rapidly and accurately detect food contaminant bacteria, while little or no sample preparation is required (Falasconi et al., 2012).

Several reasons exist for developing ENs for food control applications. The monitoring of the quality of foods is one of the most important reasons. Moreover, food aroma analysis offers a good opportunity for comparing the EN performances with those of natural olfaction. Foods contain a huge variety of different chemicals, many of which are responsible for the differences in taste and aroma as well as in edibility (Di Natale et al., 1997). ENs have been used for quality control processes, to monitor process and aging, determination of geographical origin, adulteration, contamination and spoilage. In most cases classification of samples obtained a good classification rate, but before the establishment of the technology can take place in an industrial scale, several challenges still need to be met, including for example the review of various characteristics of EN presentation, such as drift, humidity influence, redundancy of sensors, selectivity and signal to noise ratio. Although new sensor materials and designs, and improvement of algorithms, which can be used for each sensor, are continuously developed, the major drawback of the currently used MOS sensors is their authorization and selectivity. Sensors characterized by poor selectivity change adversely the discriminating authority of the array. It is commonly accepted that, at present, it is not possible to envisage a universal EN that will be able to be used for every odour type offering accurate results, while frequently the development of instrumentation is required for particular applications (Zohora et al., 2013).

1.2.16 Chemometric Methods in Food Authentication

Chemometrics can be defined as a statistical approach to the interpretation of patterns in multivariate data. Its use in the analysis of instrument data often fastens the assessment and increases the precision. For example, the composition (protein, fibre, moisture, carbohydrate) of dairy or grain products can be rapidly evaluated using NIR spectroscopy and chemometrics. Monitoring of food properties (such as taste, smell, astringency) can also be carried out on a continuous basis. In all cases, the data patterns are employed for the development of models aiming at predicting quality parameters for future data. The two basic applications of chemometrics are:

- the prediction of a property of interest (commonly conformance to a standard); and
- the classification of the sample into several categories (Riverside and Bothell, 1996).

Chemometrics focuses on the mathematical and statistical modelling of analytical data aiming at extracting relevant chemical information. It analyzes the variations of analytical data acquired in a vast range of domains. One of these domains is the verification of the authenticity of food products. Chemometrics can be successfully used for the development of decision rules with regards to the authenticity of food origin. Some of the advantages of chemometrics are that their models can be easily maintained and updated, are simple and robust and can be easily converted into a set of specifications (Vandeginste, 2013). Some examples of the use of chemometrics in combination with other techniques, for the authentication of foods are presented in Table 1.1.

Chemometrics is based on the analysis of data matrices. Measurements of different variables are taken for each sample. In contrast to some techniques that can give results based on any available data, chemometrics must take into account all data and cannot give accurate results if some values are missing. In cases when data are collected without having any specific project, the result is a "sparse" matrix, which does not take into account the values that are missing. In such cases, if the percentage of missing data is quite high, the dataset cannot be used for a multivariate analysis and, as a result, the variables and/or the objects with the lowest number of data must be removed. This will result in the loss of a large part of the experimental data. All the chemometrical software allows the import of data from ASCII files or from spreadsheets. Therefore, it is recommended that the data be organised in matrix form from the start, so that the import can be carried out in a single step. If, however, the data are spread over several files or sheets, the import procedure can be much longer. The interpretation of data with intermediate and therefore "useless" numbers is another commonly observed problem. These values are not necessary for conducting the data analysis, and should always be removed before the chemometrical elaboration, to facilitate the analysis (Leardi, 2008).

According to Van der Veer et al. (2011): "Chemical fingerprinting methods (e.g., NMR, NIR and chromatography) in combination with chemometric techniques provide a powerful tool for verifying the authenticity of food and related commodities. In this integrated approach, a multivariate classification model is "trained" to distinguish between authentic and non-authentic food samples based on chemical fingerprinting data. After training, such a model can be used to determine whether a suspect food sample is authentic or not at a certain level of confidence. These authentication methods generally have the following characteristics: (1) The test is specifically developed for a single combination of commodity/product and authenticity question. (2) The test is based on a multivariate classification model. (3) The classification model is empirically derived (i.e., based on a reference dataset). (4) The fingerprinting data used for building the classification model is often of a non-targeted or semi-targeted nature."

It is obvious that a good data collection, and therefore a complete data sheet, is essential for a successful data analysis. This is highly important especially when the data have to be elaborated by an external consultant, who can thus begin the analysis by simply and directly importing the data, without wasting time (and possibly making mistakes) with joining files or removing redundant information (Leardi, 2008).

Chemometrics is not always used to obtain new knowledge particularly in industrial applications. It is mainly used to produce data and extract information from these data. If the quality of the measurement processes and thus the quality of the data is poor, the information generated may not be correct. Quality is one of the most important pre-occupation of chemometrics. In fact, techniques that chemometricians apply for obtaining better measurement processes are also applied for obtaining better processes or better products. The measurement processes themselves are frequently

Table 1.1. Examples of chemometrics used in combination with analytical techniques in the field of food authentication.

Food Type/Target	Chemometric techniques	Method of analysis	Effectiveness	References
Gilthead sea bream (*Sparusaurata*)	Non-supervised and supervised multivariate analysis (PCA, LDA)	^{1}H NMR fingerprinting of lipids	Clear distinction between wild and farmed samples and effective classification of samples according to the geographic origin	Rezzi et al., 2007
Fishmeal produced using different fish species	PCA, DPLS, LDA	NIR spectroscopy	Very effective and rapid authentication and identification	Cozzolino et al., 2005
Detection of meat and bone meal in fishmeal	PLS discriminant analysis	Vis- and NIR spectrometry	Most effective with combined visible and NIR region, error of calibration: 0.85%, R^2: 0.94	Murray et al., 2001
Chicken, turkey, pork, beef and lamb meat	FDA, SIMCA, K-nearest neighbour analysis and PLS regression		85 and 100% correct identifications. FDA and PLS were the most accurate	McElhinney and Downey, 1999
Detection of non-milk fat in milk fat	LDA	GC	Correct classification of 94.4% of samples	Gutiérrez et al., 2009
Identification of geographical origin of lamb meat		Multi-element (H, C, N, S) stable isotope ratio analysis	78% correct classification	Camin et al., 2007
	K-means clustering technique and canonical discriminant analysis	Stable isotope ratio analysis	Very good resolution	Piasentier et al., 2003
Determination of microbial load on chicken breast fillets	PLS, LDA, SIMCA	Synchronous front-face fluorescence spectroscopy	Recoveries of 100 to 102%, $R^2 = 0.99$ and root mean squares errors between 0.1 and 0.2 log cfu/cm^2	Sahar et al., 2011
Species identification in raw meat	LDA, SIMCA	NIR	Only two classification errors on 115 validation samples	Arnalds et al., 2004

Table 1.1. contd....

Table 1.1. contd.

Food Type/Target	Chemometric techniques	Method of analysis	Effectiveness	References
Detection of chicken bone in chicken feed	PCA, orthogonal PLS	Real time ionization–mass spectrometry	Correct classification	Cajka et al., 2013
Quantification of adulterants in honey	PLS	FTIR spectroscopy with attenuated total reflectance	Standard error of prediction: 1.5–2.1 for corn syrup, 2.1–3.0 for high fructose corn syrup and 1.4–2.5 for inverted sugar	Gallardo-Velázquez et al., 2009
Discrimination between fresh and frozen–thawed fish	PCA, FDA	MIR spectroscopy	Correct classification of 75% to 100%	Karoui et al., 2007
Crabmeat authentication	PLS	UV and NIR spectroscopy	Effective for detecting adulteration	Gayo and Hale, 2007

used to assist in developing improved products or in controlling processes. Not rarely, therefore, the ultimate aim of chemometrics is the improvement, optimization or monitoring and control of the quality of a product or process (Massart et al., 1997).

1.3 Conclusions and Introduction to the Book

In earlier times the food analyst focused in detecting gross adulteration. Nowadays, foods and raw materials are produced in accordance with prescribed manufacturing formulations, and must also comply with legal or other requirements. This can be done through the standardization of the processes at each of the following stages: farm, raw material, manufacturing process and storage. The significantly increased manufacturing capacity and the complexity of modern food products has rendered essential the development of techniques, which can be used to rapidly assess and control the quality of foods. Many efforts have been made for the replacement of subjective methods of assessment of various organoleptic qualities with more accurate objective procedures. Even minor food constituents can now be detected and analyzed, a fact that demonstrates the recent advances in analytical techniques of separation, identification and measurement (Kirk and Sawyer, 1991). A vast variety of techniques have therefore been developed and improved to assist in the analysis of foods. These techniques use different technologies and can be applied for the identification, analysis and detection of different types of compounds or properties, always aiming at ensuring the authenticity of foods.

Traceability and systems related with the ability to link a finished food product with its ingredients and processing have always been an essential part of any good manufacturing practice and quality assurance scheme. The lack of such schemes would hamper the substantiation of any on-pack marketing claims, for example 'Organic'; and would render the defense of due diligence impossible. These schemes are also very important for minimizing the quantities involved in the event of a recall. There has been a substantial growth in both global sourcing and manufacturing, along with centralization of food production into specialist sites, in recent years. As a consequence, huge volumes of raw materials and ingredients are supplied from producers around the world and converted into equally huge quantities of finished products for global distribution. In these circumstances, traceability is a significant element for any manufacturer, whether international, national or regional. It is also obvious that, in order to be proactive in meeting ever-increasing consumer demands for clear labelling and transparency, traceability must be adopted by all food and food-related industries (Morrison, 2003). For GMOs specifically, for example, Directive 2001/18 requires traceability throughout the whole supply chain. Additionally, both the initial notification procedure and the written consent provisions need to comply with labelling requirements and specifically highlight that the product contains genetically modified organisms. Member States must ensure that labelling requirements are met and minimum thresholds for labeling can be established (Grossman, 2007). Universal traceability systems for all products, and companies cannot exist, due to the great complexity of the different supply chains, products, and industries. However, specialized traceability systems and services should be established and usually need to be customized to meet certain requirements and to offer some flexibility in terms of adaptation to changes in processes and procedures. The majority of current systems do not support and integrate objective and reliable acquisition and analysis of information about product origin and characteristics. Furthermore most of them are developed for individual companies, while it is now widely accepted that food traceability systems should be managed at a "supply chain level". Finally many companies are still managing traceability issues and processes manually, without the support of Information and Communications Technology (Morreale and Puccio, 2011).

The definition of food quality should take into account changes in consumer expectations, relevant legislation and new developments in food analysis. The production processes have to be monitored using modern methods that are based on physical, chemical or biochemical/immunological principles. New chromatographic, immunological and mass-spectrometric methods are now used for detecting trace amounts of residues, allergens, trans-fatty acids and other contaminants or impurities. Furthermore, analysis of new metabolic pathways in nutritional studies is of increasing interest, providing information with regards to potential positive health effects of minor food ingredients such as omega-3 fatty acids, isoflavones or conjugated linoleic acids. An increasing interest exists in the use of mass-spectrometric, biochemical and online/at-line methods for analyzing allergens, residues/contaminants, trans-fatty acid and for determining the geographic origin of foods (Müller and Steinhart, 2007). Also, since meat fraud is an illegal procedure that affects the composition of the products, and aims at increasing the profit, it needs to be controlled by legal authorities by means of robust, accurate and sensitive methodologies that can assure that fraudulent or accidental mislabelling can be identified. Common methodologies used for the assessment of meat authenticity have been based on methods such as chemometric analysis of a large set of data analysis, immunoassays, DNA analysis, etc. (Sentandreu and Sentandreu, 2011). Furthermore, according to Drivelos and Georgiou (2012): "The determination of the geographical origin of food and beverages has been a growing issue over the past decade for all countries around the world, mostly because of the concern of consumers about the authenticity of the food that they eat. An increasing number of research articles in the past five years have investigated the elemental composition and the isotope ratios as indicators to determine the origin of food and beverages."

Finally, microbial contamination detection and analysis is of paramount importance for ensuring the safety of foods. Traditional techniques are usually time consuming but are quite sensitive. Recently developed methods can be faster but less sensitive. However, no single method, to date, has been completely successful of the accurate estimation of microbial contents. The disadvantages of the different detection techniques may include long response time, low sensitivity and high cost of instrumentation. Also, lack of discrimination between viable and non-viable biomass can easily result in errors (Tothill and Magan, 2000).

This book aims at giving an insight into the various modern and traditional techniques used nowadays for food authenticity and food quality assurance. A high number of studies are described and analyzed in an effort to provide the reader with information about the current research in the field of food authentication and the advantages and disadvantages of the different techniques and methodologies.

References

Adhikary, R. 2010. Application of fluorescence spectroscopy: excited state dynamics, food-safety, and disease disgnosis. Graduate Theses and Dissertations. Iowa State University-Digital Repository @ Iowa State University. Available at: http://lib.dr.iastate.edu/cgi/viewcontent.cgi?article=2922&context=etd. Accessed on 08.03.2015.

Al-Jowder, O., E.K. Kemsley and R.H. Wilson. 1997. Mid-infrared spectroscopy and authenticity problems in selected meats: a feasibility study. Food Chemistry. 59(2): 195–221.

Arnalds, T., J. McElhinney, T. Fearn and G. Downey. 2004. A hierarchical discriminant analysis for species identification in raw meat by visible and near infrared spectroscopy. Journal of Near Infrared Spectroscopy. 12: 183–188.

Asensio, L., I. González, T. García and R. Martín. 2008. Determination of food authenticity by enzyme-linked immunosorbent assay (ELISA). Food Control. 19: 1–8.

Berna, A. 2010. Metal Oxide Sensors for Electronic Noses and Their Application to Food Analysis. Sensors. 10: 3882–3910.

Bianchi, T.S. and E.A. Canuel. 2011. Analytical chemical methods and instrumentation. pp. 49–78. *In*: Chemical Biomarkers in Aquatic Ecosystems. Princeton University Press, New Jersey.

Brooks, R. 2011. ELISA and PCR Method Allergen Testing in Food Products. Microbac Laboratory Services. Available at http://www.microbac.com/uploads/Technical%20Articles/pdf/ELISA%20and%20PCR%20Method%20Allergen%20Testing%20in%20Food%20Products.pdf.

Bruylants, G., J. Wouters and C. Michaux. 2005. Differential scanning calorimetry in life science: thermodynamics, stability, molecular recognition and application in drug design. Current Medicinal Chemistry. 12(17): 2011–2020.

Cajka, T., H. Danhelova, M. Zachariasova, K. Riddellova and J. Hajslova. 2013. Application of direct analysis in real time ionization–mass spectrometry (DART–MS) in chicken meat metabolomics aiming at the retrospective control of feed fraud. Metabolomics. 9: 545–557.

Camin, F., L. Bontempo, K. Heinrich, M. Horacek, S.D. Kelly, C. Schlicht, F. Thomas, F.J. Monahan, J. Hoogewerff and A. Rossmann. 2007. Multi-element (H,C,N,S) stable isotope characteristics of lamb meat from different European regions. Analytical and Bioanalytical Chemistry. 389: 309–320.

Capone, S., C. Distante, L. Francioso, D. Presicce, A.M. Taurino, P. Siciliano and M. Zuppa. 2005. The electronic nose applied to food analysis. Annales de la Asociación Química Argentina. 93: 1–14.

Carter, J. and V. Barwick (eds.). 2011. Good practice guide for isotope ratio mass spectrometry. Forensic Isotope Ratio Mass Spectroscopy. 1st Edition, ISBN 978-0-948926-31-0.

Cotte, J.F., H. Casabianca, S. Chardon, J. Lheritier and M.F. Grenier-Loustalot. 2003. Application of carbohydrate analysis to verify honey authenticity. Journal of Chromatography A. 1021: 145–155.

Cozzolino, D., A. Chree, J.R. Scaife and I. Murray. 2005. Usefulness of Near-Infrared Reflectance (NIR) Spectroscopy and Chemometrics to Discriminate Fishmeal Batches Made with Different Fish Species. Journal of Agricultural and Food Chemistry. 53: 4459–4463.

Deyl, Z., F. Tagliaro and I. Mikšík. 1994. Biomedical applications of capillary electrophoresis. Journal of Chromatography B. 656: 3–27.

Di Natale, C., A. Macagnano, F. Davide, A. D'Amico, R. Paolesse, T. Boschi, M. Faccio and G. Ferri. 1997. An electronic nose for food analysis. Sensors and Actuators B: Chemical. 44(1-3): 521–526.

Drivelos, S.A. and C.A. Georgiou. 2012. Multi-element and multi-isotope ratio analysis to determine the geographical origin of foods in the European Union. Trends in Analytical Chemistry. 40: 38–51.

Falasconi, M., I. Concina, E. Gobbi, V. Sberveglieri, A. Pulvirenti and G. Sberveglieri. 2012. Electronic Nose for Microbiological Quality Control of Food Products. International Journal of Electrochemistry. 2012: 1–12.

Gallardo-Velázquez, T., G. Osorio-Revilla, M. Zuñiga-de Loa and Y. Rivera-Espinoza. 2009. Application of FTIR-HATR spectroscopy and multivariate analysis to the quantification of adulterants in Mexican honeys. Food Research International. 42: 313–318.

Gayo, J. and S.A. Hale. 2007. Detection and Quantification of Species Authenticity and Adulteration in Crabmeat Using Visible and Near-Infrared Spectroscopy. Journal of Agricultural and Food Chemistry. 55: 585–592.

Gordon, M.H. and R. Macrae. 1987. Infrared spectroscopy. pp. 133–145. *In*: Instrumental Analysis in the Biological Sciences. Blackie Academic & Professional, Glasgow, UK.

Grossman, M.R. 2007. European community legislation for traceability and labeling of genetically modified crops, food and feed. pp. 32–62. *In*: P. Weirich (ed.). Labeling Genetically Modified Food. Oxford University Press, New York.

Gutiérrez, R., S. Vega, G. Díaz, J. Sánchez, M. Coronado, A. Ramírez, J. Pérez, M. González and B. Schettino. 2009. Detection of non-milk fat in milk fat by gas chromatography and linear discriminant analysis. Journal of Dairy Science. 92: 1846–1855.

Herrero, A.M. 2008. Raman spectroscopy—a promising technique for quality assessment of meat and fish: A review. Food Chemistry. 107: 1642–1651.

Hoorfar, J., R. Prugger, F. Butler and K. Jordan. 2011. Future Trends in food chain integrity. pp. 303–308. *In*: Food Chain Integrity: A Holistic Approach to Food Traceability, safety, quality and authenticity. Woodhead Publishing Limited, Cambridge, UK.

http://amrita.vlab.co.in/?sub=3&brch=186&sim=321&cnt=1. Accessed on 07 August 2014.

http://apps.who.int/phint/en/p/docf/. Accessed on 07 August 2014.

http://biotech.about.com/od/pcr/a/PCRtheory_2.htm. Accessed on 14 August 2014.

http://cdn.intechopen.com/pdfs-wm/37268.pdf. Accessed on 14 August 2014.

http://chem.ch.huji.ac.il/nmr/whatisnmr/whatisnmr.html. Accessed on 01 August 2014.

http://chemwiki.ucdavis.edu/Analytical_Chemistry/Instrumental_Analysis/Chromatography/Gas_Chromatography. Accessed on 13 July 2014.

http://chemwiki.ucdavis.edu/Analytical_Chemistry/Instrumental_Analysis/Capillary_Electrophoresis. Accessed on 14 August 2014.

http://content.piacton.com/Uploads/Princeton/Documents/Library/UpdatedLibrary/Raman_Spectroscopy_Basics.pdf. Accessed on 15 July 2014.

http://forensicscienceeducation.org/wp-content/uploads/2013/02/Theory_and_Instrumentation_Of_GC_Introduction.pdf. Accessed on 15 July 2014.

http://hiq.linde-gas.com/en/analytical_methods/Infrared_spectroscopy/infrared_spectroscopy.html. Accessed on 15 July 2014.

http://instructor.physics.lsa.umich.edu/adv-labs/Raman_Spectroscopy/Raman_spect.pdf. Accessed on 17 July 2014.

http://orgchem.colorado.edu/Spectroscopy/irtutor/IRtheory.pdf. Accessed on 18 July 2014.

http://particle.dk/methods-analytical-laboratory/dsc-differential-scanning-calorimetry/dsc-theory/. Accessed on 14 August 2014.

http://physweb.bgu.ac.il/~bogomolc/Books/An%20Introduction%20to%20Fluorescence%20Spectroscopy.pdf. Accessed on 20 July 2014.

http://ritchie.chem.ox.ac.uk/research_files/midIR.htm. Accessed on 20 July 2014.

http://teaching.shu.ac.uk/hwb/chemistry/tutorials/chrom/gaschrm.htm. Accessed on 20 July 2014.

http://teaching.shu.ac.uk/hwb/chemistry/tutorials/molspec/nmr3.htm. Accessed on 01 August 2014.

http://teaching.shu.ac.uk/hwb/chemistry/tutorials/molspec/uvvisab1.htm. Accessed on 03 August 2014.

http://www.azom.com/materials-equipment.aspx?cat=125. Accessed on 08 August 2014.

http://www.bruker.com/products/infrared-near-infrared-and-raman-spectroscopy/ft-nir.html. Accessed on 03 August 2014.

http://www.brynmawr.edu/chemistry/Chem/mnerzsto/The_Basics_Nuclear_Magnetic_Resonance%20_Spectroscopy_2.htm. Accessed on 01 August 2014.

http://www.chem.ucalgary.ca/courses/351/Carey5th/Ch13/ch13-uvvis.html. Accessed on 03 August 2014.

http://www.chemguide.co.uk/analysis/chromatography/hplc.html. Accessed on 07 August 2014.

http://www.clinchem.org/content/51/12/2415.full. Accessed on 03 August 2014.

http://www.elisa-antibody.com/ELISA-applications/food-industry. Accessed on 03 August 2014.

http://www.foodprocessing-technology.com/projects/nmr-technology/. Accessed on 06 August 2014.

http://www.foodsafetycentre.com.au/docs/Applications%20of%20Near%20Infrared%20Spectroscopic%20Analysis%20in%20the%20Food%20Industry%20and%20Research.pdf. Accessed on 03 August 2014.

http://www.geniaphotonics.com/business-markets/industrial/i-mid-ir-spectroscopy/. Accessed on 03 August 2014.

http://www.highveld.com/pcr/pcr-basics.html. Accessed on 14 August 2014.

http://www.infometrix.com. Accessed on 15 August 2014.

http://www.kdanalytical.com/instruments/technology/pcr.aspx. Accessed on 14 August 2014.

http://www.medterms.com/script/main/art.asp?articlekey=9100. Accessed on 12 August 2014.

http://www.metrohmsiam.com/processtalk/PT_04/PT04_MEP_Monograph_NIRS_81085026EN.pdf. Accessed on 12 August 2014.

http://www.ncbi.nlm.nih.gov/projects/genome/probe/doc/TechPCR.shtml. Accessed on 07 August 2014.

http://www.piercenet.com/method/overview-elisa. Accessed on 13 August 2014.

http://www.princeton.edu/~achaney/tmve/wiki100k/docs/Ultraviolet-visible_spectroscopy.html. Accessed on 13 August 2014.

http://www.spectroscopyonline.com/spectroscopy/data/articlestandard/spectroscopy/442001/836/article. pdf. Accessed on 12 August 2014.

http://www.unityscientific.com.au/page.asp?id=28. Accessed on 08 August 2014.

http://www2.chemistry.msu.edu/faculty/reusch/VirtTxtJml/Spectrpy/UV-Vis/uvspec.htm#uv1. Accessed on 08 August 2014.

http://www2.warwick.ac.uk/services/rss/business/analyticalguide/fluorescence/. Accessed on 08 August 2014.

https://static.thermoscientific.com/images/D14181~.pdf. Accessed on 08 August 2014.

Karoui, R., B. Lefur, C. Grondin, E. Thomas, C. Demeulemester, J. De Baerdemaeker and A.-S. Guillard. 2007. Mid-infrared spectroscopy as a new tool for the evaluation of fish freshness. International Journal of Food Science & Technology. 42: 57–64.

Kirk, R.S. and R. Sawyer. 1991. Introduction. Legislation, Standards and Nutrition. pp. 1–7. *In*: Pearson's Composition and Analysis of Foods. Longman Scientific & Technical, Essex, UK.

Lakowicz, J.R. 2006. Instrumentation for Fluorescence Spectroscopy. Available at http://www.google.co.uk/url?sa=t&rct=j&q=&esrc=s&source=web&cd=1&ved=0CEAQ FjAA&url=http%3A%2F%2Fwww.springer.com%2Fcda%2Fcontent%2Fdocument%2 Fcda_downloaddocument%2F9780387312781-c2.pdf%3FSGWID%3D0-0-45-351714- p134266323&ei=4QfxU5iTJ-ON7Abku4CoBA&usg=AFQjCNE-KwKqnpj_LlcHw- 9ULUgcl7keXA&bvm=bv.73231344,d.ZGU.

Leardi, R. 2008. Chemometric Methods in Food Authentication. pp. 585–616. *In*: D.-W. Sun (ed.). Modern Techniques for Food Authentication. Elsevier.

Lehotay, S.J. and J. Hajšlová. 2002. Application of gas chromatography in food analysis. Trends in Analytical Chemistry. 21: 686–697.

Li-Chan, E.C.Y. 1996. The applications of Raman Spectroscopy in food science. Trends in Food Science & Technology. 7(11): 361–379.

Lüthy, J. 1999. Detection strategies for food authenticity and genetically modified foods. Food Control. 10: 359–361.

Luykx, D.M.A.M. and S.M. Van Ruth. 2008. An overview of analytical methods for determining the geographical origin of food products. Food Chemistry. 107: 897–911.

Martin, A.J.P. and R.L.M. Synge. 1941. A new form of chromatography employing two liquid phases. Biochemical Journal. 35: 1358–1368.

Martinez, I., M. Aursand, U. Erikson, T.E. Singtad, E. Veliyulin and C. van der Zwagg. 2003. Destructive and non-destructive analytical techniques for authentication and composition analyses of foodstuffs. Trends in Food Science & Technology. 14: 489–498.

Massart, D.L., B.G.M. Vandeginste, L.M.C. Buydens, S.D.E. Jong, P.J. Lewi and J. Smeyers-Verbeke. 1997. Handbook of Chemometrics and Qualimetrics: Part A. pp. 1–887. *In*: B.G.M. Vandeginste and S.C. Rutan (eds.). Elsevier, Netherlands.

McElhinney, J. and G. Downey. 1999. Chemometric processing of visible and near infrared reflectance spectra for species identification in selected raw homogenised meats. Journal of Near Infrared Spectroscopy. 7(3): 145–154.

Morreale, V. and M. Puccio. 2011. The role of service orientation in future web-based food traceability systems. pp. 3–22. *In*: J. Hoorfar, K. Jordan, F. Butler and R. Prugger (eds.). Food Chain Integrity. A holistic approach to food traceability, safety, quality and authenticity. Woodhead Publishing, USA.

Morrison, C. 2003. Traceability in food processing: an introduction. pp. 459–472. *In*: M. Lees (ed.). Food Authenticity and Traceability. CRC Press, USA, Woodhead Publishing Ltd, UK.

Muccio, Z. and G.P. Jackson. 2009. Isotope ratio mass spectrometry. Analyst. 134: 213–222.

Müller, A. and H. Steinhart. 2007. Recent developments in instrumental analysis for food quality. Food Chemistry. 102: 436–444.

Murray, I., L.S. Aucott and I.H. Pike. 2001. Use of discriminant analysis on visible and near infrared reflectance spectra to detect adulteration of fishmeal with meat and bone meal. Journal of Near Infrared Spectroscopy. 9: 297–311.

Nakyinsige, K., Y.B.C. Man and A.Q. Sazili. 2012. Halal authenticity issues in meat and meat products. Meat Science. 91: 207–214.

Nilsson, R. 2014. Applications of Near Infrared Spectroscopic Analysis in the Food Industry and Research. pp. 1–12. Food Safety Centre, Tasmanian Institute of Agricultural Research, University of Tasmania, Australia. Available at: http://www.foodsafetycentre.com.au/docs/Applications%20of%20Near%20 Infrared%20Spectroscopic%20Analysis%20in%20the%20Food%20Industry%20and%20Research.pdf.

Ozaki, Y., R. Cho, K. Ikegaya, S. Muraishi and K. Kawauchi. 1992. Potential of near-infrared Fourier transform Raman spectroscopy in Food Analysis. Applied Spectroscopy. 46(10): 1503–1507.

Peris, M. and L. Escuder-Gilabert. 2009. A 21st century technique for food control: Electronic noses. Analytical Chimica Acta. 638: 1–15.

Piasentier, E., R. Valusso, F. Camin and G. Versini. 2003. Stable isotope ratio analysis for authentication of lamb meat. Meat Science. 64: 239–247.

Pojić, M., J. Mastilović and N. Majcen. 2012. The Application of Near Infrared Spectroscopy in Wheat Quality Control. pp. 167–184. Available at http://cdn.intechopen.com/pdfs-wm/36048.pdf.

Primrose, S., M. Woolfe and S. Rollinson. 2010. Food forensics: methods for determining the authenticity of foodstuffs. Trends in Food Science & Technology. 21: 582–590.

Puckett, R.P. 2004. Food Service Manual for Health Case Institutions. Jossey-Bass. San Francisco. pp. 1–785.

Reilly, C. 2002. Metal Contamination of Food. Its significance for Food Quality and Human Health. Blackwell, UK. pp. 1–280.

Rezzi, S., I. Giani, K. Héberger, D.E. Axelson, V.M. Moretti, F. Reniero and C. Guillou. 2007. Classification of gilthead sea bream (*Sparusaurata*) from ^{1}H NMR lipid profiling combined with principal component and linear discriminant analysis. Journal of Agricultural and Food Chemistry. 55: 9963–9968.

Riverside, E. and W.A. Bothell. 1996. Chemometrics in Food and Beverage. Chemometrics Applications Overview. Infometrix, Inc. Available at: http://www.chemometrix.net/apps/15-1096_FoodBevAO.pdf.

Sahar, A., T. Boubellouta and É. Dufour. 2011. Synchronous front-face fluorescence spectroscopy as a promising tool for the rapid determination of spoilage bacteria on chicken breast fillet. Food Research International. 44: 471–480.

Scheer, F.P. 2006. Optimising supply chains using traceability systems. pp. 52–64. *In*: I. Smith and A. Furness (eds.). Improving traceability in food processing and distribution. Woodhead Publishing Ltd, Cambridge, UK.

Schmidt, O., J.M. Quilter, B. Bahar, A.P. Moloney, C.M. Scrimgeour, I.S. Begley and F.J. Monahan. 2005. Inferring the origin and dietary history of beef from C, N and S stable isotope ratio analysis. Food Chemistry. 91: 545–549.

Schubbert, R., W. Hell, T. Brendel, S. Rittler, S. Schneider and K. Klöpper. 2008. Food forensics: Analysis of food, raw and processed materials with molecular biological methods. Forensic Science International: Genetics Supplement Series. 1: 616–619.

Scotter, C.N.G. 1997. Non-destructive spectroscopic techniques for the measurement of food quality. Trends in Food Science & Technology. 8: 285–292.

Sentandreu, M.A. and E. Sentandreu. 2011. Peptide biomarkers as a way to determine meat authenticity. Meat Science. 89: 280–285.

Ševčík, J. 1976. Detectors in gas chromatography. Journal of Chromatography Library. 4: 1–192.

Stuart, B. and E. Prichard. 2003. Practical Laboratory Skills Training Guides. Gas Chromatography. LGC (Teddiigton) Limited, Cambridge. pp. 1–22.

Sulzman, E.W. 2007. Stable isotope chemistry and measurement: a primer. pp. 1–21. *In*: R. Michener and K. Lajtha (eds.). Stable Isotopes in Ecology and Environmental Science. Blackwell Publishing.

Tothill, I.E. and N. Magan. 2000. Rapid detection methods for microbial contamination. pp. 136–160. *In*: I.E. Tothill (ed.). Rapid and On-line Instrumentation for Food Quality Assurance. Woodhead Publishing Ltd, England.

Tudor, A.M., C.D. Melia, J.S. Binns, P.J. Hendra, S. Church and M.C. Davies. 1990. The application of Fourier-transform Raman spectroscopy to the analysis of pharmaceuticals and biomaterials. Journal of Pharmaceutical and Biomedical Analysis. 8(8-12): 717–720.

Vandeginste, B. 2013. Chemometrics in studies of food origin. pp. 117–145. *In*: New Analytical Approaches for Verifying the Origin of Food. Woodhead Publishing Limited, Cambridge, UK.

Van der Veer, G., S.M. Van Ruth and W. Akkermans. 2011. Guidelines for validation of chemometric models for food authentication. RIKILT—Institute of Food Safety, Netherlands. pp. 1–26.

Vandegineste, B. 2013. Chemometrics in studies of food origin. pp. 117–145. *In*: P. Brereton (ed.). New Analytical Approaches for Verifying the Origin of Food. Woodhead Publishing, Cambridge.

Woolfe, M. and S. Primrose. 2004. Food forensics: using DNA technology to combat misdescription and fraud. Trends in Biotechnology. 22: 222–226.

Xu, Y. 1996. Tutorial: Capillary Electrophoresis. The Chemical Educator. 1: 1–14.

Zohora, S.E., A.M. Khan, A.K. Srivastava and N. Hundewale. 2013. Electronic Noses Application to Food Analysis Using Metal Oxide Sensors: A Review. International Journal of Soft Computing and Engineering (IJSCE). 3: 199–205.

2

Traceability of Foods

Ioannis S. Arvanitoyannis* and *Konstantinos V. Kotsanopoulos*

2.1 Introduction

Traceability in the food industry is described as the ability to track food products through all stages of production, processing and distribution (including importation and at retail). Traceability should allow all movements to be traced one step backwards and one step forward at any point of the food supply chain (http://www.foodstandards.gov. au/industry/safetystandards/traceability/pages/default.aspx). Therefore, the concept of traceability includes information on the history of a product with regards to the direct properties of that product and/or properties that are associated with that product once it has been processed/treated by any value-adding processes using associated production means and under associated environmental conditions. The information regarding relationships at origin may be used upstream in the supply chain (e.g., for defining the requirements of an ordered product), or downstream (e.g., at the delivery stage for determining the specification of a food material). Moreover, the information may be used for reporting purposes, both in the supply chain and for any third parties (Regattieri et al., 2007). Traceability and procedures related to the ability to trace a finished food product back to its ingredients and processing steps are essential parts of good manufacturing practices and quality assurance schemes/systems. The absence of traceability can hinder the substantiation of any on-pack marketing claims, such as 'Organic', or the defense of the due diligence in the case of a Public Enforcement challenge. Furthermore, lack of a traceability system would make the minimization of the quantities involved in recalls impossible. The modern food industry is characterized by both global sourcing and manufacturing, as well as by centralization of food production into specialist sites. As a result, huge amounts of raw materials and ingredients are sourced from suppliers around the world and converted into equally large volumes of finished product that can again be distributed around the world.

School of Agricultural Sciences, Department of Agriculture, Ichthyology and Aquatic Environment, University of Thessaly, Fytoko St., 38446 Nea Ionia Magnesias, Volos, Hellas, Greece.
* Corresponding author

Under these conditions, traceability is of paramount importance to the manufacturer. It is also obvious that increasing demands of consumers towards the establishment of clear labelling and transparency can only be achieved if robust traceability systems are in place (Morrison, 2003). According to Martinez and Epelbaum (2011): "Although consumers show little understanding of the mechanism underpinning food traceability, they appreciate its potential to improve food safety as well as to promote consumer protection. It is important to develop effective communication strategies to inform consumers about the benefits traceability systems can provide in addressing their concerns if consumer trust in food safety monitoring systems is to be developed and maintained".

Various identification techniques exist for individual animals, batches of similar sources and types of products. Each species is characterized by unique harvesting, processing, storage and distribution characteristics that highly affect any identification methods and subsequent systems employed for obtaining information about the supply chain. Record keeping and data management are essential characteristics of an effective traceability system. Radio frequency identification (RFID) is commonly employed for pallet, case and individual package identification and tracing, while bar coding and quick response technologies are extensively used for some production and marketing schemes for inventory control and the conveying of information to both processors and consumers. Systems employing multiple flexible reading and recording technologies as well as centralized Internet data access facilitate the detection of any linkages between different components in the meat, poultry and seafood industries (McMillin, 2012).

Traceability systems are developed to strengthen processes through which information on food origins, attributes and processing technologies is gathered and made available to the relevant parties of a supply chain. In general, these systems are based on data recording and management systems developed for tracking information forwards and backwards, respectively, following not only product/material flows, but also specific product characteristics. Traceability is one of the most significant elements used to demonstrate the transparency of the supply chain using verifiable data and labelling. The traceability systems aim at preventing the introduction of an unwanted agent, such as a pathogenic microorganism, to the food chain. Should this occur, though, it can minimize the magnitude and impact of such an event by facilitating the identification of the product(s) and/or batches affected and by unravelling the circumstances under which the incident occurred, determining the place and time of the occurrence of the lapse in food integrity and indicating who have the relevant responsibilities. Traceability operations can be internal systems established by individual companies (in-house traceability), or systems allowing information flow through the whole supply chain (chain traceability). The speed of response and the reliability of information are the most significant elements for ensuring that a robust traceability system is in place (Morreale et al., 2011).

The need for the development of traceability systems has arisen primarily as a result of consumer and government concerns over food safety, hygiene and authenticity. These concerns range from health risks related to animal-borne pathogens, such as salmonella, listeria, clostridium and *Escherichia coli* O157 through agents causing bovine spongiform encephalitis (BSE) to topics associated with Genetically Modified

Organisms (GMOs). As a result of national and international needs and industry responses for traceability, triggered by developments in global trade and consumer demands, the development of numerous traceability guidelines has been reported. An increasing number of food-related companies adopt and implement traceability systems. Covering different parts of the supply chain and emphasizing on particular foodstuffs such as fish, meat and wine, these systems can satisfy some traceability needs for these products but cannot be easily adjusted to encompass different types of products. Since more and more supply chains are working towards the development of sector-specific traceability systems, different problems will inevitably appear where the connections between different supply chain boundaries are essential for tracing multi-ingredient food products (Furness and Osman, 2006).

Specifically, food scandals such as the BSE scandal in the UK have made quite clear to food business that a lapse in the integrity of traceability can detrimentally affect the relevant parties, as the companies may not be able to determine which batches have been affected and which batches can be safely used. This will inevitably result in the recall of all products from the market (Rasmussen, 2012). Presently in Europe and in the International FAO/WHO Codex Alimentarius Commission, new regulations have been established on "traceability" and labelling. The need for these regulations was triggered by issues arisen as a result of the weak European control over foods, animal feed, and animal diseases. Furthermore, in many European and other countries there is a trend to restrict the production or use of foods or feed materials derived from the use of biotechnology. When such products are used, they need to be labelled correspondingly. Moreover, due to the higher production costs, local products that may not be otherwise competitive in open markets, should be accompanied by labelling to include a statement that indicates what the country of origin is or even the specific production area (Lupien, 2005). According to 2003/89/EC directive as amended by 2000/13/EC, to guarantee a high level of health protection for consumers and cover their right to have access to information, it is essential that sufficient information is provided to consumers in relation to foodstuffs, inter alia, through the listing of all ingredients on labels. In the text of regulation (EC) No. 1830/2003 concerning the traceability and labelling of GMOs and the traceability of food and feed products produced using biotechnology and amending Directive 2001/18/EC, it is stated that differences between national laws, regulations and administrative provisions in relation to traceability and labelling of GMOs as products or in products as well as traceability of food and feed produced from GMOs do not facilitate their free movement, thus leading to phenomena of unequal and unfair competition.

A harmonized community framework for traceability and labelling of GMOs should allow the internal market to function effectively. Traceability requirements for GMOs should facilitate the withdrawal of products that can potentially be harmful for the human and animal health or the environment, including the ecosystems, and should target at monitoring any potential environmental effects in particular. Traceability should also contribute to the facilitation of the establishment of risk management measures in accordance with the precautionary principle. Traceability requirements for food and feed products derived from GMOs are of paramount importance for facilitating the accurate labelling of such products, in order to ensure that correct information is provided to operators and consumers thus enabling them to effectively exercise their freedom of choice and enable control and verification of labelling claims.

According to Bosona and Gebresenbet (2013): "The major driving forces behind the development and implementation of food traceability systems are food safety and quality, regulatory, social, economic, and technological concerns. Similarly, the major benefits of food traceability systems can be categorized as: increase in customer satisfaction, improvement in food crises management, improvement in food supply chain management, enhanced company competence, enriched technological and scientific contribution and contribution to agricultural sustainability. The identified barriers to implementation of food traceability systems also have been presented as limitations in: resource, information, standard, capacity, and awareness".

The effectiveness of the traceability system chosen depends on the structure of the food supply chain under consideration; connections between the members of the chain; capacity (human or technological) for the management of transactions, quality and production processes; and packaging materials and methods. When long food supply chains are considered, implementation of effective traceability system requires integration of different parts of the supply chain. The implementation of traceability systems that fully cover the whole length of the supply chain leads to better results than focusing on partial improvement of traceability (Bosona and Gebresenbet, 2013). As regards automated traceability systems, traceability relations support stakeholders in understanding the dependencies between artifacts developed during manufacturing software systems, thus facilitating many development-related tasks. To ensure that the anticipated benefits of these tasks can be realized, it is necessary to have an up-to-date set of traceability relations between the established artifacts. Therefore, the creation of traceability relations is required during the initial development process. Moreover, the continuous maintenance of traceability relations is also required as the software system evolves for preventing their decay (Mäder and Gotel, 2012).

In recent years automatic identification procedures (Auto-ID) have gained popularity in many service industries, purchasing and distribution logistics, food manufacturing industries and material flow systems. Automatic identification procedures have been developed to support the information availability with regards to people, animals, goods and products in transit. The omnipresent barcode labels was the first step towards a revolution in identification systems a few years ago, but are no more adequate in an increasing number of cases. Barcodes can be characterized by low cost, but their low storage capacity and the fact that they cannot be reprogrammed are their main disadvantages (Finkenzeller, 2010). RFID technology has various potential applications within a high number of industrial sectors. One of the most significant applications is their use for complete traceability of a specific product, offering at the same time the additional advantage of being able to verify that quality controls have been passed (Azuara et al., 2010). The results of several studies examining the application of this technology on foods of animal origin are detailed in this chapter.

Stakeholders in the beef industry, including government agencies, retailers and consumers, require information with regards to the whole-chain traceability from farm to plate. There are numerous benefits from clearly defining the objectives of an effective traceability system including protection against animal disease issues, food safety and premiums for marketing of specialty products. Despite the fact that technology has been developed to trace ground beef back to a specific animal that provided the trim, it is currently applicable only in very small-sized operations where carcass by carcass processing is carried out. It is highly possible that the identification of the specific animals used for the production of the commercial packaged ground beef, would not be feasible, apart from lot of animals being slaughtered or lots of trim being ground to adjust fat content of a particular order during a recorded time-of-day (Crandall et al., 2013).

DNA barcoding can also be employed as a universal tool for food traceability. Although it has only found applications during the last few years, it has already become commonly used. A number of factors have attributed to that, including: (i) the dropping cost of molecular analyses; (ii) the fact that the availability of equipped laboratories and skilled personnel has increased; (iii) the existence of freely available web-based resources; and (iv) the increasing number of well-informed consumers who demand food of very high quality. Therefore, a requirement has been generated for the development of a technique built around molecularization, standardization and computerization. Considering that, DNA barcoding is the result of this requirement of the twenty-first century. Numerous case studies and technical advancements clearly indicate that DNA barcoding can sensitively, rapidly and reliably identify and track a wide range of raw materials and final foods and detect allergens or poisonous substances potentially occurring in food, while eliminates the need for expensive tools. Due to its universality, DNA barcoding can find applications in different contexts, and be used by different operators. International agencies or institutions, having the responsibility of quality control of raw materials or food products, can cooperate and exchange their data, thus facilitating the development of population reference databases, the absence of which is the only major limitation of the method. In reality, although some groups of organisms, such as fish, are efficiently represented, a lot of work is still needed to generate a reliable source of reference DNA barcoding data for sectors that have not been extensively investigated. As a result, in the near future DNA barcoding could become part of the routine testing in many fields, and specifically in food quality control and traceability (Galimberti et al., 2013).

The establishment of recall systems can be carried out even if a minimum of traceability information (e.g., production date) is available, but the more sub-descriptors that are available (e.g., production time, batch number, production conditions) the more specific the product recall can be, thus minimizing any financial costs and reputation damages. The particular aims (or limitations) of one or more of the steps in a food processing chain establish the demands (or limits) for the extent of information included in the chain traceability system (Moe, 1998).

2.2 Food Labelling

2.2.1 Identification Systems (EAN, SSCC, GTN, and GLN)

EAN (European Article Numbering) International and the Uniform Code Council Inc. (UCC), its partner in the USA and Canada, are moving towards the development of a common traceability system for the global market through leadership, innovation, technology support and the development of multi-industry standards for product identification and related electronic communications. Through the use of robust systems and logistic identification tools, standard bar codes and electronic commerce activities, EAN and UCC provide all trading partners with the tools required for the effective implementation of a global supply chain management (EAN International, 2002). The EAN.UCC System originated in the United States and was established in 1973 by the Uniform Product Code Council, now commonly known as the Uniform Code Council, Inc.® (UCC™). The UCC system is based on a 12-digit identification number, and the first ID numbers and bar code symbols in open trade were being scanned in 1974. The successful implementation of this UPC System led to the establishment of the EAN Association (currently known as EAN International) in 1977 that would aim at developing a compatible system for use outside North America. The EAN System was developed as a superset of the UCC System and uses principally 13-digit numbers. As a result of using certain bar code symbols and data structures, the EAN.UCC System has expanded. Nowadays, full global compatibility has been successfully established using the GTIN (Global Trade Item Number) Format, a 14-digit reference field in computer files that can be used for the storage of data structures ensuring that any trade item identification number will be unique in the global market.

The EAN.UCC System is based on the use of unambiguous numbers for the identification of goods, services, assets, and locations worldwide. The representation of these numbers is made using bar code symbols thus enabling their electronic reading wherever demanded in business processes. The system is developed to overcome the limitations faced when company, organisation, or sector specific coding systems are used, and to make trading much more efficient and responsive to customers (EAN International and Uniform Code Council, Inc., 2005). The fundamental principle of the EAN.UCC System is an unambiguous numbering schema employed for the identification of goods or services throughout any supply chain. The use of automatic data capture techniques allows the successful application of this numbering system at every stage of production and distribution. Although the immediate and most obvious application of the EAN.UCC System is a bar code (an UPC or EAN-13 symbol), it is of high importance to mention that a bar code is simply a machine-readable representation of its associated number. The use of the EAN.UCC System guarantees that the associated number used for the identification of the product, to which it is assigned, will be unique. In general, each traded product, such as a meat package intended for retail sale, or an aggregate of tradable products (for example, a crate of several meat packages transferred from the storage warehouse to the retail outlet) is labelled with a global unique EAN.UCC number, known as the Global Trade Item Number (GTIN). The GTIN does not provide any information about the product. It is just a world wide unique and unambiguous identification number. The GTIN can be encoded into a bar code (EAN International, 2002).

The development of the EAN.UCC System has been carried out following an "open architecture" approach. It has been carefully designed for modular expansion while minimally disrupting the existing applications and focuses on: (1) Open Standards philosophy—targeting at the development of a single, open, business led, integrated system of identification and information flow, thus enabling an effective supply chain management in any company, in any type of industry, all over the world. (2) Differentiation—the system has been developed based on standards that, when followed, ensure globally unique and discrete identification of products, handling units, assets, and locations. The system incorporates the tools for transferring EAN.UCC System identification numbers as well as any relevant data related to these numbers. (3) Transparency—EAN.UCC System identification numbers must be relevant and applicable to any supply chain, regardless of who assigns, receives, and processes the standards. This should indicate a unique way of performing any given function. The introduction of new features to the standard should only be performed if they are to follow the establishment of new applications or improve ways of performing existing functions. (4) Non-Significance—Global uniqueness of EAN.UCC System identification is secured only when the standard number is taken as a whole and processed in its entirety. Fixed information on characteristics of a product/material or service should be available through a computer or other data source by entering the item or service's EAN.UCC System identification number and thus using it as a reference. Multilateral instead of time-consuming bilateral agreements and regulations are essential for the members of any supply chain, to rapidly, efficiently and productively interact in an open environment (TRACE-I Guideline, 2003).

UCC stands for Uniform Code Council, Inc., an organisation that until 1972 was named the Uniform Grocery Product Code Council. When established, UCC was aiming at taking a global leadership role in the establishment and promotion of multi-industry standards for product identification, including the Universal Product Code (U.P.C.), and related electronic communications. The goal is to strengthen the efficiency of the supply chain management, contributing added value to the customer (http://www.upccode.net/upc-guide/uniform-code-council.html).

GS1 stands for Global Standards One and as the name suggests, the organisation functions mainly as the standards organisation managing the assignment of different numbering schemes upon which global commerce has come to rely. Furthermore, though interacting with different levels of success, the organisation aims at promoting a uniform and consistent way of coding distributable goods. GS1 was formed in 2005 by the merger of the UCC and EAN International. On the 7th of June 2005 the UCC became the official GS1 member organisation for the United States of America and was named GS1 US (http://www.upccode.net/upc-guide/gs1.html). GS1, thereby, was successfully established as the single worldwide origination point for UPC and EAN numbers (now known as GTIN-12 and GTIN-13, respectively) (http://www.upccode.net/upc-guide/uniform-code-council.html). GS1 US is one of numerous member organisations (MO) functioning under the GS1 umbrella. The majority of countries nowadays, possess their own GS1MOs. The country-specific MOs can present significant differences as regards pricing, policies, and procedures. GS1 is now the single organisation in the world with the authority to provide UPC and EAN numbers (now known as GTIN-12 and GTIN-13, respectively) (http://www.upccode.net/upc-guide/gs1.html).

According to Bennet (2010): "For more than 30 years GS1 has been a leading global organisation dedicated to the design and implementation of standards and solutions to improve the efficiency and visibility of global and sector supply and demand chains by offering a diverse range of products, services, and solutions. GS1 is a neutral, not-for-profit standards (and related services) organisation. GS1 operates in more than 20 industry sectors ranging from retail, food, and fast-moving consumer goods to health care, logistics, and military defense. Formed from the joining together of EAN International and the UCC, GS1 is truly global, with a presence in more than 150 countries driven by more than one million companies that execute more than five billion transactions each day using GS1 standards, solutions, and services. The GS1 Traceability Standard was developed by the GS1 Global Standards Management Process Team. This group was composed of 73 experts from 18 countries. The global GS1 Traceability Standard has been developed to meet important business needs, including regulatory compliance. It addresses the entire supply chain and can be applied to any product. The GS1 Traceability Standard is based on current business practices used by a large majority of supply chain Partners".

The use of barcodes by companies for identification, capture, and sharing information within the supply chain can be carried out only after the unique numbers, used for formatting the barcode, have been created. These numbers are used to link information about a material with the flow of business transactions to the physical flow of those products. The first step in establishing an identification number is to obtain a Company Prefix from GS1 US. The Company Prefix is a constant in every GS1 identification number structure, and is the basis of unique identification in the supply chain. Company Prefixes vary in length depending on the quantity of identification numbers that the company may need for its system. After receiving a Company Prefix, a company can assign identification numbers to their trade items, locations, and logistics units. Each type of identification number is characterized by a specific format, which is a combination of the company's prefix and reference numbers that have been assigned to develop globally unique identification numbers. There are various methods of assigning identification numbers, which may be based on numerous factors such as the organisational strategy, current business practices, and customer requirements (http://www.gs1us.org/DesktopModules/Bring2mind/DMX/Download. aspx?Command=Core_Download&EntryId=224&PortalId=0&TabId=785).

The GS1 Traceability System mainly focuses on factors affecting the materials, covering whole supply chains consisting of mainly distinct partners. It is relied on the assumption that traceability is a tool that should be used in various predetermined objectives and can be considered as one of numerous elements developed for improving security, control quality, combat fraud, and for assisting in the management of complex logistical chains. The existence of pre-established tracking and tracing capabilities can significantly facilitate the establishment of a traceability system. Unique identification is achieved using the GS1 globally unique identifiers functioning as keys that allow access to all available information with regards to the product's history, application, or location. Unique identification of locations is ensured through allocating a GS1 Global Location Number (GLN) to each location and functional entity, while unique product identifications are ensured through allocating a GS1GTIN to each individual product (consumer unit). Considering traceability, the GTIN has to be used in combination with a serial number or batch number, which is used for the identification of the particular item. Traceability of series is ensured through allocating a GS1GTIN and serial number to each product, while traceability of lots/batches is ensured through allocating a GS1GTIN and lot/batch number to each product. Furthermore, a GTIN needs to be established for each of the three levels of a trade unit: consumer unit, traded unit, and pallet. Finally, the pallets can be effectively identified and traced through the allocation of a GS1 Serial Shipping Container Code (SSCC) (Bennet, 2010).

A barcode is the data carrier or symbology employed to allow the automatic electronic capture of data about a product. The GTIN is the primary data element encoded within barcode symbologies for food and food-related industries. There are several symbologies or data carriers employed nowadays to enable the bar coding of food products and packaging materials. The level of information encoded into the barcodes can vary significantly and depends on the barcode symbology used. All barcode formats apart from the SSCC contain a GTIN. Barcodes currently used for food products include the SSCC, GSI-128, ITF-14, UPC-A, UPC-E, EAN-8, EAN-13, UPC-Type 2 and GS1DataBar Expanded Stacked for meat (http://www.gs1us.org/DesktopModules/Bring2mind/DMX/Download.aspx?Command=Core_Download&EntryId=224&PortalId=0&TabId=785). Some representative examples of barcodes used for the traceability of food products are shown in Figs. 2.1 and 2.2.

A GTIN/UPC/EAN is the number allocated to a product sold in a retail store, while the bar code is the graphical interpretation of a GTIN/UPC/EAN number into bars and spaces of varying widths. A bar code is developed to allow its scanning to be performed by an optical reader that translates the bar and space widths into the number represented by the bars since reading straight lines is easier than reading numerical digits. Although a GTIN/UPC/EAN can be used without a bar code, no barcode can be used if it is not based on a GTIN/UPC/EAN. A bar code is needed to render the GTIN/UPC/EAN scannable by bar code readers in retail stores. However, no scannable bar code is needed in the case of online sales—and a GTIN/UPC/EAN is sufficient.

Figure 2.1. Barcode used for traceability purposes of final products.

Figure 2.2. Barcode used for traceability purposes of products in retail market.

The bar codes generated by a GTIN-13 (EAN) are identical to their GTIN-12 (UPC) counterpart. They only differ on the numbers along the bottom of the bars and spaces. No differences exist in the actual bar code of a UPC and an EAN and bar code scanners cannot recognize differences between a GTIN-13/EAN and a GTIN-12/UPC (http:// mybarcodegraphics.com/index.php?main_page=page&id=7&chapter=1).

Currently, GTIN is used exclusively within bar codes, but it can also be applied in other data carriers such as RFID. The GTIN is only a term and does not affect any pre-existing standards, nor does it place any additional requirements on scanning hardware. As regards the North American companies, the UPC is an existing type of the GTIN. The groups of data structures (not symbologies) that comprise GTIN include: GTIN-12 (UPC-A)—a 12-digit number used mainly in North America, GTIN-8 (EAN/ UCC-8)—an 8-digit number used principally outside North America, GTIN-13 (EAN/ UCC-13)—a 13-digit number used mainly outside North America and GTIN-14 (EAN/ UCC-14 or ITF-14)—a 14-digit number used for the identification of trade items at all packaging levels (http://www.gtin.info/). Parameters, such as the position where the symbol will be scanned, the type of data encoded in the symbol, and the way the symbol will be printed will determine the selection of the correct barcode for a given business application. Attention is also required when considering the scanning environment of the company, when deciding on the barcode selection process (http://www.gs1us.org/ DesktopModules/Bring2mind/DMX/Download.aspx?Command=Core_Download& EntryId=224&PortalId=0&TabId=785).

The majority of products in the market place are bar-coded only with the GTIN. Nevertheless, there are cases where additional data needs to be bar-coded. The use of multiple information segments in one barcode does not facilitate the determination of the point where one information segment ends and the next segment begins. For example, the expression of the batch 43 of the product 90614141000869 can potentially be either 9061414100086943 or 4390614141000869. Thus, the scanner could not realize which string of numbers should be used for identifying the product and/or the batch. The EAN.UCC system uses a bar code that can carry special prefixes for the identification and separation of multiple data elements. These two-, three-, or four-digit numbers, named Application Identifiers (AI) can only appear within the EAN.UCC symbols. When a scanner detects this special type of bar code, it automatically searches for AIs aiming at separating and interpreting ID numbers properly. The information that follows the AIs can contain numeric (n) or alphanumeric (an) data characters. More than 100 Application Identifiers can be used by the EAN.UCC system. The heart of the AI system is the AI Table and the related AI definitions are included in the General EAN.UCC Specifications. These application identifiers are recognized at a global scale. By definition an AI should be interpreted in the same way in all countries (http://www.google.gr/url?sa=t&rct=j&q=&esrc=s&source=web&cd=1&ved=0CCs QFjAA&url=http%3A%2F%2Fwww.fda.gov%2Fohrms%2Fdockets%2Fdailys%2F 02%2FAug02%2F082902%2F80025b72.doc&ei=jHxuU4zbAuHL0AW374GgAw& usg=AFQjCNGBqmv76poAo7uwum4UNGJ4eYOAA&bvm=bv.66330100,d.bGQ).

When products are traded between several organisations, information about their movements remain hidden in separate databases hindering the determination of their history and authenticity, or track the final sale. If the information about the product movements is not available, companies do not possess the tools to effectively implement cost saving measures, streamline operations, and avoid risks. Even when product and logistics data is communicated between companies, it is essential to be done carefully and securely. Each company needs to secure the confidentiality of their proprietary data and control the access to this information. TraceTracker's Global Traceability Network (GTNet) is a cloud-based discovery service that allows the establishment of a higher level of supply chain visibility by providing a look up mechanism. It connects disparate Electronic Product Code Information Services (EPCIS) repositories in a secure network, enabling supply chain visibility, product traceability, asset tracking, and business intelligence applications. TraceTracker's GTNet allows the effective establishment of global traceability, based on a set of data sharing standards called EPCIS. EPCIS constitutes an easy and secure way for companies to access data regardless of the type of database employed for data storage. GTNet functions similarly to a web search engine. Following a user query about a unique ID, such as an electronic product code (EPC), the GTNet searches through all of the available databases that can be accessed by the user and provides a list of related items. The user can then shortlist the information to identify the data needed about the item, such as the history of the item's movements, temperature logs, documentation or other properties (http://www.tracctracker.com/LiteratureRetrieve.aspx?ID=4353 9&A=SearchResult&SearchID=1223254&ObjectID=43539&ObjectType=6). The following GS1 standards are either currently used by industries or being supported by GTNet and the GTNet applications: (a) The low level RFID tag standards (UHF, HF), (b) The EPC Tag Data Standard (TDS 1.7), (c) The EPC Tag Data Translation Standard (TDT 1.6), (d) The Object Name Service (ONS 2.0.1), (e) EPCIS 1.0.1 (The EPCIS Capture Interface and the EPCIS Query and Control Interface), (f) The Core Business Vocabulary specification (CBV 1.1 Working Draft), (g) The GS1 Traceability Standard (GTS 1.2.2), and (h) The GS1 Global Data Dictionary (GDD).

GTNet is probably also one of the most representative examples of the efficiency and sophistication of GS1EPC global Architecture Framework—where a "loosely connected" set of independent EPCIS services can provide information to EPCIS (http://www.gs1.no/sfiles/5/45/50/6/file/traceability-and-supply-chain-visualisation-with-tracetracker-gtnet-and-gs1-standards.pdf). The GTNet is a service for internal and external electronic traceability. The GTNet is a subscription-based service that allows a sophisticated sharing of data with partners. GTNet has numerous user-friendly mechanisms to connect to the network, enabling external applications access. The framework is quite generic, thus enabling different solutions for different businesses (http://tracker.oukej.cz/wiki/images/Njord_report.pdf). According to Olson and Criddle (2005), traceability solutions for the salmon industry can be provided by different third-party providers. Maritech, Akvsmart, TraceTracker, Intentia, C-Trace, etc. can provide various IT, ERP, and traceability solutions covering fish from either vessel or farm through the chain. Being touted as the first automated global traceability chain, a new tracing project, TELOP (Technology Development for Profitable Fish Farming), should provide accurate and sufficient information for traceability in the salmon supply chain. TELOP employs the GTN (Global Traceability Network) provided by TraceTracker. This decentralized, platform-independent network links internal traceability systems to an on-line hub from where accredited users can obtain information. GTN subscribers have full control over their own data with security certificates and access-controlled systems. The use of automatic information exchange will significantly limit the costs of demonstrating traceability and related certifications. By using TELOP, labelling and information exchange can be easily carried out by retailers to fully cover the requirements of regulation or consumers.

A product traceability system should be able to identify all the physical entities (and locations) from which the product originates, including the location where it is processed, packaged, and stocked, thus including the involvement of every agent in the supply chain. Nowadays technical and operative resources are primarily available in the form of alphanumerical code, bar code, and RFID. Alphanumerical codes are defined as a sequence of numbers and letters of different sizes located on labels, which in turn are attached to the products or packaging. Clearly, when food products are considered, the labels are usually attached on the packages. The design phase of this system can be very easily and cost effectively carried out, but its management requires significant human resources (and thus costs) since code writing and code reading cannot be automatically performed. Moreover, performance is not particularly good—there are several problems related to the large amount of manually managed data. Also, the risk of data integrity corruption is very high. No standards have been established for alphanumerical codes, and they are commonly "owners" codes, and therefore there is a unique and not general tie between the different parts (raw material suppliers, manufacturers and distributors) of the supply chain. The EAN association has made some effort towards standardization through the introduction of several codes such as the EAN/UCC Global Location Numbers (GLN) in the EAN/UCC-13 version (Regattieri et al., 2007). The GLN is a 13-digit number used for the unique identification of any legal entity, functional entity, or physical location. It is basically composed of a GS1 Company Prefix, Location Reference and Check Digit (GS1 US, 2013). The GLN (Global Location Number) offers some standardization towards the way of identifying legal entities, trading parties and locations in order to cover the requirements of the electronic commerce. The GLN aims at improving the efficiency of integrated logistics while contributing added value to both the parties involved and to customers. Examples of parties and locations that can be identified using GLNs are:

- Functional entities, such as a purchasing department within a legal entity, an accounting department, a returns department, a nursing station, a ward, a customer number within a legal entity, etc.
- Physical entities, such as a specific room in a building, warehouse, warehouse gate, loading dock, delivery point, cabinet, cabinet shelf, housing circuit boards, room within a building, hospital wing, etc.
- Legal entities/Trading Partner, such as buyers, suppliers, organisations, subsidiaries or divisions, financial services companies, freight forwarders, etc. (UCC, 2012).

The use of the GLN would lead to standardisation of account/location numbers allowing each food service location to only possess one, standards-based identifier in a standardised data format based on standardised assignment and change management guidelines. This same identifier could then be applied by all trading partners for the identification of this location in all supply chain transactions, supply chain communications, and internal systems. This could resolve all problems related to the current approach of account/location identification, thus improving foodservice business processes such as contract management, procure-to-pay and recall (GS1 US, 2012). The GLN can be used for the unique identification of any location or legal entity, thus ensuring that it is always identified correctly. Some companies choose to use one GLN for all their processes, while others use an individual GLN for each location that needs to be identified (e.g., a warehouse goods-in door). The GLN is completely flexible and can be applied to all levels of location identification. As a GS1 Identification Key, the GLN will always be unique, thus allowing trading parties to exchange GLN data with regards to location identification without risking number duplications (GS1 US, 2013).

2.2.2 Radio Frequency Identification (RFID): Theory/Functioning Principles and Case Studies

The most commonly used type of technology adopted by electronic data-carrying devices nowadays is the smart card based upon a contact field (e.g., bank cards). However, the need for mechanical contact in the smart card is not always practical. The flexibility of a technology that would allow the contactless transfer of data between a data-carrying device and its reader would be much higher. Ideally, the power needed for the operation of the electronic data-carrying device would also be obtained through contactless transfer from the reader. Due to the procedures employed for transferring power and data, contactless ID systems are called RFID systems (Finkenzeller, 2010).

According to Subhashrahul and Kumar (2012): "RFID is the term used for technologies that leverage radio waves to identify items automatically. Typically, this happens through a stored serial number on the tag that identifies a product. The tag may also have other information, such as where the item was made, manufacture date and other information stored in that item like temperature, vibration, contact status, current, voltage, speed, pressure, flow, etc. RFID has two main components namely RFID reader and RFID tags. A reader (RFID interrogator) is basically a radio frequency (RF) transmitter and receiver, controlled by a microprocessor or digital signal processor. The reader, using an attached antenna, captures data from tags, then passes the data to a computer for processing. As with tags, readers come in a wide range of sizes and offer different features. Readers can be affixed in a stationary position, portable or even embedded in electronic equipment such as print-on-demand label printers".

The advantages of RFID technology can create a lot of value to an organisation. Increased levels of information exchange between companies can be extremely beneficial to all parties, leading to better performance metrics and creating stronger business relationships. Increased trust and reliability between companies can increase business opportunities. By maintaining product visibility to its end location, and through knowledge of expiration dates, and reduction of stock levels and lost products significant financial benefits can be seen. The use of RFID systems can also directly reduce the costs associated under the need for more employees. However, the quantification of the challenges and costs related to RFID can much easier be performed than the advantages of its implementation and the expected financial benefits (Kumar et al., 2011).

RFID is of increasing importance in the field of the meat industry due to the increasing need for managing information with regards to animal products and the progress of these products through the value chain. The use of RFID technology has been promoted by retailers such as Wal-Mart as well as US government initiatives from the Department of Defense and US Department of Agriculture. All these organisations have suggested RFID as valuable technology for allowing the facilitation of product traceability, location intelligence and product tracking. Due to the increasing importance that is technology gains in many levels of the meat supply chain, it is highly important that stakeholders in this industry fully explore the opportunities arising by its use as well as its limitations (Townsend and Mennecke, 2008). In the review study of Mc Carthy et al. (2011), it was noted that with regards to the use of RFID in the field of bovine and beef products traceability, UHF RFID can add significant value to the supply network of organisations due to the inherent high levels of automation offered to the end users. It has been demonstrated that UHF RFID technology can add value to the meat supply chain as well as the meat production cycle, and facilitates automation, storage and transport. The fact that this technology can be used to monitor environmental conditions of the product strengthens product safety and consumer confidence factors. The full potential of this technology could only be estimated after its widespread adoption across an extended network of organisations, an opportunity that is not currently being commercially explored. The effective implementation and maintenance of a traceability system can only be achieved through considering all relevant aspects, such as any legislative requirements, technical aspects of data capture, storage and transfer along the supply chain, and relevant costs. Thus, the widespread adoption will help to increase chances of a successful adoption of RFID technology in the meat industry and demonstrate the full potential of the technology. It is therefore time that the industry moves towards the adoption of this technology. It is also stated by Tóth et al. (2010) that RFID technology is the most up to date technology that exists in the poultry sector for marking. The Passive Integrated Transponder tags are very effective on marking these animals. Since they are passive tags, they have no energy source and therefore their size is small enough to be implantable. The only disadvantage is the risk of loosing the tag, during its use and therefore attention is required. Further, costs would however unavoidably arise as a result of this process.

Furthermore, the use of RFID will assist farmers in the maximization of their productivity—which is of high importance in the modern competitive dairy industry. It is expected that the modern farm management practices supported by the RFID technology will allow farmers to significantly increase the volume and improve the quality of milk output from their herd. This could be achieved by (1) using improved practices for monitoring the health of the animals, targeting at the minimization of illnesses (and thus reduced production of cows), (2) speeding up the milking process—which will enable the animals to return to the paddocks quicker, and (3) by focusing on the optimization of feed to suit each cow production and stage of lactation cycle, etc. The use of RFID towards the automation of the process will also facilitate the minimization of labour input, thus allowing each farmer to cater for more cows, or allowing farmers to have more time to spend on other activities (Singh et al., 2014).

According to Aung and Chang (2014), the Internet can act as an important tool for food traceability. Traceability systems relied on web applications could provide information on traceability chains for products to personal computers and smart phones of consumers depending on the access control level of the consumer identification system. This will enable the delivery of real-time information to consumers regarding the quality and safety status of products and will also allow rapid information exchange and recalls when quality and safety standards are breached. The technology of the future is expected to be based on the convergence of smart phones with the Internet of Things (i.e., Internet-connected real world objects). Devices such as smart phones can be used as sensors and RFID readers, allowing the interaction of consumers with real world objects in a much more detailed manner. Furthermore, most current research with regards to food traceability examines the development of systems that focus on traceability until the retail point of the food chain thus not taking into account the consumer part of the food chain. As regards food safety, the consumer part of the supply chain is also important and thus the traceability should be extended to also cover this segment. Traceability obviously comes at a cost. However, the costs that could potentially arise by not having a robust traceability system in place may be much higher for governments, consumers, individual companies and the industry. All in all, complete food traceability from "farm to fork" will only become a reality if market forces, consumer demand and government regulations move towards a totally new level of supply chain visibility.

As regards the dairy industry, it has been addressed by Wamba and Wicks (2010) that further research is required for assessing the impact of RFID technology on the interdependency of the dairy value chain activities and the main technical and business challenges that are faced during the process of RFID integration within the whole dairy value chain. The consideration of key facilitating factors which will increase RFID adoption within the dairy value chain should also be taken into account in future research.

According to Abad et al. (2009), a RFID smart tag developed for real-time traceability and cold chain monitoring for food applications was developed. This RFID based system is based on the use of a smart tag and a commercial reader/writer. The smart tag is attached on the product of interest, integrating light, temperature and humidity sensors, a microcontroller, a memory chip, low power electronics and an antenna for RFID communications. The storage of these sensor logged data in the memory is performed together with traceability data. A commercial reader/writer was employed to read and write data on the smart tag in real time, and was characterized by a wireless reading distance potential of 10 cm. The results of the use of the system along an intercontinental fresh fish logistic chain were presented in this study. The smart tag developed offers significant advantages regarding these conventional tools. The main advantage derives from the automation of the system that integrates online traceability data and chill chain conditions monitoring. Furthermore, a key factor related to this RFID system is that the data can be read-out at any time of the logistic chain and there is no need to open the polystyrene boxes that contain the fish and the tags. Multiple tags can be read simultaneously as they pass through a reader in a fully automated way. Moreover, the system is cost effective enough when high added value products are considered, since there would be only one tag per box. These developed RFID tags can also be used for measuring and recording temperatures below 0°C, and can thus be effectively used for monitoring frozen foods logistic chains. The integration of a humidity sensor increases significantly the sensitivity to changes in the storage conditions. In the study of Trebar et al. (2013), a temperature monitoring system employed for each package in the fish supply chain was presented as part of the traceability system implemented with RFID technology. RFID data loggers are placed inside the box to measure the temperature of the product and on the box for measuring ambient temperature. RFID portal performed identification of boxes and RFID data loggers passing through the door. It was shown that the system could be very useful during the stages of storage and transportation of fish increasing the quality control level that can be implemented. The sensor data is available immediately at the delivery to be checked on the mobile RFID reader and are then stored in the traceability systems database in order to be available on a web to stakeholders and private consumers.

In another study, a novel traceability system architecture relying on web services was developed. The web services were employed for integrating traceability data captured through RFID systems with environmental data collected using Wireless Sensor Networks (WSN) infrastructure. This system can be applied in Small to Medium Enterprises (SMEs), and is based on the integration of information collected along the entire food supply chain, providing full traceability information from the farm to the consumer. The results of the use of this method in two pilots in the aquaculture business are also described, demonstrating how business processes in the aquaculture supply chain can be improved using this flexible system, since the two companies under consideration were of very different sizes. Furthermore, an analysis of the advantages gained by introducing the system in the companies based on predefined objectives and the evaluation of KPIs is presented. It was proved that by using this system, the efficiency of the companies can be improved by 89–95% (Parreño-Marchante et al., 2014).

In the study of Azuara et al. (2010), a food traceability system was described. The system is based on the use of RFID tags with contents guaranteed secure by using public-key cryptography. Nevertheless, the system comes at an affordable cost while no substantial investment in infrastructure is required. Aggregate signatures are employed so that all the steps can be signed in a reduced memory space. This type of signature is described as a cryptographic primitive that combines several signatures into one allowing the signatures of any users to be grouped into one single signature. The RFID tags in the system suggested can be used to obtain evidence that all the controls in a production process have been complied with. This allows the rapid verification of the fact that all products have correctly passed through the production cycle with the guarantee of trust provided under the responsibility of the signer. As the signatures are controlled by computer, the tag is freed from this task, and that offers a major saving both in terms of cost and processing time.

Azzalin et al. (2008) examined how feasible a RFID system is for the traceability of carcass and beef, since computers are increasingly used in agricultural business, RFID can be an efficient method of herd and individual animal management. However, the development of such traceability system cannot be currently achieved during slaughtering, where products are identified by lots. RFID is a well-rounded technology if correctly implemented. Thus, this study was relied on the know-how of slaughtering process in Italy. It was shown that there are different levels of industrialization and productivity in slaughterhouse, which can be summarized to two principal situations:

(A) high industrialized and high productivity abattoir; and (B) small abattoir. The simulation of a slaughter chain was carried out and an animal database recalling the official National Cattle Database was compiled. Several RFID systems were developed and evaluated from hardware and software point of view. These systems could be integrated in current processes, combining Low Frequency and High Frequency device and barcode. The implementation of RFID technologies can be effectively carried out in the slaughterhouses. The implementation of an RFID system in the abattoir removes the possibility of human errors, favours automation of the process and allows traceability of single carcasses. It is however essential to customize the system in accordance with the specific process implemented by each single slaughterhouse. Although the integration of High Frequency can be effectively performed in labels at the end of the process, Low Frequency assures best performances along the slaughter chain, due to the fact that it is not as disturbed by electromagnetic interferences.

The use of an item-level RFID traceability system was suggested by Barge et al. (2014) for a high-value, pressed, long-ripened cheese. Specifically, several different techniques were tested for fixing tags to the cheese and for the automation of the identification. An automatic recording of all item movements was performed during the production, handling in the maturing room and warehouse, delivery, packing and selling phases. Both fixed and mobile RF devices that operate at low, high or ultra-high frequency bands were evaluated for the identification of single/multiple cheese wheels. The tags resulted were not affected by the environmental conditions or any operations commonly carried out in ripening rooms and the quality of the product was not affected by the insertion of the tags. Cheese presence highly affected the reading zone, especially at higher frequencies (UHF band) and during the initial processing stages when cheese water content is still high. Thus, just before the introducing of the RFID system all technical solutions should be carefully evaluated and comparison should be made with regards to frequency band and tag/antenna coupling for tracking the cheese in different situations. It was shown that LF systems were unsuitable for dynamic and multiple cheese wheels tracking in this study. UHF systems cannot be effectively used for identifying cheese wheels during the cheese production process since the transfer of the signal from tag to reader can only be carried out during ripening, warehousing and distribution. It was therefore concluded that the successful integration of an RFID system in a food production process can be affected by various factors related not only to the RFID characteristics of the devices, but also to each production process layout.

According to Chou et al. (2011): "The information including breeding, performance test, screening, mating, collection of mated eggs, fertilization rate, incubation rate, feeding amount before laying period and bodyweight, are very important to management of the poultry breeders. The collection, storage and analysis of these useful data are highly related to the successful management of the breeder farm and thus a study was carried out to develop an egg-collecting system integrated with the RFID for floor-raised poultry housing. RFID system was installed integrated with laying cage. The antenna was placed on laying board and the tag-rings were put on legs of each duck. When ducks enter the laying cage, tags will be detected and the data will be transmitted to PC by the wireless sensing network. An egg collection system has been developed and could be started and stopped automatically, and a PLC counter was installed to count the number of eggs produced for different laying cages. Through the interpretation of the data collected, the laying efficiency for each testing duck could be obtained. Based on the results shown, we may conclude the integrated RFID and egg collection system is able to improve the production management successfully."

According to Costa et al. (2011), RFID can be used for the attainment of massive amounts of data. It can be applied for traceability of products during their flow along the production chain, as well as in behavioural sciences for tracking different animals within a group. Therefore, an application of RFID technology on food supply chain was described. It was based on the use of tags along the high commercial value supply chain of fresh fish. A second example of the use of RFID was however also given, this time for tracking the behavioural locomotor activity in commercially important crustaceans. Specifically, a microcosm tank was endowed with set of RFID controllers below it, each handling group readers in order to contemporarily track four Norway lobsters (*Nephrop snorvegicus*). It was proved that the use of the technology was cost-effective enough for high commercial value fish. Also, the RFID technology allowed tracking behavioural rhythms at a resolution of centimetres. Its distributed topology design rendered it highly flexible offering the potential of an easily surface expansion, should this is required, through a single connection to the USB ports new controllers with antennas.

In another study, a method was suggested by Livestock Research Institute, Council of Agriculture, Executive Yuan that is based on the use of the "poultry foot ring" combined radio frequency identification (RFID) and barcode identification for enhancing the traceability systems of the poultry industry. The Personal Digital Assistants (PDAs) records all the data of the native chicken production traceability. The development of a system was also reported that could enable chicken farmers to implement individual chicken farm standard operating procedures. Digitizing applications implemented for assisting in everyday feeding and management procedures help not only to strengthen the control of internal operations, but also to increase the farm's operational efficiency. The PDA can be used to save production traceability information onto the poultry farm commercial poultry production flow operating system, and complete a variety of production traceability records required for the broiler chicken in the system. Upon reaching a suitable age, native chickens are delivered to the abattoir by the shipper. Even following depilation, the foot ring remains at the foot of the chickens for identification purposes to avoid any confusion with native chickens from other chicken farms. This offers an important identifying mark allowing consumers to recognize the brand of native chicken they wish to buy. Therefore, both targets of ensuring reasonable profit for producers and providing consumers with information about the products they buy, are met (Liu et al., 2012).

A system that can be employed for the identification of all aspects of beef traceability from farm to slaughter has been developed, taking into account the relevant European Union law and global standards. An integrated traceability system that involves all stakeholders that form part of the supply chain can be used to considerably enhance consumer confidence in beef products by simply giving consumers access to traceability data. The use of RFID for identifying individual cattle, as well as biometric identifiers for verifying cattle identity was suggested. A BioTrack database was proposed to be used for storing retinal images. A data structure for RFID tags could be used in accordance with ISO 11784 and a middleware to convert animal ID data to the EPC (electronic product code) data structure, facilitating the use of EPC global Network for the exchange of traceability data. At the time of the publication of this study, DAFF operated a Calf Birth Registration system for keeping records of all the births of cattle and for authorizing and tracking the movement of bovines using the CMMS. When combined, these systems can accurately depict the herd characteristics in Ireland. Nevertheless, it is important to mention that herd-keepers are not required to retain electronic records of their current herd status, which is probably the biggest disadvantage of the system, since it renders vulnerable to loss, inaccessibility of traceability records or even vulnerable to fraudulent activities. This situation also exists where no records of the animal identification number are taken directly from the animals and the passport accompanying the animal is only used. As a result, the use of an identity verification system that employs biometric identifiers was proposed in combination with RFIDs that contain animal identification numbers in ISO 11784 compliant format, the electronic storage and transfer of traceability data, and the use of global standards to enable uniformity, transparency and precision. Retinal scanning is currently considered to be the most viable biometric method of cattle identity verification (Shanahan et al., 2010).

2.3 Protected Geographical Origin and Indication

It is generally accepted that the demands of the consumers have changed gradually. Their requirements have now been developed to not only much higher dietary, hygienic and health standards in the products they consume, but they also take into account the certification and reassurance of products' origins and production methods. 'Quality' is the most important term defining consumers' choices. This increased consumer awareness is reflected in the demand for products with individual characteristics, due to particular production methods, composition or origin. The increased freedom of movement for goods that has been established nowadays has helped to increase the availability of a significantly higher variety of products from all over Europe, creating at the same time a need for providing consumers with more information about the products. Two regulations were adopted in 1992, namely Regulation (EEC) No. 2081/92 on the protection of geographical indications and designations of origin for agricultural products and foodstuffs, and Regulation (EEC) No. 2082/92 on certificates of specific character for agricultural products and foodstuffs, which both reflect a requirement for adaptation to the changing demands of both producers and consumers (European Commission, 2004).

It has been stated that: "The abundant supply of goods from all over Europe, which used to be disconcertingly diverse, is now a precious asset. The development of specific and traditional products will lead to further diversity, which is just what the consumer wants (European Commission, 2004). Three EU schemes known as PDO (protected designation of origin), PGI (protected geographical indication) and TSG (traditional speciality guaranteed) promote and protect names of quality agricultural products and foodstuffs. These EU schemes encourage diverse agricultural production, protect product names from misuse and imitation and help consumers by giving them information concerning the specific character of the products. Specifically, Protected Designation of Origin (PDO) covers agricultural products and foodstuffs, which are produced, processed and prepared in a given geographical area using recognised know-how. Protected Geographical Indication (PGI) covers agricultural products and foodstuffs closely linked to the geographical area. At least one of the stages of production, processing or preparation takes place in the area and finally, Traditional Speciality Guaranteed (TSG) highlights traditional character, either in the composition or means of production" (http://ec.europa.eu/agriculture/quality/schemes/index_en.htm). The indication or designation must be a name that can only be used to define a product that originates from a specific region, or—only as an exception for Designations of Origin—in a specific country. Specific traditional geographical or non-geographical names can also be regarded as designations of origin or geographical indications. These are denominations that are no longer widely used, or indirect references to geographical indications (e.g., "feta"). In both cases the products can be protected if there is a link between product characteristics and its production in the region of origin (http://www.deutsches-patentamt.de/docs/service/formulare_eng/marke_eng/w7729_1.pdf).

There is an increasing demand by consumers for high quality food with a clear local identity. Consumers are now interested in the production method and they want to support local producers. Producers can add value and often secure a better price by linking Geographical Indication products with this consumer demand. From the point of view of the producers the main benefit to consider is the opportunity to negotiating a better price. Nevertheless, the added value of a PDO/PGI label may not be obvious to consumers without investment in a campaign of increasing public awareness with regards to these products. Geographical Indications can open market opportunities for producers to markets where competition is not only based on price. Internationally, in trade negotiations between the European Union and other countries over Free Trade agreements, the EU constantly promotes the recognition and protection of the Geographical Indication System (http://www.google.com/search?client=safari&rls=en&q=infofoodnames.com&ie=UTF-8&oe=UTF-8).

2.3.1 (Protected) Geographical Origin [(P)GO]

The protected designation of origin is related to products, which are closely associated with the region whose name they bear. To be eligible for a protected designation of origin (PDO), the quality or attributes of the product must be mainly or exclusively attributed to them due to the effects of the specific geographical environment of the place of origin. The geographical environment includes both inherent natural and human factors, such as climate, soil quality, and local expertise. Also, the production and processing of the raw materials, and final product, must be carried out in the defined geographical area whose name the product bears. It is therefore obvious that there must be a very close link between the characteristics of the product and its geographical origin. The regulation, however, provides for exceptions to cases in which the name of the product should designate the defined area from which the product originates, in order in particular that a non-geographical product name may be allowed to be registered, if it is traditionally related to a specific geographical area. For example, the PDO '*Reblochon*' is based upon a traditional name, which designates a French cheese, associated with its geographical area of production. However, the regulation clearly requires that all other requirements are met. Particularly, the production area must be accurately defined, production, processing and preparation steps must be carried out in that area, and there must still be the close and clear link between the characteristics of the product and its place of origin. Furthermore, certain geographical designations can be registered as protected designations of origin although the raw materials used for manufacturing the product derive from a geographical area larger or different from the processing area. As regards this case, the regulation provides for a limited period to register these names, in order to consider some particular situations previously covered by national laws. Examples include "*Prosciutto di Parma*" and "*Roquefort*" (European Commission, 2004). Some representative examples of foods of animal origin with protected designation of origin are given below:

The "Beacon Fell Traditional Lancashire Cheese" is a product manufactured using full fat cows' milk. It must contain a minimum of 48% butter fat in the moisture-free substance and moisture of 48% maximum. The cheese is made in traditional cylindrical form. The geographical area where the product is manufactured is the Fylde area of Lancashire, north of the River Ribble and including the Preston and Blackpool district of Lancashire. The cheese has a buttery texture and mellowness due to the sandstone bedrock of this part of Lancashire and thus its soft water and lush grazing. This geographical area is also characterized by moderate climate and therefore the fat constituents of the milk are relatively consistent. The high level of rainfall of this area contributes to the lushness of the grazing and the cleanliness of the grass. Also, the area of Lancashire is a traditional dairying area as it is characterized by one of the most consistently moderate climates in Europe. The traditional cheese making process has been carefully preserved by the local cheesemakers and is significantly different from other cheesemaking methods. Traditional methods of production are used, for protecting the delicate curd, which must be properly handled in order for the required flavour and texture to be developed, while ensuring that the fat is not broken out nor lost in the whey. The long production time of Beacon Fell Traditional Lancashire Cheese is essential for the particular microorganisms present in the dairies involved to be absorbed and allowed and contribute to the development of the necessary flavour characteristics (https://www.gov.uk/government/uploads/system/uploads/attachment_data/file/271080/pfn-beaconfell-lancashire-cheese.pdf).

Bonchester cheese is a white coated full fat soft cheese that contains a minimum of 20% milk fat and a maximum of 60% water. It is manufactured from unpasteurized whole milk from Jersey cows. The geographical area in which Jersey cows are grazed and where the product is produced and matured is defined as the border lands of England and Scotland, including the river systems of the Tyne, the Tweed and the Solway Firth. The designated area falls within a radius of 90 kilometres from the summit of Peel Fell on the Border and in the middle of the Cheviot Hills. The milk used for manufacturing the cheese is the morning and evening milk of Jersey cows, which must not be subjected to any form of heat treatment (http://ec.europa.eu/agriculture/quality/door/registeredName.html?denominationId=144).

"Buxton Blue" is a coloured, blue veined cheese made from cows' milk treated with pasteurization. This product is produced in a 30-mile radius of Buxton. The processes that must determine the production of Buxton Blue cheese include specific time and temperature conditions, specific salt level and time of ripening, and final deep knowledge and expertise with regards to the manufacturing process. It has been proved that since Buxton Blue cheeses have been made in this specific area and thus any cheese made under the same conditions in a different geographical area could not be named Buxton Blue. The milk used for the production of Buxton Blue originates from Hartington in the county of Derbyshire. The designated area is a high limestone area with a thin topsoil, and high average rainfall where high quality grazing can be produced with a characteristic sward. During the winter months the cows are fed upon grass conserved from the same pastures ensuring the production of a consistent type of milk. At peak times, however, it may be required to source milk from the neighbouring counties of Shropshire and Cheshire. Comparison of this milk to the local milk and blending is carried out to produce milk, which is directly comparable to milk used at all other times. Finally, it should be mentioned that the production of Buxton Blue highly depends on the expertise of the local cheese makers which often passes from generation to generation (https://www.GOV.UK/government/uploads/system/uploads/attachment_data/file/271204/pfn-buxton-blue.pdf).

The name "Fal Oyster" is a protected name associated with oysters caught within the Truro Port Fishery area by sailing and rowing vessels using traditional methods (http://cornishnativeoysters.co.uk/2011Oysters-Fal.html). The Fal Oyster is a product derived from the oyster species *Ostrea edulis* known as a flat or native oyster. It is characterized by a less than round or uneven round shaped shell with a rough scaly surface. The colour of its shell is brown or cream with light brown or bluish concentric bands on the outer surfaces. The inner surfaces are quite smooth and pearly and white or bluish-grey, frequently with darker blue areas (https://www.gov.uk/government/uploads/system/uploads/attachment_data/file/271313/pfn-fal-oyster-pdo.pdf). There is a specific legislation in force which controls the fishing methods used in the fishery of this species. These fishing methods distinguish the Fal Oyster from other native oysters. The environmental factors of the area where the oyster is caught determine the taste of the product. The rivers of the Fal area are fed from steep sided valleys and are characterized by high mineral and biological content. The estuary is very deep and there is high water circulation. These characteristics of the environment lead to increased plankton generation, which functions as feed for the oysters. The taste of the final product is salty, metallic, creamy and sweeter than other species or the same species in other areas. It is believed that the biology and mineral content of the water is responsible for the mix of plankton on which the oysters feed and affects the taste of the product at a great extent (http://cornishnativeoysters.co.uk/2011Oysters-Fal.html).

Isle of Man Queenies is a protected name that refers to queen scallops caught in Isle of Man waters. The queen scallop (*Aequi pectenopercularisis*) is a medium sized species of scallop that belongs to the family Pectinidae. The shell can have various colours such as yellow, orange, red, brown and purple and grows to a maximum of 90 mm in diameter. Manx Queenies, as are locally known, are caught in Isle of Man territorial waters by Manx registered vessels or other registered fishing vessels that possess the appropriate license. All vessels are landed on the Isle of Man and the products are processed in the same area. The Isle of Man Queenie is fished in extensive beds within the Manx Territorial Sea. The area covers 3,917 square kilometres and extends to a total of 12 nautical miles (http://archive.defra.gov.uk/foodfarm/food/industry/regional/foodname/products/documents/isle-of-man-queenie-pdo.pdf).

"Orkney Lamb" is the name given to a meat product, which is derived from lambs born and reared in the Orkney Isles and which were slaughtered and dressed in Orkney. Following slaughter and dressing the lamb may be marketed as a whole carcass, or the carcass may be broken down into cuts. The geographical area where rearing of the animals must be carried out is the group of islands in the North Atlantic off the North Coast of Scotland known as the Orkney Islands. The Orkney Islands have always been known for the production of Quality Lamb. It is essential that traditional breeds are used for the production of a product of very high quality. The unique "North Ronaldsay Sheep" is a sheep breed found only in Orkney. The animals feed on a diet of seaweed. These sheep are protected by the Rare Breed Survival Trust. These lambs spend their entire life grazing freely on farms or crofts before being selected for slaughter. Processing is performed exclusively in Orkney. The production and preparation of the product is carried out exclusively in the Orkney Isles and this meat is considered to be distinctly different with regards to texture and flavour from lamb produced in other parts of the UK (https://www.gov.uk/government/uploads/system/uploads/attachment_data/file/271263/pfn-orkney-lamb.pdf).

Champagne is a sparkling (or carbonated) wine. Its production is carried out in the Champagne region of France. Only a few specific varieties of grapes, such as Pinot noir, Chardonnay and Pinot Meunier can be used for producing Champagne. Although not only white grapes are used, champagne is usually a white wine because of the extraction methods that minimize the contact between the juice and skin. Pink champagnes are produced when the skin is in contact with the juice for longer periods or when small amounts of red wine are added into the champagne (http://foodreference.about.com/od/bar_beverage/a/What-Is-Champagne.htm). Champagne is a tightly regulated PDO product. Several parameters such as the pruning method, the maximum permitted yields per hectare, the maximum permitted press yield, the minimum potential alcohol content of newly harvested grapes and the secondary fermentation in the bottle, determine whether a product can be named Champagne. A minimum of 15 months is required for the maturation on lees for non-vintage Champagne, while three years are required for vintage Champagne. These guidelines are continuously revised and updated. Some of the last measures taken for improving the qualitative characteristics of the product include regulations (in effect in 1978) governing the training and pruning of vines, their height, spacing and planting density. These regulations targeted at optimizing fruit quality through high-density (8000 plants per hectare) low-yield vineyards. Furthermore, rules have been established that forbid the bottling of wines until the second day of January following the harvest and approval of all press operations. Finally there are rules that came in effect in 1993 defining that press yields have to be set at 102 litres of must per 160 kg grapes (up from 150 kilos) (http://www.champagne.fr/en/terroir-appellation/appellation/appellation-origine-controlee-aoc).

Greek authorities submitted on 21 January 1994 the name 'Feta' for registration as a protected designation of origin (PDO). After examining the request, Commission concluded that the name 'Feta' met the criteria of the Regulation and should therefore be protected (COM 314 final, 2002). Feta is made primarily from sheep's milk or a mixture of sheep and goat milk (maximum 30% goat's milk). Milk is collected by farmers and brought to the cheese dairy in large kegs. The milk coagulation has to be completed two days after the collection. Traditionally unpasteurized milk was used for the production of Feta, but modern industries use milk that has been subjected to pasteurization to avoid any problems with regards to public health and export regulations. The pasteurization of the milk is performed at a minimum of 72°C for 15 seconds or any equivalent time-temperature combination. When pasteurized milk is used, lactic acid starters cultures are necessary, while calcium chloride is used to obtain the ideal taste profile. Both the addition of the micro-starters and calcium chloride is performed after pasteurization, while the milk is being held at 34–36°C. These are the only additives that can be added during the production of the product. Following refrigeration of milk for 20 minutes, rennet is added causing coagulation. After that the curds are cut in small cubes and left for 10 minutes. They are then placed into molds, which allow effective draining and shaping of small cuts. The filled molds are stored at 16–18°C for 18–24 hours and turned periodically ensuring that drainage occurs evenly and that all the curds are subject to the same level of pressure. Following drainage, the molds are removed and the product is placed temporarily either in wooden barrels or metal containers, where salting in layers is carried out, to achieve a final concentration of salt in the product of approximately 3%. After salting, the cheese is subjected to maturation in brine for 14–20 days. This stage of the process is carried out at 16–18°C, under conditions of high relative humidity. Following this stage the humidity level of the product is approximately 56% and the pH is 4.4–4.6. Following that, a second maturation is performed during which the product remains refrigerated (1–4°C) in brine (http://www.greek-feta-cheese.com/production.html).

Also, another type of cheese, named Galotiri, is one of the oldest traditional cheeses of Greece with PDO. It is manufactured in the Greek regions of Epirus and Thessalia using ewe's or goat's milk or mixtures. It is characterized by a soft and spreadable texture with sour and very pleasant refreshing taste and is consumed as table cheese. Galotiri is composed of approximately 70.8% moisture, 13.8% fat, 9.8% protein, 2.7% salt and pH 4.1 (http://expoaid.gr/resources/cheese-encyclopedia/cheese-overview). Galotiri is mainly produced for household consumption rather than at an industrial scale. The fresh curds from successive days are placed in barrels and sealing is carried out using fat. Alternatively, they are hung from the rafters in sacks to drain. Mould growth can be treated by scraping it off, thus allowing the whey to escape and the cheese to develop its unique characteristics (Aspinwall et al., 2009).

The designation of origin 'Parma Ham' can only be used for ham that is branded in a specific identifiable way and is derived from the fresh legs of pigs that are born, bred and slaughtered in any one of the Italian Regions indicated in Art. 3 of Ministerial Decree No. 253 dated 15th February 1983. Since 1970 the designation of origin "*Prosciutto di Parma*" has been protected by law (No. 506 dated 4th July) at a national level. This law was subsequently replaced by Law No. 26 dated 13th February 1990 (Protection of the 'Parma Ham' designation of origin), which determines the current situation. These general provisions have been supplemented and completed by the respective Execution Regulations, that came in force by Presidential Decree No. 83 dated 3rd January 1978 and by Ministerial Decree No. 253 dated 15th February 1993 (Prosciutto di Parma—Parma Ham—Protected Designation of Origin, Specifications and Dossier, 1992). The product "*Prosciutto di Parma*" is characterized by:

(1) Rounded external form which is attributed to the lack of the distal part of the leg (foot), lack of any external imperfections and by determining that the permissible extension of the uncoated muscular part placed over the head of the femur has to be shorter or equal to six centimetres; (2) Weight of eight to ten kilograms and not less than seven; (3) Uniform colour of pink to red, streaked with pure white of the fat; (4) Sweet flavour and slightly salty taste and a characteristic and fragrant aroma; and (5) Compliance with characteristics as defined through specific analytical parameters (http://www.monicaegrossi.it/en/prosciutto-di-parma-dop.html).

Generally, protection is not given to generic names that are now used as the common names of agricultural products or foodstuffs and are no longer used for the identification of a product that originates from a specific geographical origin, for example, "Emmentaler" or "Pils". Similarly, names conflicting with the name of a plant variety or an animal breed or with previously registered homonymous names and well-known trademarks can be excluded from registration (http://www.dpma.de/docs/service/formulare_eng/marke_eng/w7729_1.pdf).

2.3.2 (Protected) Geographical Indication [(P)GI]

The protected geographical indication (PGI) also designates products associated with the region whose name they bear; but the link differs to that of the product with a PDO and its geographical area or origin. To be allowed to use a protected geographical indication, the production of the product must have been carried out in the geographical area whose name it bears. Unlike the protected designation of origin, it is sufficient that only one of the stages of production has been carried out in the defined area. Therefore, the raw materials used for producing the product may originate from another region. Also, a link must exist between the product and the area, which gives its name. Nevertheless, this link is not essential (as in the case of the protected designation of origin) and there is a degree of flexibility. It is sufficient, for example, that a specific quality, reputation or other characteristic is attributed to the geographical origin. Under the rules for protected geographical indications, the link may be only associated with the reputation of the product, if it owes its reputation to its geographical origin. In this case, the attributes of the product itself are not the determining factor for registration. It is sufficient for the name of the product to enjoy a reputation based specifically on its origin. These rules are based on the idea that a geographical indication should be protected even if it cannot be proved that the product owes its special features to its region of origin. The indication is a very important factor that can determine the prices of products and can open new customer opportunities. Based on the distinct characteristics of each product, producers choose whether to apply for a PDO or a PGI (European Commission, 2004). From the point of view of the consumers, GI markings and labels are information mechanisms with the power to guarantee a consistent quality and could therefore highly affect the consumers' choices. It is believed that some consumers value the food quality characteristics that GIs offer, but this has to be balanced against the fact that GI marks are not widely recognized in many European countries. Since GIs are not the only quality mark or certification that consumers should consider, there may be some confusion over the meaning of different logos. Also, in some cases, consumers have a well-established attachment to a particular brand/mark or type of product (e.g., Camembert Cheese) and may thus not perceive the use of GI marks to add any additional value to the products (http://www.google.com/search?client=safari&rls=en&q=infofoodnames.com&ie=UTF-8&oe=UTF-8). Some representative examples of animal-derived products with PGI are given below.

"Irish Salmon" is a name associated with the farmed fresh fish of the anadromous species *Salmo salar*, commonly known as Atlantic salmon. This PGI does not cover the wild salmon. The "Irish Salmon" has characteristic bluish coloured scales and body shape identical to the body shape of the wild Atlantic Salmon. The name "Irish Salmon" can only be used for superior grade salmon, certified to the Irish Quality Salmon (IQS) Scheme, or equivalent. In 2012, a national objection procedure started where anyone with a legitimate interest could examine and lodge an objection to the application of "Irish Salmon". A period of 12 weeks was given to all interested parties (ended on 9 October 2012) to submit their objections in writing to DAFM. The application was examined and modified with the aim to reflect the results of the consultation and was submitted to the Commission on the 30 August 2013 (https://www.agriculture.gov.ie/gi/pdopgitsg-protectedfoodnames/products/). A few representative examples of Protected Geographical Indication animal-origin foods are given below.

Another PGI food product is "Dorset Blue Vinny Cheese". It is a lightly pressed cheese, made from unpasteurised milk. The cheese is firm and of uniform colour with irregular blue/green veining and a rough, dry brown mould coating. Its flavour is piquant, peppery mild to strong. It is produced in the county of Dorset. Dorset Blue Vinny Cheese has been traditionally produced in the area. Traditionally, cream was made into butter or clotted, and there was therefore a need to produce a high value cheese from semi skimmed milk. Also, the sale of fresh milk was quite limited due to the low population level and the poor transport links to the more heavily populated areas. The flavour of the cheese is attributed to the quality of the milk, which is obtained from cattle grazing on the pastures of the Blackmore Vale. Low lying permanent pasture on top of Oxford clay produces lush sward contributions to the flavour of the milk. The production of the cheese is performed in accordance with a traditional and local recipe using the local expertise. All production of the cheese is carried out at present on only one farm in Dorset (https://www.gov.uk/government/uploads/system/uploads/attachment_data/file/271184/pfn-dorset-blue-vinny-cheese.pdf).

"Scottish Wild Salmon" is a name associated with the pelagic fish of the species *Salmo salar*, which are caught up to 1500 metres off the Scottish Coast and landed by nets men at fisheries in Scotland. Fresh Scottish Wild Salmon has a bright silver colour, with a dark blue back and is firm, with scales intact. They are characterized by a distinctive fresh smell and are in excellent condition throughout the season. The geographical area where the fish is caught and processed is the whole of Scotland, including an area up to 1500 metres off the coast. Atlantic salmon is the symbol of Scotland since ancient times. The animal begins its life in freshwater, heads downstream to the ocean and, when fully grown, returns to the same river to give birth. The rivers of Scotland have one of the largest and most diverse population of Atlantic salmon in Europe. Specifically, a system of 400 rivers supports many hundreds of populations, each of which is genetically distinct. The characteristics of the Scottish Wild Salmon are directly linked to this geographical area as regards the traditional ways of catching and processing. These practices are well-documented (http://scotland.gov.uk/Resource/Doc/305128/0095733.pdf).

"Scotch Beef" is a PGI product derived from cattle born, reared, slaughtered and dressed in the designated geographical area of the mainland of Scotland, which covers the islands off the West Coast, Orkney and the Shetland Isles. Since the 19th century Scotch Beef has been known for this superior quality foodstuff produced through traditional feeding systems and it has established a high reputation in the UK meat market and in Europe. The quality and characteristics of the product are mainly attributed to the extensive grazing on the pastures of Scotland (https://www.gov.uk/government/uploads/system/uploads/attachment_data/file/271297/pfn-scotchbeef-pgi.pdf).

Two other PGI products are the Welsh Beef and Welsh Lamb. Their protection status was awarded in November 2002 and July 2003 respectively (http://hccmpw.org.uk/index.php/tools/required/files/download?fID=2663). "Welsh beef" is the name given to both carcasses and cuts of meat taken from cattles that have not bred and are born and reared in Wales. Welsh beef producers aim at meeting carcass classification of R conformation or better and 4L fat content or leaner. Historically the traditional cattle breeds of Wales were mainly the Welsh Black and Hereford. These breeds are the main ones used by the Welsh beef industry today. Welsh beef is derived from Welsh breeds or any other recognized breed. The efficient production and use of grass are the main parameters affecting the well being of Welsh beef production. The grass found in various regions of Wales together with heathers and indigenous fragrant wild herbs contribute to the distinctive flavour of the beef. The farms of this region are mainly family farms that may have both sheep and cattle (https://www.gov.uk/government/uploads/system/uploads/attachment_data/file/278748/welsh-beef-pgi.pdf). On the other hand, Welsh lamb is the name given to meat obtained from lambs, which are born and reared in Wales. Similar to the producers of beef, producers of Welsh lamb target at meeting a carcass classification of R conformation or better and 3H fat content or leaner. Welsh lamb is derived from the sheep breeds of Wales, mainly the Welsh Mountain, Welsh Mules, Welsh Halfbreds, Beulah, Welsh Hill Speckled Face, Lleyn Sheep, Llanwennog, and Radnor. Crossing of the above mentioned breeds may be carried out with Texel, or Suffolk rams, or any other terminal sire breed for prime lamb production. Welsh lamb is produced using traditional extensive farming practices and the local knowledge of the associated processes. Welsh lamb is derived entirely from lambs born, reared, slaughtered and processed in Wales. Processing can only be carried out using abattoirs/processors approved by a HCC verification scheme. The distinctive characteristics of the product are attributed to the traditional hardy Welsh breeds that dominate the Welsh flock as well as by the lambs feeding on the Welsh grassland (https://www.gov.uk/government/uploads/system/uploads/attachment_data/file/271300/pfn-welsh-lamb.pdf).

It is important to mention that a geographical name cannot be registered under the designations PDO/PGI if it is:

1. A name that has already been given to a plant variety or animal breed.
2. The name of an EU Member State.

3. A generic name (excluding cases when the generic term is only a part of the name)—the term "generic" may refer to the name of an agricultural product or food which, although it is associated with the geographical area where the product was originally produced or marketed, it has become the common name of the product or foodstuff.
4. A name that is similar or identical to a trademark used for a similar product and which if registered, could lead to misleading of the consumers as to the true identity of the product (http://www.fsai.ie/uploadedFiles/About_Us/forums/artisan/PDO_PGI_TSG_Info_Note.pdf).

Finally, the registration of indications of geographical origin could not influence the rights conferred by the earlier establishment of trademarks acquired in good faith. The co-existence of the mark and of a designation of origin or geographical indication should be allowed. On the other hand, protection cannot be given to the later trademark if the application has been submitted for comparable products or in cases where the use of the trademark could counteract the Article 13 of the Regulation (Article 14 of the Regulation) (http://www.dpma.de/docs/service/formulare_eng/marke_eng/w7729_1.pdf).

2.4 Conclusions

It has been stated by Bosona and Gebresenbet (2013): "Experts are often inclined toward technical aspect of food traceability while consumers consider that traceability alone is less important unless it addresses well issues of food quality and safety as well as sustainability of food production. Some actors of food supply chain consider traceability as bureaucratic burden and are less willing to implement it. There exist also societal concerns about the potential impacts of traceability technologies (e.g., RFID) on consumers' health and on data privacy. In general, the implementation of food traceability should be associated with effective way of communicating traceability information to the consumers and other stakeholders."

The development and implementation of a full chain food traceability system can be very complex. The development of modern systems is supported by effective and efficient traceability technologies and innovative FSCM, while trainings provided to improve employee's skill, awareness, and motivation together with the essential senior management commitment are all important factors that can enable researchers to develop more user-friendly traceability tools and data processing software. Furthermore, the integration of traceability activities and food logistics activities and the development of advanced information connectivity systems as well as communication among partners, consideration of traceability issues at early stage of designing food logistics network, organisation of the increasing interests from society, government agencies, and researchers in the security of FSC and use of the latter to attract financial funding and finally the preparation of clearer traceability guidelines can all enhance the robustness of a newly developed modern traceability system (Bosona and Gebresenbet, 2013). Also, it seems important to consider the traceability needs of each company on a case by case basis since according to the study of Resende-Filho and Hurley (2012), the use of a traceability system by itself is neither a necessary nor sufficient solution for triggering further food safety effort when there is some form of traceability even without the full adoption of a traceability system. But in cases when there is complete anonymity in the food supply chain, traceability adoption is essential and sufficient for ensuring safer food since the elimination of complete anonymity, incentive contracts can be used based on contingent payments. Also, to trigger a certain level of food safety effort, more intensive contingent payments can substitute for higher traceability precision. It is therefore clear that the precision of traceability should not be considered as an unequivocal mark for food safety. Thus low precision traceability systems may finally lead to safer food than high precision systems.

Traceability should keep providing logistical support and function as a tool for removing potentially hazardous products from the market through the establishment of accounting systems, which can be both audited and verified. Increased tightening of rules and monitoring are causing this evolution. The use of many of the same tools as traceability can aim at keeping track of physical products or ingredients, and this could lead to an improvement of product claims and thus increase customer satisfaction. A future trend could be based on the accommodation by the food system of a spectrum of foods and consumer tastes, with appropriate levels of oversight and auditing. Where adequate mechanisms to provide what consumers want with sufficient safety do not exist, the government should be responsible for providing necessary guidance through its regulatory tools. There may always be a disconnection between the market place and governments; particularly in the way each sector reacts to the continuously changing needs of the consumers (Bennet, 2010). The need to integrate more and more information in food production management and the increasing demands for information along the food processing chain will unavoidably set higher requirements for robust traceability systems in the future. There is a need to analyze the fundamental and practical aspects of traceability in food manufacture. Having that into account, guidelines that can assist organisations in assessing their particular need for the degree of detail in their internal and chain traceability systems can be established. In the future, the information flowing in the food manufacturing chain could offer a competitive advantage since it can be added to the price of the product and promoted as quality parameters (Moe, 1998). It is important to mention, however, that according to Karlsen et al. (2013): "No common understanding of the definitions and principles of traceability exists, nor is there a sound common theoretical framework with respect to implementation of food traceability. When no common theoretical framework exists, this affects the implementation process of traceability in the food industry. With a common theoretical framework, all traceability studies could have been more similar, and the implementation processes could have been more goal-oriented and efficient. Based on the review, it is clear that traceability is an interdisciplinary research field, and it spans the natural sciences as well as the social sciences. Further theoretical developments on implementation of food traceability are needed."

References

Abad, E., F. Palacio, M. Nuin, A.G. de Zárate, A. Juarros, J.M. Gómez and S. Marco. 2009. RFID smart tag for traceability and cold chain monitoring of foods: Demonstration in an intercontinental fresh fish logistic chain. Journal of Food Engineering. 93: 394–399.

Aspinwall, M., S. Blohorn, V. Bozzatti, K.J. Broome, R. Buck, S. Cooper, D. Curtin, J. Davies, S. Davis, A. Gray, R. Hijikata, R. Honma, K. Jarvis, M. Linton, G. Pretty, H. Renz, R. Sutton, W. Studd, J. Warwick and A. Vernooij. 2009. The Finest Selection. World Cheese Book. pp. 1–352. *In*: J. Harbutt (ed.), Dorling Kindersley Limited, New York.

Aung, M.M. and Y.C. Chang. 2014. Traceability in a food supply chain: Safety and quality perspectives. Food Control. 39: 172–184.

Azuara, G., J.L. Salazar, J.L. Tornos and J.J. Piles. 2010. Reliable Food Traceability Using RFID Tagging. IFCA/Springer-Verlag Berlin Heidelberg. pp. 57–67.

Azzalin, T., C.U. Biader and C.M. Medaglia. 2008. TraceMEAT, RFID technology in the service of meat traceability. International Conference: September 15–17, 2008, Ragusa, Italy.

Barge, P., P. Gay, V. Merlino and C. Tortia. 2014. Item-level Radio-Frequency Identification for the traceability of food products: Application on a dairy product. Journal of Food Engineering. 125: 119–130.

Bennet, G.S. 2010. IPT Software Providers. In: Food Identity Preservation and Traceability. Safer Grains. CRC Press. Taylor and Francis Group, LLC., Boca Raton, USA

Bosona, T. and G. Gebresenbet. 2013. Food traceability as an integral part of logistics management in food and agricultural supply chain. Food Control. 33: 32–48.

Chou, C.-Y., Y.-N. Jiang, J.-J. Chou, C.-Y. Chen, A.-C. Liu, J.-F. Huang and H.-C. Liu. 2011. Application of RFID System in Production Management for Floor-Raised Poultry Breeders. Animal Industry Department, Council of Agriculture, Executive Yuan. pp. 1–7.

Commission of the European Communities. Brussels, 14.6.2002. COM(2002) 314 final. Proposal for a Council Regulation amending the Annex to Commission Regulation (EC) No. 1107/96 with regard to the name 'Feta'. pp. 1–12.

Costa, C., J. Aguzzi, P. Menesatti, A. Mànuel, C. Boglione, D. Sarriá, J.A. García, F. Sardà, J. del Río, F. Antonucci, V. Sbragaglia, M. Rampacci, R. D'Anbra and S. Cataudella. 2011. Versatile application of RFID technology to commercial and laboratory research contexts: fresh fish supply-chain and behavioural tests. pp. 1–3.

Crandall, P.G., C.A. O'Bryan, D. Babu, N. Jarvis, M.L. Davis, M. Buser, B. Adam, J. Marcy and S.C. Ricke. 2013. Whole-chain traceability, is it possible to trace your hamburger to a particular steer, a U.S. perspective. Meat Science. 95: 137–144.

Directive 2003/89/EC of the European Parliament and of the Council of 10 November 2003 amending Directive 2000/13/EC as regards indication of the ingredients present in foodstuffs (Text with EEA relevance). Official Journal of the European Union. 25.11.2003.

European Commission Directorate-General for Agriculture—Food Quality Policy in the European Union. 2004. Protection of Geographical Indications, Designations of Origin and Certificates of Specific Character for Agricultural Products and Foodstuffs—Working Document of the Commission Services. Guide to Community Regulations. pp. 1–46.

Finkenzeller, K. 2010. RFID Handbook. Fundamentals and Applications in Contactless Smart Cards, Radio Frequency Identification and Near-Field Communication, Third Edition. pp. 1–462.

Furness, A., UK, AIM and K.A. Osman. 2006. Developing traceability systems across the food supply chain: an overview. pp. 3–25. *In*: I. Smith and A. Furness (eds.). Improving Traceability in Food Processing and Distribution. Woodhead Publishing in Food Science, Technology and Nutrition, Cambridge, England.

Galimberti, A., F. De Mattia, A. Losa, I. Bruni, S. Federici, M. Casiraghi, S. Martellos and M. Labra. 2013. DNA barcoding as a new tool for food traceability. Food Research International. 50: 55–63.

General EAN.UCC Specifications. Basics and Principles of the EAN.UCC System. EAN International and Uniform Code Council, Inc., January 2005.

GS1 US. 2012. GS1 Global Location Number (GLN). Opportunities in the Foodservice Supply Chain. Foodservice GS1 US Standards Initiative Work group White Paper. pp. 1–14.

GS1 US. 2013. An Introduction to the Global Location Number (GLN). pp. 1–10.

http://archive.defra.gov.uk/foodfarm/food/industry/regional/foodname/products/documents/isle-of-man-queenie-pdo.pdf. Accessed 13/06/2014.

http://cornishnativeoysters.co.uk/2011Oysters-Fal.html. Accessed 13/06/2014.

http://ec.europa.eu/agriculture/quality/door/registeredName.html?denominationId=144. Accessed 13/06/2014.

http://ec.europa.eu/agriculture/quality/schemes/index_en.htm. Accessed 13/06/2014.

http://expoaid.gr/resources/cheese-encyclopedia/cheese-overview. Accessed 15/06/2014.

http://foodreference.about.com/od/bar_beverage/a/What-Is-Champagne.htm. 06/06/2014.

http://hccmpw.org.uk/index.php/tools/required/files/download?fID=2663.Accessed 13/06/2014.

http://mybarcodegraphics.com/index.php?main_page=page&id=7&chapter=1. Accessed 13/06/2014.

http://scotland.gov.uk/Resource/Doc/305128/0095733.pdf. Accessed 13/06/2014.

http://tracker.oukej.cz/wiki/images/Njord_report.pdf. Accessed 13/06/2014.

http://www.champagne.fr/en/terroir-appellation/appellation/appellation-origine-controlee-aoc. Accessed 10/06/2014.

http://www.deutsches-patentamt.de/docs/service/formulare_eng/marke_eng/w7729_1.pdf. Accessed 13/06/2014.

http://www.dpma.de/docs/service/formulare_eng/marke_eng/w7729_1.pdf. Accessed 13/06/2014.

http://www.foodstandards.gov.au/industry/safetystandards/traceability/pages/default.aspx. Accessed 13/06/2014.

http://www.fsai.ie/uploadedFiles/About_Us/forums/artisan/PDO_PGI_TSG_Info_Note.pdf. Accessed 13/06/2014.

http://www.google.com/search?client=safari&rls=en&q=infofoodnames.com&ie=UTF-8&oe=UTF-8. Accessed 04/06/2014.

http://www.google.gr/url?sa=t&rct=j&q=&esrc=s&source=web&cd=1&ved=0CCsQFjAA&url=http%3A%2F%2Fwww.fda.gov%2Fohrms%2Fdockets%2Fdailys%2F02%2FAug02%2F082902%2F80025b72.doc&ei=jHxuU4zbAuHL0AW374GgAw&usg=AFQjCNGBqmv76poAo7uwum4UNGJ4eYOAA&bvm=bv.66330100,d.bGQ. Accessed 10/05/2014.

http://www.greek-feta-cheese.com/production.html. Accessed on 18/06/2014.

http://www.gs1.no/sfiles/5/45/50/6/file/traceability-and-supply-chain-visualisation-with-tracetracker-gtnet-and-gs1-standards.pdf. Accessed 13/06/2014.

http://www.gs1us.org/DesktopModules/Bring2mind/DMX/Download.aspx?Command=Core_Download&EntryId=224&PortalId=0&TabId=785. Accessed 13/06/2014.

http://www.gtin.info/. Accessed 13/06/2014.

http://www.monicaegrossi.it/en/prosciutto-di-parma-dop.html. Accessed 15/06/2014.

http://www.tracetracker.com/LiteratureRetrieve.aspx?ID=43539&A=SearchResult&SearchID=1223254&ObjectID=43539&ObjectType=6.pdf. Accessed 13/06/2014.

http://www.upccode.net/upc-guide/gs1.html. Accessed 10/05/2014.

http://www.upccode.net/upc-guide/uniform-code-council.html. Accessed 10/05/2014.

https://www.agriculture.gov.ie/gi/pdopgitsg-protectedfoodnames/products/. Accessed 13/06/2014.

https://www.gov.uk/government/uploads/system/uploads/attachment_data/file/271080/pfn-beaconfell-lancashire-cheese.pdf. Accessed 13/06/2014.

https://www.gov.uk/government/uploads/system/uploads/attachment_data/file/271184/pfn-dorset-blue-vinny-cheese.pdf. Accessed 13/06/2014.

https://www.GOV.UK/government/uploads/system/uploads/attachment_data/file/271204/pfn-buxton-blue.pdf. Accessed 13/06/2014.

https://www.gov.uk/government/uploads/system/uploads/attachment_data/file/271263/pfn-orkney-lamb.pdf. Accessed 13/06/2014.

https://www.gov.uk/government/uploads/system/uploads/attachment_data/file/271297/pfn-scotchbeef-pgi.pdf. Accessed 13/06/2014.

https://www.gov.uk/government/uploads/system/uploads/attachment_data/file/271300/pfn-welsh-lamb.pdf. Accessed 13/06/2014.

https://www.gov.uk/government/uploads/system/uploads/attachment_data/file/271313/pfn-fal-oyster-pdo.pdf. Accessed 13/06/2014.

https://www.gov.uk/government/uploads/system/uploads/attachment_data/file/278748/welsh-beef-pgi.pdf. Accessed 13/06/2014.

Karlsen, K.M., B. Dreyer, P. Olsen and E.O. Elvevoll. 2013. Literature review: Does a common theoretical framework to implement food traceability exist? Food Control. 32: 409–417.

Kumar, S., B.B. Kadow and M.K. Lamkin. 2011. Challenges with the introduction of radio-frequency identification systems into a manufacturer's supply chain—a pilot study. Enterprise Information Systems. 5(2): 235–253.

Liu, H.-l., T.-l. Liu, Y.-f. Lin, C.-h. Hung, Y.-s.Cheng and L.-y. Wang. 2012. Application of RFID in the Production Traceability of Native Chicken. pp. 1–6.

Lupien, J.R. 2005. Food Quality and Safety: Traceability and Labeling. Critical Reviews in Food Science and Nutrition. 45: 119–123.

Mäder, P. and O. Gotel. 2012. Towards automated traceability maintenance. The Journal of Systems and Software. 85: 2205–2227.

Martinez, M.G. and F.M.B. Epelbaum. 2011. The role of traceability in restoring consumer trust in food chains. pp. 294–302. *In*: J. Hoorfar, K. Jordan, F. Butler and R. Prugger (eds.). Food Chain Integrity. A Holistic Approach to Food Traceability, Safety, Quality and Authenticity. Woodhead Publishing Limited, Cambridge, UK.

Mc Carthy, U., G. Ayalew, F. Butler, K. McDonnell and S. Ward. 2011. The case for UHF RFID application in the meat supply chain in the Irish context: a review perspective. Agricultural Engineering International. 13: 1–11.

McMillin, K.W. 2012. Traceability in the meat, poultry and seafood industries. pp. 565–595. *In*: J.P. Kerry (ed.). Advances in Meat, Poultry and Seafood Packaging, Woodhead Publishing Limited, Cambridge, UK.

Moe, T. 1998. Perspectives on traceability in food manufacture. Trends in Food Science & Technology. 9(5): 211–214.

Morreale, V., M. Puccio, N. Maiden, J. Molina and F.R. Garcia. 2011. The role of service orientation in future web-based food traceability systems. pp. 3–22. *In*: J. Hoorfar, K. Jordan, F. Butler and R. Prugger (eds.). Food Chain Integrity. A Holistic Approach to Food Traceability, Safety, Quality and Authenticity. Woodhead Publishing Limited, Cambridge, UK.

Morrison, C. 2003. Traceability in food processing: an introduction. pp. 459–472. *In*: M. Lees (ed.). Food Authenticity and Traceability. Woodhead Publishing Limited, Cambridge, UK.

Olson, T.K. and K. Criddle. 2005. Industrial Evolution in Response to Changes in the Demand for Traceability and Assurance: A Case Study of Chilean Salmon Aquaculture. Economics Research Institute Study Paper. 12: 1–21.

Parma Ham. Designation of Origin Specifications and Dossier. 1992. Prosciutto di Parma (Parma Ham) Protected Designation of Origin. Pursuant to Article 4 of Council Regulation (EEC) No. 2081/92 dated 14th July 1992. pp. 1–83.

Parreño-Marchante, A., A. Alvarez-Melcon, M. Trebar and P. Filippin. 2014. Advanced traceability system in aquaculture supply chain. Journal of Food Engineering. 122: 99–109.

Rasmussen, M.-L.R. 2012. A study of traceability and quality assurance in fish supply chains. PhD Thesis. DTU Food. National Food Institute. pp. 1–154.

Regattieri, A., M. Gamberi and R. Manzini. 2007. Traceability of food products: General framework and experimental evidence. Journal of Food Engineering. 81: 347–356.

Regulation (EC) No. 1830/2003 of the European Parliament and of the Council of 22 September 2003 concerning the traceability and labelling of genetically modified organisms and the traceability of food and feed products produced from genetically modified organisms and amending Directive 2001/18/EC. Official Journal of the European Union. 18.10.2003.

Resende-Filho, M.A., T.M. Hurley. 2012. Information asymmetry and traceability incentives for food safety. International Journal of Production Economics. 139: 596–603.

Shanahan, C., G. Ayalew, F. Butler, S. Ward and K. McDonnell. 2010. Implementation Protocol Utilising Radio Frequency Identification (RFID) and Biometric Identifiers; in the Context of Irish Bovine Traceability. pp. 105–120. *In*: C. Turcu (ed.). Sustainable Radio Frequency Identification Solutions, INTECH, Croatia.

Singh, A.K., S. Ghosh, B. Roy, D.K. Tiwari and R.P.S. Baghel. 2014. Application of Radio Frequency Identification (RFID) Technology in Dairy Herd Management. International Journal of Livestock Research. 4(1): 10–19.

Subhashrahul, S. and P.S. Kumar. 2012. RFID Technology & its Application for Detection of Removal of Fish Plate in Rail Tracks. IOSR Journal of Engineering. 2(2): 229–233.

The case for UHF RFID application in the meat supply chain in the Irish context: a review perspective. Agricultural Engineering International: CIGR Journal. 13(3): 1–11.

Tóth, Á., K.K. Gaál, Z. Turcsán, N. Ásványi-Molnár, B. Ásványi, J. Szigeti and H. Fébel. 2010. Tracking possibilities in the poultry sector—a review. ArchivTierzucht. 53: 328–336.

Townsend, A. and B. Mennecke. 2008. Applications of radio frequency identification (RFID) in meat production: two case studies. CAB Reviews: Perspectives in Agriculture, Veterinary Science, Nutrition and Natural Resources. 3: 1–10.

Traceability of Beef. Application of EAN.UCC Standards in implementing. Regulation (EC) 1760/2000. EAN International, 2002. pp. 1–24.

TRACE-I Guideline. 2003. EAN.UCC Traceability Implementation. TRACE-I Project Deliverable. pp. 1–106.

Traceability of Beef. Application of EAN.UCC Standards in implementing. Regulation (EC) 1760/2000. EAN International, 2002. pp. 1–24.

Trebar, M., M. Lotrič, I. Fonda, A. Pleteršek and K. Kovačič. 2013. RFID Data Loggers in Fish Supply Chain Traceability. International Journal of Antennas and Propagation. 2013: 1–9.

UCC. 2012. Global Location Number (GLN) Implementation Guide. Available at http://www.naesb.org/pdf3/weq_jiswg052108w1.pdf. Accessed 15/06/2014.

Wamba, S.F. and A. Wicks. 2010. RFID Deployment and Use in the Dairy Value Chain: Applications, Current Issues and Future. Research Directions. IEEE International Symposium on Technology and Society, 7–9 June 2010, Piscataway, New Jersey, USA. pp. 172–179.

PART B
Authenticity of Foods of Animal Origin

3

Meat Authenticity

Ioannis S. Arvanitoyannis, * *Konstantinos V. Kotsanopoulos*
and *Antonios Vlachos*

3.1 Introduction

Quality improvement is becoming more and more important nowadays. Different factors that can affect quality, such as origin of food, labelling and ingredients are now of high significance. Food authenticity is among the most important factors for consumers since they usually want to be sure of what they consume. It now becomes obvious that the high number of national and international laws for labelling and trade of foods, including meat, cannot effectively prevent adulteration. During the last years, a number of modern methods have been proposed for determining food authenticity. Food processors and retailers are also very interested in food authenticity in order to avoid the exposure to unfair competition from unscrupulous suppliers who would benefit through misenterpreting the food they are selling (Reid et al., 2006).

The majority of meat authenticity tests are mainly based on lipid or protein analysis, and have both advantages and disadvantages. Several factors such as the type, quantity, part, condition and processing of meat will be taken under consideration when deciding which analytical technique will be used (Rodriguez-Ramirez et al., 2011). Recently, food scandals, such as the bovine spongiform encephalopathy (BSE) scandal, resulted in the extensive use of DNA analysis. The polymerase chain reaction (PCR) technique is one of the most commonly used methods for assessing the authenticity of meat and meat products (Reid et al., 2006). The review of Mafra et al. (2008) describes the application of PCR-based techniques on foods of animal origin. Vallejo-Cordoba et al. (2005) concluded that the use of PCR-based capillary electrophoresis (CE) has been significantly increased during the last 10 years. It is important to mention that CE was used in the field of food analysis and authenticity

School of Agricultural Sciences, Department of Agriculture, Ichthyology and Aquatic Environment, University of Thessaly, Fytoko St., Nea Ionia Magnesias, 38446 Volos, Hellas, Greece.
* Corresponding author

assessment, but its application on meat authenticity was quite limited. During the last few years, PCR-based CE is applied for assessing meat authenticity (Prado et al., 2007; La Neve et al., 2008; Fajardo et al., 2009; Pun et al., 2009; Wang et al., 2010). The effectiveness of DNA-based techniques may be compromised as a result of the effect of storage, cooking and processing. It has been proved that the thermal processing of meat under high-temperature conditions significantly reduce the overall DNA yield (Aslan et al., 2009).

Luykx and van Ruth (2008) presented an overview of the main analytical methods as well as an insight on several recent studies and chemometric techniques. The chemometric analysis was substituted by analytical instrumentation with the potential of determining more than one ingredient in a sample, at each test. Chemometric techniques can be effectively applied for the identification of characteristic patterns and become more and more useful nowadays since the number of ingredients necessary for the discrimination of samples of different geographical origin increases. The application of multivariate analysis (MAV) on the assessment of authenticity is also mentioned. The organoleptic characteristics are statistically analysed. MAV (e.g., discriminant analysis) can be used to differentiate between different kinds of meat products, based on these characteristics.

A Swiss study reported that the food origin highly affected the purchase decision of 82% of the consumers, while the origin of meat was found to be of high importance for 71% of them (Franke et al., 2005). The most important parameters that affected this purchase decision were: health issues, media attention, particular gastronomic or sensory characteristics connected with regional commodities, friendly-to-the-environment production and reduced confidence in the quality and safety of food products produced outside the local region (Gilg and Battershill, 1998; Ilbery and Kneafsey, 1998; Luykx and van Ruth, 2008).

3.2 The Role of Specific Compounds as Meat Authenticity Markers and the Use of Instrumentation for Their Determination

PCR technique is one of the most important DNA-based methods, and a simple, specific and highly sensitive technique for assessing the authentication of species in foodstuffs (Fei et al., 1996; Ballin et al., 2009; Montowska and Pospiech, 2011). PCR has been used for the examination of various food products and is frequently applied by food authorities to reveal commercial frauds usually represented by incorrect labelling (Mafra et al., 2008). PCR, and especially real-time PCR (RT-PCR), is a very effective technique, which gives the potential to confirm the absence of specific allergens, genetically modified organisms (GMO), bovine meat and adulterations. Pork adulteration is a critical issue nowadays as concerns Halal verification. The contamination of Halal materials with pork involves the detection of a minute amount of porcine DNA, thus a highly specific and sensitive method is demanded for these levels of porcine. Real-time PCR is a technique commonly applied for identifying and quantifying the targeted species. Incorporation of a highly sensitive and specific probe can render the assay more specific and sensitive. However, derivation of PCR primers, either from nuclear DNA (nDNA) or mitochondrial DNA (mtDNA can have some effects on the sensitivity and specificity of the reaction as well as the quantification result (Mohamad et al., 2013). Cammà et al. (2012) achieved the development, standardisation and validation of a five species-specific Real-Time PCR protocols for identifying beef, pork and sheep meat in complex food products.

Specific primers and probes relied on 16S rRNA and cyt-b target genes located in the mitochondrial DNA were designed for the assays. The limit of detection of the Real-Time PCR methods ranged between 0.02 pg and 0.80 pg of template DNA, with an efficiency of 95% to 100%. All methods effectively detected the target species when added at a level of 1% in any other species, while no relevant difference was observed between the Ct values of raw and cooked samples. An Internal Amplification Control was applied for the detection of possible false negatives due to the presence of inhibitory substances in the sample matrix. Tests were performed on meat mixtures to assess the diagnostic sensitivity and specificity of the assays. Drummond et al. (2013) reported the development of a method for quantifying the bovine and buffalo content in milk- and meat-derived food products. Real-time PCR using primers designed to specifically amplify bovine or buffalo DNA were used. The assessment of the amplification efficiencies of both primer sets was performed using the TaqMan and SYBR Green systems. Both sets of primers provided satisfactory results. The process involves the amplification of a sample with both primer sets and the subsequent normalization of the total DNA using the total non-normalized bovine and buffalo DNA. The correction of potential deviations between the real and measured DNA quantity as a result of biological differences between species, can be performed using calibration curves generated from each analyzed matrix. These curves include a set of controlled admixtures of bovine and buffalo material. The use of this method for dairy samples containing known quantities of bovine and buffalo meat proved that the calibration curve always approximated the expected results. Further tests were performed proving that the method is very reliable and can be used to routinely analyse these products.

Furthermore, the PCR—restriction fragment length polymorphism (RFLP) method has been applied for generating DNA fingerprints of a high number of animal species (22), through amplification of particular regions within the cytochrome b gene and digestion of the amplified material by Hae III and Hinf I (Meyer et al., 1994b; Partis et al., 2000). The two restriction enzymes mentioned above led to effective discrimination, with the exception of kangaroo and buffalo, while thermal processing of the meat samples prior to examination did not affect the effectiveness of the enzymes. In the tested mixtures, pig presented an elevated degree of amplification, dominating over all species examined, even at very low levels (1%). Furthermore, quantitative competitive PCR (QC-PCR) has been used for detecting and quantifying porcine DNA, through the development of a new porcine specific PCR method, which uses the growth hormone gene of *Sus scrofa* and differentiates between contaminated and admixture products (Wolf and Lüthy, 2001).

In the study of Mane et al. (2012), a species-specific polymerase chain reaction (PCR) assay was developed to specifically detect of beef using self-designed primer pair based on D-loop region of mitochondrial gene for amplification of 513 bp DNA fragments from raw, processed and autoclaved meat and meat products. The beef-specific primer pair was self-designed based on the available gene sequences on NCBI nucleotide database. The optimization of the primer pair was individually performed in order to amplify 513 bp DNA fragments from isolated DNA of fresh beef. After successfully amplifying the desired DNA fragments using this primer pair, the evaluation of the PCR assay was performed, assessing its efficiency to amplify DNA obtained from cooked and autoclaved meat and meat emulsion. The level of detection of this beef-specific primer pair was less than 1% using PCR assay, and remained at this level even in admixtures of meat products that contained meat of beef, buffalo meat, pork, chevon and mutton.

The aim of the study of Natonek-Wiśniewska et al. (2013) was the development of a universal method for identifying bovine, porcine and ovine materials using PCR. The primers used generated short amplicons of 90, 85 and 67 bp for cattle, pigs and sheep respectively within the gene encoding COX1 in the case of ovine and porcine tissues and 12S rRNA for cattle. These primers only amplified the DNA of targeted species while no cross-reactivity was observed with the DNA of other species of animals and plants. The use of short amplification products for the indicators resulted in effectively identifying the species, both in raw and thermally processed samples. This method was proved to be effective for various animal products such as lard, animal meals, pet foods, plasma, whey, milk substitute, etc. According to the authors "the specificity and high sensitivity of the indicators, as well as the universality and usefulness of the method regardless of the degree of processing, type and form of the source material are its greatest advantages".

A polymerase chain reaction followed by a High Resolution Melting (HRM) analysis was employed for the detection of buffalo meat in six commercial meat products. A pair of specific 12S and universal 18S rRNA primers was used yielding DNA fragments of 220 bp and 77 bp, respectively. It was found that all samples contained buffalo meat and melting curves with at least two visible inflection points were obtained, deriving from the amplicons of the 12S specific and 18S universal primers. The existence of buffalo meat in the samples and the detection of adulterated buffalo meat with unknown species were established down to a level of 0.1%. The method was rapid and precise and could be successfully applied for performing authentication tests on meat products (Sakaridis et al., 2013).

Regarding mitochondrial genes, the cytochrome b (Matsunaga et al., 1998; 1999; Verkaar et al., 2002; Wolf et al., 1999), the 12S and 16S ribosomal RNA subunits (Bottero et al., 2003; Fajardo et al., 2006; Rodríguez et al., 2003a,b,c; 2004), and the displacement loop region (D-loop) (Fei et al., 1996; Gao et al., 2004; Montiel-Sosa et al., 2000) are among the most important markers that have been used for developing DNA-based techniques for species identification (Fajardo et al., 2007a,b). In addition, the use of 12S RNA mitochondrial gene has effectively been used for the identification of duck species in meat mixtures (Martin et al., 2007a).

Guoli et al. (1999) presented a PCR assay, which can be used to identify beef by amplifying bovine 1.709 satellite DNA. The selected sequence amplification of the 218 bp DNA fragment gave the potential to effectively amplify all raw, cooked and autoclaved DNA. Furthermore, pair of synthetically produced oligonucleotides flanked in this sequence, and genomic DNA derived from beef, were used as primers and templates, respectively. Following a 33-cycle process in a Taq DNA polymerase mixture, the resulted amplified materials were quickly electrophorized in 3% agarose gel and visualisation was performed due to the UV illumination of the ethidium bromide staining. The limit of detection was 33.6 fg of DNA from raw beef product and 0.32 pg of DNA from cooked or autoclaved beef, while all samples (n = 103) tested were correctly identified.

Ebbehoj and Thomsen (1991a) suggested the use of a quantification assay for pork in thermally processed meat samples, which was based on isolating DNA from meat and applying agarose gel electrophoresis for determining the average size of DNA fragments. After immobilising the DNA on nylon membranes, it was hybridized with a 32P-labelled probe derived from genomic porcine DNA. The method effectively detected a 0.1% and 0.5% of pork in beef, in raw and thermally-processed products, respectively.

The specificity of genomic DNA probes for species differentiation by slot blot hybridization has been applied for the examination of the differentiation between monkey and human and between cattle, goat and sheep (Ebbehoj and Thomsen, 1991b). In this study, the cross hybridization between probe and DNA sequences from closely related species was limited by adding unlabelled DNA from the cross hybridizing species, targeting at quantitatively differentiating between the species. The limits of detection for differentiating between cattle and sheep or goat and the closely related species sheep and goat were proved to be less than 0.01% and about 10%, respectively.

Koh et al. (1998) made an effort to establish a library of reference fingerprint patterns for a number of meat products from different species, using the random amplified polymorphic DNA (RAPD) technique. The generation of fingerprint patterns of 10 species (wild boar, pig, horse, buffalo, beef, venison, dog, cat, rabbit and kangaroo) was performed, while the most appropriate primer sets for characterising specific species were found and reported. The extraction of high molecular weight DNA was effectively performed with a satisfactory A260 : A280 ratio range from 1.6 to 2.0. As a result, a total of 29 10-nucleotide primers, with GC contents ranging from 50 to 80%, were assessed, including 50% GC (six different sequences), 60% (eight sequences), 70% (five sequences) and 80% (10 sequences). The resulted fingerprint patterns were proved to be species-specific, thus allowing the differentiation between species.

Matsunaga et al. (1999) used the PCR technique to identify the meat of cattle, pig, sheep, goat and horse, which are used as raw materials for foodstuffs. The correct ratios presented as a result of mixing seven primers showed the presence of species-specific DNA fragments after performing only one multiplex PCR. A conserved DNA sequence in the mitochondrial cytochrome b gene was the basis of a forward primer. Moreover reverse primers were based on species-specific fragments. The PCR primers led to differentiation between the length of the fragments from the different species. Specifically, the DNA fragments for each species were as follows: 157 bp for goat, 274 bp for cattle, 331 bp for sheep, 398 bp for pig and 439 bp for horse. Where the preparation of the samples was performed at high-temperature conditions, multiplex PCRs were used, while the DNA of those samples was amplified effectively for all species apart from the horse. The latter was not amplified effectively, due to its high length (439 bp). The limit of DNA detection for all species was 0.25 ng.

In the same way, red and sika deer were discriminated by digesting the PCR products using a restriction enzyme (EcoRI, BamHI or ScaI) and analysing the samples using 4% agarose gel electrophoresis. Although successful digestion was observed for the first species fragment when EcoRI was used (resulting in 67/127 bp fragments), the other two restriction enzymes did not digest the material. The digestion of the second species fragment led to 481/46 bp and 49/145 bp fragments by using the BamHI and ScaI, respectively, thus effectively differentiating between the two animal species (Matsunaga et al., 1998).

Pascoal et al. (2005) presented an innovative pair of primers for detecting bovine meat: CYTbos1 (forward) and CYTbos2 (reverse). The above primers can be used to specifically amplify a 115 base pair fragment of the bovine cytochrome b gene (cyt b) between nt 844 (mitochondrial site 15,590) and nt 958 (mitochondrial site 15,704), without significantly cross-reacting with DNA from twelve different meat species. The confirmation validity of the PCR technique was highly increased through the specific cleavage by the endonucleases ScaI and TspE1, while the sensitivity reached 0.025%.

Pascoal et al. (2004) used PCR coupled to RFLP to evaluate the incorrect labelling issues associated with foodstuffs containing meat from one or more species. The experiment was performed by using the primers CYT b1/CYT b2, in order to amplify a variable region of the mit cyt b of vertebrates, and the endonucleases PalI, MboI, HinfI and AluI. A quite extensive variety of meat products, both raw and/or cured, were examined revealing incorrect labelling in about 30% of samples, while 11.1% of the raw/cured products, and 34.2% of the thermally-processed foodstuffs were labelled incorrectly.

The PCR-RFLP technique was applied on the mitochondrial 12S rRNA gene aiming at identifying different species such as cattle, buffalo, sheep (mutton) and goat (chevon). The resulted PCR-amplified yield was a 456-bp fragment in each of the above-mentioned species. The fragments were digested using AluI, HhaI, ApoI and BspTI restriction enzymes resulting in a pattern capable of identifying and differentiating the different animal species. The restriction enzymes HhaI, ApoI and BspTI are related to unique sites in buffalo, sheep and goat sequences, respectively; while AluI led to 97 and 359 bp fragments in cattle and 246 and 210 bp fragments in sheep and goat. As a consequence, the interpretation of the results was very straightforward and the technique can be equally used for both fresh and processed meat products, but with limit efficiency when used in meat mixtures.

In the study of Fajardo et al. (2006), the PCR-RFLP method was applied to discriminate between red deer (*Cervus elaphus*), fallow deer (*Dama dama*), roe deer (*Capreolus capreolus*), cattle (*Bos taurus*), sheep (*Ovis aries*), and goat (*Capra hircus*). The experiment was performed by amplifying a conserved fragment from the mitochondrial 12S rRNA gene of ~712 base pairs. The MseI endonuclease was selected and collaboratively used with combined MboII, BslI and ApoI enzymes, allowing for the effective discrimination between the examined species. The MboII endonuclease led to fragmentation of the 12S rRNA gene materials of red deer and fallow deer into two DNA fragments of 384/328 and 489/223 bp, respectively. Conserved DNA sequences of 384, 223 and 105 bp were produced for roe deer, goat, cattle, and sheep, thanks to two identical restriction sites of the above endonuclease. Following the similar pattern, five DNA fragments were obtained due to the exploitation of four restriction sites for

BslI endonuclease in red deer and fallow deer 12S rRNA gene sequences. Four DNA fragments were obtained due to the presence of three restriction sites in the same enzyme, in the group of roe deer, cattle and sheep PCR-derived materials, while five DNA fragments were obtained from goat. All in all, when ApoI endonuclease was used to digest the samples, three DNA fragments of 553, 96 and 63 bp in red deer and fallow deer were obtained, while a unique restriction site of the roe deer sequence led to the generation of two DNA fragments of 649 and 63 bp. The identification of two restriction sites for this enzyme in sheep produced three DNA fragments of 412, 204 and 96 bp, while cattle and goat samples, with one ApoI restriction site, yielded two DNA fragments of 616 and 96 bp. The method led to effective differentiation between the different game meats and the meat of bovine, caprine, and ovine species.

In another study, Murugaiah et al. (2009) used the same method in the mitochondrial gene for differentiating between beef (*Bos taurus*), pork (*Sus scrofa*), buffalo (*Bubalus bubali*), goat (*Capra hircus*) and rabbit (*Oryctolagus cuniculus*) meat, as well as for Halal authentication. Smaller bands were effectively detected by using CE since it is characterised by higher resolution than agarose gel. As a result, the identification of PCR products of 359-bp was performed from the cyt b gene of the above-mentioned species using the AluI, BsaJI, RsaI, MseI and BstUI enzymes as potential restriction endonucleases. The genetic differentiation within the cyt b gene was carried out using PCR-RFLP indicating that the method can be used with high efficiency for detecting genetic differences. The resulted bands of beef were significantly less intense in comparison to those of pork, when presented at 1%, 3% or 5% in a meat mixture, but they could still be clearly detected on the agarose gel.

Kesmen et al. (2007) used a species-specific PCR assay for detecting small quantities of pork, horse and donkey meat in cooked sausages. The preparation of two series of binary meat mixtures was performed through addition of 0.0%, 0.1%, 0.5%, 1.0% and 5.0% of horse, donkey and porcine in minced meat. The determination of the detection limits of the particular PCR assay was carried out with PCR amplification of mitochondrial DNA after extraction of the latter at levels ranged between 0.01 and 100 ng of DNA in water. The determination of the sensitivity of the technique was defined as 0.01 ng DNA per species. By implementing the assay to DNA extracted from the samples, it was possible to detect each species when added in another species meat at a level as low as 0.1%.

Ilhak and Arslan (2007) used PCR to determine the origin of horse, dog, cat, bovine, sheep, porcine and goat meat. The preparation of the samples was performed through addition of 5%, 2.5%, 1%, 0.5% and 0.1% levels of pork, horse, cat or dog meat to beef, sheep and goat meat. After amplification, the species-specific fragments of the mtDNA were as follows: 439 bp for horse, 322 bp for dog, 274 bp for cat, 271 bp for bovine, 225 bp for sheep, 212 bp for porcine, and 157 bp for goat meat. For lowering the limit of detection below 1%, the PCR cycles were increased from 30 to 35 as previously reported by Meyer et al. (1994a), Hopwood et al. (1999) and Partis et al. (2000).

Dalmasso et al. (2004) used the multiplex PCR technique for analysing foodstuffs. In this experiment, a variety of species, including pork, was identified. Primer binding sites (PBS) were selected for the generation of particular amplimers shorter than 300 bp in length, offering the opportunity for application of the assay to DNA samples degraded by heat, following the treatment indicated by the EU Law for meat materials

(134.4–141.90°C and 3.03–4.03 bar for 24 min). The primers formed were adjusted to different regions of mitochondrial DNA and had alternate well-conserved regions (12S rRNA-tRNA Val), and a length of 290 bp. The detection limit was 0.002%.

In a similar experiment species-specific primers, based on the nucleotide sequence variation in the 12S rRNA mitochondrial gene, were generated by Martin et al. (2007b) to quantitatively detect and identify cat, dog and rat or mouse in both food and feedstuffs. The detection limit of the method in meat/oats mixtures was 0.1%, while the extensive heat-treatment (up to 133°C for 20 min at 300 kPa) had no effect on the effectiveness of the method, offering an opportunity for its use in origin verification of denatured raw materials of food and feedstuffs.

In the study of Montiel-Sosa et al. (2000) a straightforward, species-specific technique for detecting pork meat and fat in meat foodstuff was reported. The basis of the method was the species-specific primers of pork D-loop mtDNA. It could be characterised more effective than the nuclear 18S ribosomal RNA and growth hormone genes or Y chromosome PCR amplification, reported by Meyer et al. (1994a) and Meer and Eddinger (1996), as well as the porcine mtDNA described by Ghivizzani et al. (1993). As a result, the generation of the above-mentioned primers in combination with restrictive PCR amplification was used for identifying both raw and thermally processed pork meat and fat, using the presence of an amplified band of 531 bp. Moreover, the lack of this band in samples of bovine, ovine, and human, subjected to PCR amplification; and the ease discrimination among wild boar and pork amplified DNA, through digesting the samples with AvaII restriction enzyme (production of two bands of 286 and 245 bp vs none, respectively), make it quite efficient.

Fajardo et al. (2007a) evolved a PCR assay for discriminating between three caprinae species. The method was based on the selective amplification of mt D-loop sequences from chamois, pyrenean ibex, and mouflon by employing oligonucleotides, which targeted mitochondrial D-loop sequences. The application of mitochondrial DNA (mtDNA) sequences for identifying different species using PCR is commonly preferred against other genetic markers, i.e., the cell nucleus DNA; however its detection level may not be very high. On the other hand, the use of mtDNA renders the PCR amplification more sensitive due to the increased number of mtDNA copies per cell. Furthermore, the high number of mtDNA forms can lead to effective discrimination and identification of specific species in food mixtures (Girish et al., 2004). The D-loop region (~700–1000 bp) was initially amplified and sequenced from various game and domestic meat DNAs, and was used as a matrix for subsequently designing three primer sets. Therefore, for the PCR-amplified D-loop fragments from chamois (88 bp), pyrenean ibex (178 bp), and mouflon (155 bp) meats, a high degree of specificity and reproducibility were observed in comparison to numerous game and domestic meats. Sufficient positive results were observed as regards the amplification carried out during analysing meats treated under pasteurisation (72°C for 30 min) and sterilization conditions (121°C for 20 min), with a detection limit of ~0.1% for each one of the species examined.

The increased consumer demand for low-fat products, unavoidably led to increased consumption of game meat. Venison meat, widely consumed in Europe, is derived from red deer (*Cervus elaphus*), fallow deer (*Dama dama*), and roe deer (*Capreolus capreolus*). The different kinds of meat in this product can be identified using a selective

PCR amplification of DNA fragments on the mitochondrial 12S rRNA gene (Fajardo et al., 2007b). Using a common reverse primer, in combination with forward specific primers, the selective amplification of the desired cervid sequences was carried out. Amplification and sequencing of a conserved 12S rRNA gene fragment (~720 bp) from red deer, fallow deer, roe deer, chamois, mouflon, pyrenean ibex, cattle, sheep, goat, and swine meats was performed using the oligonucleotide primers 12S-FW and 12S-REV (Fajardo et al., 2006). After aligning the 12S rRNA gene sequences, specific primers were formed by analysing and comparing aligned 12S rRNA sequences. They consisted of a reverse primer (12SCERV-REV), common to all species of interest, and three forward species-specific primers: 12SCE-FW for red deer, 12SDD-FW for fallow deer and 12SCC-FW for roe deer. By combining each forward primer, with the common reverse oligonucleotide, DNA fragments of ~170–175 bp were generated in the 12S rRNA gene from each deer species. Having that in mind, PCR of species-specific primers targeting short DNA fragments could be successfully used for meat authentication, through identifying targeted species from a pool of different DNAs, thus eliminating the need for further sequencing or digestion of the PCR products using restriction enzymes.

According to Dooley et al. (2004), the development of the assays described below was accomplished around short (amplicons < 150 base pairs) regions of the mitochondrial cytochrome b (cytb) gene, their high speciation was connected with species-specific primers. As a consequence, the introduction of species-specific real-time PCR (TaqMan) assays for detecting beef, pork, and lamb was performed in combination with a TaqMan probe, specific to the these mammalian species. PCR was limited to 30 cycles under normal end-point TaqMan PCR conditions to DNA extracts from raw meat admixtures, offering the advantage of detecting each species when introduced in any other species at a level as low as 0.5% level. The limits of detection for beef and lamb were below 0.1%.

Ali et al. (2012) developed a method for detecting pork adulteration in meatballs, using TaqMan probe real-time polymerase chain reaction. The assay used a combination of porcine-specific primers and TaqMan probe for detecting a 109 bp fragment of porcine cytochrome b gene. Specificity test with 10 ng DNA of eleven different species yielded a threshold cycle (Ct) of 15.5 ± 0.20 for the pork and no positive results for the others. By analysing beef meatballs with spiked pork, it was proved that the assay could effectively determine 100–0.01% contaminated pork with 102% PCR efficiency, high linear regression ($r^2 = 0.994$) and ≤ 6% relative errors. Residuals analysis showed that the method was very precise in all determinations. Random analysis of commercial meatballs from pork, beef, mutton and goat, yielded a Ct between 15.89 ± 0.16 and 16.37 ± 0.22 from pork meatballs and no positive results from the others, thus confirming that the assay was suitable for determining pork in commercial meatballs both accurately and precisely.

Santos et al. (2012) employed a DNA-based technique for identifying hare meat. Species-specific primers were designed using mitochondrial cytochrome b gene and they were characterized by high specificity to Lepus species. The limit of detection was about 0.01% of hare meat in pork meat using polymerase chain reaction (PCR). A real-time PCR assay with the new intercalating EvaGreen dye was then used to specifically and rapidly identify hare meat. The assay was more sensitive (1 pg) than the end-point PCR (10 pg). It was proved that the new primers generated can be used by both species-specific end-point PCR or real time PCR to provide an accurate authentication of hare meat.

A Duplex PCR assay was employed by Di Pinto et al. (2005) for the determination of the adulteration of horse fresh sausages with pork meat. The process was performed by extracting DNA following a procedure in which total DNA was bound on a silica membrane followed by amplification of a fragment of the cytochrome b gene of mitochondrial DNA (mtDNA). This revealed the presence of two—439 and 398 bp— specific amplicons. Thus, it was proved that both the primer specificity and detection limit agreed with the values found by Matsunaga et al. (1999), while all bovine DNA samples gave negative results. This assay revealed traced pork meat in 6/30 while the absence of horsemeat was reported for 1/30 of the examined horse-sausage samples.

Targeting at restoring consumers' distrust to beef or beef-containing products due to the recent cases of BSE and variant Creutzfeld Jacob Disease (vCJD), Brodman and Moor (2003) employed the following specific beef detection system: TaqMan Universal PCR Master Mix, 0.3 m M of the primers BtaGH-1 and BtaGH-2 (Microsynth, Balgach, Schweiz), 0.1 mM of the probe BtaGH-S (Applied Biosystems). The reagents were set to the appropriate concentration with oligo storage buffer. General mammal detection system was: TaqMan Universal PCR Master Mix, 0.45 mM of the primers GH-1 and GH-4, 0.1 mM of the probe GH-S, oligo storage buffer for dilution of the reagents to the appropriate concentration. Amplified fragments, as short as 66 and 76 bp, were determined even in meat and bone meal (MBM) samples with a detection limit of 0.02 which corresponds to about 5 or 6 genome copies.

The detection of low levels of beef and pork in processed food samples was performed in the study of Laube et al. (2003) using two TaqMan TM-PCR systems. The method was based on the amplification of the phosphodiesterase gene (104 bp) (with cattle-specific primers) and the ryanodin gene (108 bp) (with swine-specific primers). The above-mentioned assay was further enforced using a third system, which effectively excluded false-negative results through detection of meat from sixteen different mammals or poultries in the samples examined. Nevertheless, it was proved that the method was not capable of effectively discriminating between cattle and deer species.

The quantitative detection of bovine, porcine and lamb, DNA in complex samples using six TaqMan real-time PCR systems with minor groove binding (MGB) probes was described by López-Andreo et al. (2005). This cost-effective assay was accomplished using a combination of only two fluorogenic probes and 10 oligonucleotide primers, which targeted mitochondrial sequences, demonstrating limits of detection from 0.03 to 0.80 pg of template DNA. It was proved that the above methodology could be used for the effective detection of more than 1% of pork and more than 5% of cattle or lamb in mixtures containing between two and four species.

Calvo et al. (2001) reported the development of a rapid, specific and even more sensitive PCR method for detecting pork in thermally and non-thermally processed meat, sausages, canned food, cured products and pâtés. According to this study, a new DNA-specific porcine repetitive element was isolated by non-specific PCR, was analysed and a pair of primers was synthesized. Through testing fifty-five pig blood DNA samples (from different breeds), it was proved that the method was both effective and highly specific; however no positive results were collected after analysing 200 samples from various species. The number of the PCR amplification cycles significantly affected the detection limit; 0.005% pork in beef and 1% pork in duck pâté after 30 and 20 cycles, respectively. It was therefore concluded that the identification of pork was effectively performed, while the method was very satisfactory in terms of cost, speed, simplicity and reliability.

The development of a PCR assay for the quantitative detection of pork adulteration in ground beef and pâté, in both thermally and non-thermally treated samples, using densitometry and a specific and sensitive repetitive DNA element was reported by Calvo et al. (2002). This is known as the Short Interspersed Nuclear Element (SINE) repetitive element, and is highly specific and sensitive when used for detecting pork. The detection limits of the method were identified at levels of up to 0.005%, 0.1% and 1% pork in beef (raw and heated) and pork in duck pâté, after 30, 25 and 20 PCR cycles, respectively. The optimal standard curve and correlation between pork content and band intensity was produced after completing twenty-five PCR cycles. As a result, the determination of fraud was effectively performed in commercial pâtés, and known samples containing 0% to 100% pork in beef after autoclaving at 50, 80 and 120°C for 30 min.

The importance of cytochrome b gene was also shown when used for detecting minimal quantities of horse or donkey meat in commercial foodstuffs. Chisholm et al. (2005) managed to develop real-time PCR assays specific for these species by employing primers, which were based on the mitochondrial cyt b gene, and their 3' position mismatched to closely related and other commercial species. There was therefore the potential to avoid using probes common to many mammalian species and specific primers or multiprimer assays with sense and species-specific antisense primers (Brodmann and Moor, 2003; Dooley et al., 2004; Walker et al., 2003). The prevention of amplification of non-target species DNA was achieved through truncation of primers at the 5' position, making them completely specific. Independent primer and probe sets for horse and donkey allowed an increase in specificity and led to amplicons with less than 150 base pairs. Both assays were highly sensitive and the limits of detection were 1 pg and 25 pg for donkey and horse template DNA diluted in water, respectively. The assays were tested in samples of horse or donkey muscle and commercial products containing horse, with high success.

Sawyer et al. (2003) developed a method, which was proved to be of high importance for the development of quantitative PCR assays, using universal and species-specific PCR primer pairs. By comparing the cycle number at which universal and species-specific PCR products were initially detected, together with using reference standards of pre-determined species content, a high variety of mixed samples (0.1–100%) of both beef and lamb admixtures were determined.

Yman et al. (1987) used starch gel electrophoresis and appropriate staining for examining the carcass residual blood containing both serum albumin and esterase, simplifying the examination of genetic variants of these proteins in meat derived from horse, donkey and their hybrids, mule/hinny. Due to the fact that no donkey albumin variants C and D and horse variants A, B and I were found in horses and donkeys, respectively, the differentiation between the two species and between their hybrids, was possible. Furthermore, the serum enzyme carboxylesterase was successfully used for segregating fresh meat from horses and donkeys due to the fact that it is not present in donkey but it can be found in horses and mules/hinnies. By examining the similarities in electrophoresis zones of serum albumin of the species of interest, it was proposed that the immune-diffusion or protein staining of IEF gels could be used as a starting point for identifying unknown samples that may contain horse, donkey or their hybrids. Then, discrimination of these species from other animals such as beef, swine, etc. must be carried out. Finally, determination of the genetic variants of serum albumin by stage and final confirmation by staining the other slice of the starch gel for serum carboxylesterase can take place.

Polyacrylamide gel electrophoresis provides specific protein patterns for a large variety of meat products, e.g., beef, pork, mutton, venison, and reindeer, while disc electrophoretograms can be used for the differentiation between different species and can be employed to quantitatively determinate beef and pork. Muscle extracts from beef, pork, venison, reindeer, mutton, and admixtures of the above, were examined by Skrökki and Hormi (1994) using electrophoresis. The quantities of beef and pork in commercial minced-meat mixtures were calculated using standard electrophoretograms, thus offering a valuable tool against meat adulteration. Specifically, a set of 49 minced-beef and 49 minced-beef-and-pork samples electrophoretograms were compared optically with standards obtained under the same conditions (potential 280 V, run time 1 h 30 min, amount of sample 30 μL). It was proved that it was quite easy to discriminate between pork, venison, and reindeer, but quite difficult to do so for beef, reindeer, and mutton due to the high similarity of their main bands. For these species, the problem of the high similarity of their bands can be solved by applying longer run times.

Saez et al. (2004) employed a single step DNA-based test to simultaneously identify multiple meat species. The test was based on the formation of species-specific fingerprintings by two different arbitrary DNA amplification approaches (RAPD- and arbitrarily primed PCR, AP-PCR) using the primers OPL-4 and OPL-5. The high variety of representative samples indicated the high applicability of the techniques. As a result, RAPD-PCR fingerprintings were used to easier discriminate among pork, beef and lamb, leading to a satisfactory species cluster at similarity levels ≥ 75%. The use of AP-PCR with the primer M13 was also effective for detecting five species-specific fingerprintings but also for identifying five tested species in every sample (although the generation of more complex patterns, including some low intensity bands was unavoidable). These difficulties were overcome through the introduction of a ramp time between annealing and extension temperatures, rendering the method suitable for meat authentication in routine analysis.

An important characteristic of a real-time PCR assay is the ability to directly monitor amplification products during each amplification cycle. Due to that, a quantification can be made early in the PCR process, providing more accurate results in comparison to the end point analysis. Moreover, using DNA binding dye like SYBR green, which adheres to the minor groove of the double-stranded DNA in a sequence-independent way, the technique becomes more flexible without the need for individual probe design and optimization stages. This technique was described by Fajardo et al. (2008) who used it to quantify red deer, fallow deer, and roe deer DNA in meat mixtures by combining cervid-specific primers—amplifying a 134, 169, and 120 bp of the 12S rRNA gene fragment, respectively—and universal primers—amplifying a 140 bp fragment on the nuclear 18S rRNA gene—from eukaryotic DNA. The Ct (threshold cycle) values obtained from the latter primers were applied in an effort to normalize those obtained from each of the former, functioning as endogenous control for the total content of PCR-amplifiable DNA in the sample. The resulted and quantified target cervid DNA ranged from 0.1 to 0.8%, depending on the animal species and the preceding treatment.

The real-time quantitative PCR, used by Rodriguez et al. (2005), for quantifying pork (*Sus scrofa*) in binary pork/beef muscle mixtures was mainly carried out by amplifying a fragment of the mitochondrial 12S ribosomal RNA gene (rRNA). The pork-specific primers were used for the amplification of a 411 bp fragment from pork DNA while mammalian-specific primers were used for the amplification of a 425–428 bp fragment from mammalian species DNA, which were used as endogenous control. The internal fluorogenic probe (TaqMan) was employed to monitor the amplification process of the target gene. As a result, by comparing the cycle number (Ct) at which the initial detection of mammalian and pork-specific PCR products took place, in combination with using reference standards of known pork content, it was possible to determine the percentage of pork in a mixed sample. The analysis revealed that the assay was highly specific and sensitive when used for detecting and quantifying pork in the range 0.5–5%. The detection limit for pork-specific PCR was 0.01 ng DNA, which equals to 0.1% pork DNA.

The possibility of adulterating of meat products with seagull meat is an issue addressed mainly in coastal cities. For eliminating any suspicions, a method was developed by Kesmen et al. (2013) for identifying and quantifying seagull meat in meat mixtures. For this purpose, a real-time polymerase chain reaction (PCR) assay, using species-specific primers and a TaqMan probe was designed on the mitochondrial NADH dehydrogenase subunit 2 gene. It was proved that the detection of the template DNA of seagull could be achieved at the level of 100 pg without detecting any cross-reactivity with non-target species (bovine, ovine, donkey, pork, horse). Through this method, the detection of seagull meat at the level of 0.1% in raw and heat-treated test mixtures was performed, by adding seagull meat in beef at different levels (0.01–10%). It was finally proved that the real-time PCR assay applied is a rapid and sensitive method for routinely identifying seagull meat in raw or cooked meat products.

Real-time uniplex and duplex PCR assays with a SYBR Green I post-PCR melting curve analysis was used by López-Andreo et al. (2006) in an effort to identify and quantify bovine, porcine, horse and wallaroo DNA in foodstuffs. The threshold-cycle (Ct) data, resulted from serial dilutions of purified DNA, permitted the quantification of these species, with the following limits of detection in uniplex reactions: 0.04 pg for porcine and wallaroo DNA and 0.4 pg for cattle and horse DNA. The species-specificity test of the PCR products was possible through identifying peaks in DNA melting curves, shown as a decrease of SYBR Green I fluorescence at the dissociation temperature (74.3, 73.2, 78.7 and 76.0°C, respectively). The duplex assay, which used either single-species DNA or DNA admixtures which contained different combinations of two species, finally allowed the following minimum proportions of each DNA species though the resolution of Tm peaks: 5% (cattle or wallaroo) in cattle/wallaroo mixtures, 5% porcine and 1% horse in porcine/horse mixtures, 60% porcine and 1% wallaroo in porcine/wallaroo mixtures, and 1% cattle and 5% horse in cattle/horse mixtures. The data extracted from SYBR Green I uniplex and duplex reactions with single-species DNA could be easily compared to those obtained from species-specific TaqMan probes.

According to Rastogi et al. (2007), mitochondrial markers were more effective than nuclear markers when used for identifying different species. As a result, the mitochondrial markers viz. 16S rDNA and nicotinamide adenine dinucleotide (NADH) dehydrogenase subunit 4 (ND4) were used together with a nuclear marker viz. the actin gene for the identification of specimens of animal origin for different reasons (forensic identification, food regulatory control, illegal trading prevention, poaching and conservation of endangered species). It was concluded that only one of the five tested primers (Operon I.D. C4, D5, F4, F7 and F5), the primer F4, could be used for RAPD-PCR fingerprinting, based on the number, intensity and distribution of bands, towards the clear discrimination between several species, such as buffalo, cow, goat and pig. Species identification based on intron sequences, present between exon 6 and 7 of actin gene, was not effective since, by using this nuclear marker, only the identification of chicken and pig samples was possible up to species level.

The identification and authentication of different meat muscle species, such as beef (n = 100), lamb (n = 140) and pork (n = 44), were carried out using visible infrared spectroscopy (VIS) and near infrared reflectance spectroscopy (NIRS) (Cozzolino and Murray, 2004). By scanning the homogenised materials in the region of 400 to 2500 nm, effective discrimination of muscle species was achieved in conjunction with both Principal Component Analysis (PCA) and dummy partial least-squares regression (PLS) models. As a result, more than 85% of the studied samples were correctly classified. The PLS models resulted in good classification when three wavelength segments (400–700, 1100–2500 and 400–2500 nm) were used.

Morsy and Sun (2013) evaluated the use of near infrared spectroscopy (NIRS) as a both rapid and non-destructive technique for the detection and quantification of different adulterants in fresh and frozen-thawed minced beef. Partial least squares regression (PLSR) models were generated under cross validation and tested with different independent data sets, providing determination coefficients (RP 2) of 0.96, 0.94 and 0.95 with standard error of prediction (SEP) of 5.39, 5.12 and 2.08% (w/w) for minced beef with small quantities of added pork, fat trimming and offal, respectively. It was shown that the developed models were less effective when the samples were frozen-thawed, yielding RP 2 of 0.93, 0.82 and 0.95 with simultaneous augments in the SEP of 7.11, 9.10 and 2.38% (w/w), respectively. Linear discriminant analysis (LDA), partial least squares-discriminant analysis (PLS-DA) and non-linear regression models (logistic, probit and exponential regression) were generated at the most relevant wavelengths towards the discrimination of pure and adulterated product. Both models classified the samples quite accurately, while the LDA, PLS-DA and exponential regression models were 100% accurate. It was thus proved that the VIS-NIR spectroscopy can be successfully used for the detection and quantification of adulterants added to minced beef giving precise and accurate results.

Alamprese et al. (2013) used UV-visible (UV-vis), near infrared (NIR) and mid infrared (MIR) spectroscopy, coupled with chemometric techniques for the detection of minced beef adulteration with turkey meat. Forty-four minced meat samples of pure bovine, forty-four of pure turkey and 154 mixtures of minced beef adulterated with turkey meat in the range 5–50% (w/w) were used for the analysis. Following standardisation of the spectral data using different pre-treatments, they were processed, separately or fused, using Principal Component Analysis (PCA), Linear Discriminant Analysis (LDA), and Partial Least Squares (PLS) regression. Furthermore, a variable selection method was carried out before classifying and applying regression analysis. By comparing the PLS models in terms of errors in prediction (RMSEP), it was found that the best results were obtained using NIR and MIR spectroscopy, whereas the UV-vis results were less satisfactory. Finally, the use of combined data from UV-vis, NIR and MIR spectroscopy led to a significant improvement of the overall results. According to Barbin et al. (2012), a hyperspectral imaging technique was designed to rapidly, accurately and objectively determine pork quality grades. Hyperspectral images were acquired in the near-infrared (NIR) range from 900 to 1700 nm for 75 pork cuts of *longissimus dorsi* muscle from three different quality grades (PSE, RFN and DFD). The spectral information obtained from the samples and six significant wavelengths explained of the variation among pork classes as shown from the 2nd derivative spectra. Obvious reflectance differences among the three quality grades were found mainly at wavelengths 960, 1074, 1124, 1147, 1207 and 1341 nm. Principal component analysis (PCA) was performed using the above-mentioned wavelengths and it was proved that the discrimination of pork classes could be accurately performed with overall accuracy of 96%. An algorithm was developed for the production of classification maps of the tested samples relied on score images obtained from PCA, and these results were compared with the results obtained from the ordinary classification method.

Investigation of the misclassified samples was carried out indicating that hyperspectral based classification can be useful in class determination, by showing the spatial location of classes within the samples. Barbin et al. (2013) evaluated the use of a pushbroom hyperspectral imaging system in the near-infrared (NIR) range (900–1700 nm), as a rapid and non-destructive technique for assessing meat freshness. Partial least squares discriminant analysis (PLS-DA) models were applied for the discrimination between fresh and frozen–thawed samples of longissimus dorsi. Optimal wavelengths were selected and used for sample discrimination with reduced spectral data and image processing. Classification models with limited spectral data allowed for an overall correct classification of 100% for an independent set of samples. The development of an image processing algorithm was also performed towards the visualisation of the classification results. The optimum classification model designed was applied to the images and effectively produced highly accurate classification maps. It was shown that a fast and reliable system for discriminating between fresh and frozen–thawed pork could be developed based on reflectance in the NIR wavelength range. Foca et al. (2013) used three techniques, i.e., tristimulus colorimetry, FT-NIR spectroscopy and NIR hyperspectral imaging, for discriminating fat samples derived from the two different layers (inner and outer subcutaneous layers). A variety of multivariate classification methods were employed including signal processing and feature selection techniques. The classification prediction obtained using colorimetric data was about 78.1%, while NIR-based spectroscopic methods were proved to be much more effective, reaching a prediction efficiency higher than 95%. In general, the samples of the outer were more variable than the samples of the inner layer. This is probably because of a higher variability of the outer samples as regards their fatty acid composition and water amount.

According to Gaitán-Jurado et al. (2008), the easiness of applying quality controls to typical Spanish sausages through the use of proximate analysis (fat, moisture and protein) on the finished food (intact and homogenized) is mainly relied to NIRS technology (diode array instrument). The method could be used to obtain quality measurements during storage, distribution and marketing of the final product. The selected models were calibrated and evaluated through cross and external validation. It was finally proved that, for end products, the coefficients of determination for calibration (r^2) for fat, moisture and protein were 0.98, 0.93 and 0.97, respectively, whereas for homogenised products the corresponding coefficients were 0.99, 0.98 and 0.97, respectively. It is important to mention that homogenisation led to lower values (0.71%, 0.41% and 0.95%, for fat, moisture and protein respectively).

NIRS was also used by Ortiz-Somovilla et al. (2005) for detecting meat mixture ingredients in Iberian sausages. After homogenisation, each of the five meat mixture treatments of Iberian (I) and/or Standard (S) pork was analyzed by NIRS [both fresh products (N = 75) and dry-cured sausages (N = 75)]. After selecting the most representative absorption peaks and bands, a discriminant analysis procedure was used to reveal the mixture prediction equations. The fresh products led to better results in comparison to the dry-cured sausages, with 98.3% and 91.7% (calibration) and 60% and 80% (validation) correct classification, respectively. The comparison of two instrumental modes of analysis ("Down-view" and "Up-view") was also performed by the same authors. The first was based on a spinning circular capsule, which was used to obtain spectral information at different points of the sample, whereas for the second, the instrument was inverted and the sample was placed directly over the quartz window. The results obtained from proximate analysis (fat, moisture and protein) were put into a calibration model with a diode array NIR spectrometer—spectral range 515–1650 nm—through multi-way partial least-squares regression (MPLS) and PCR (Ortiz-Somovilla et al., 2007). The values found by Kang et al. (2001) for fat, moisture and protein in ground pork sausages using NIRS of greater scanning range (400–2500 nm) were very similar to those reported by Ortiz-Somovilla et al. (2007) –0.98, 0.98 and 0.93 (R^2) and 1.38%, 1%, 0.83% (standard error of validation, SEP), respectively.

In the study of Chan et al. (2002), Vis-NIR spectroscopy was employed to non-destructively evaluate the qualitative characteristics of fresh pork loin. The qualitative attributes of the samples were used for the development of calibration models for a diode-array Vis/NIR reflectance spectrophotometer. Partial least squares (PLS) calibration models were developed for the quality attributes of interest. The prediction of XYZ tristimulus values was successful with $R^2 > 0.91$ and SECV < 0.96 for calibration, and $R^2 > 0.88$ and SEP ≤ 1.2 for validation. Less successful was the prediction of moisture, fat and protein, with $R^2 > 0.96$ and SECV < 0.007 for calibration, and $R^2 > 0.61$ and SECV < 0.007 for validation. The calibration for the other parameters (shear force, cook yield, pH, WHC and scores for colour, marbling and firmness) varied significantly providing a range of R^2 values from 0.584 to 0.002.

The differentiation between various types of bovine meat was performed by Alomar et al. (2003), using NIRS. The method was used to predict main chemical fractions on samples from two breeds, three muscles and six grading (Chilean system) categories namely V, A, C, U, N and O. After scanning (400–2500 nm) and analysing the samples for dry matter, crude protein, ether extract, total ash and collagen content, after freeze drying, it was proved that optimum calibrations for protein and fat were obtained with a second order derivative (2-20-20 and 2-10-10, respectively) and without any scatter correction of the spectral data. As a result, the discriminant analysis in conjunction with PLS technique and cross validation effectively differentiated between breed and muscle type for most samples. The results indicate that NIRS is a valuable technique that can be used for the identification of breed and muscle type in beef meat on an objective, composition-related basis and to rapidly and relatively accurately determine meat composition, particularly for protein and fat content.

NIRS was also used by Ding and Xu (2000) to identify beef hamburgers adulterated with 5–25% mutton, pork, skim milk powder, or wheat flour (accur. ~92.7%, increasing with the adulteration level). When a sample was positive, further prediction of the adulteration level was performed using calibration equations, especially as regards mutton, pork, skim milk powder, and wheat flour with standard errors of cross-validation of 3.33, 2.99, 0.92 and 0.57% and coefficients of variance of 0.87, 0.89, 0.99 and 1.00, respectively. Discrimination has also been successfully achieved for fresh and frozen-then-thawed beef (Thyholt and Isaksson, 1997). The study of Ballin and Lametsch (2008) on whole meat from *Bos taurus* (cow) and *Sus scrofus* (pig) revealed that the samples could be most effectively discriminated using a combination of analytical methods.

Furthermore, the differentiation between beef and kangaroo meat could also be effectively performed by comparing the NIR spectra. The obtained results from differences in chromatic values ($P < 0.001$) were due to the lower fat and higher polyunsaturated fatty acids concentrations in kangaroo meat in comparison to beef (Sinclair et al., 1982). Thus, selected wavelengths in the near-infrared region (2384, 2278, 2236, 2210, 2096, 1742, 1680, 1308, 1246, 1204 and 926 nm) were strongly related to C-H bonds and were used to explain the noticed differences. Furthermore, spectral bands at 2236, 2210 and 1680 nm were connected with unsaturated =C-H groups thereby improving the differentiation potential of the polyunsaturated fatty acids towards the classification (Ding and Xu, 1999).

Prieto et al. (2008) employed NIRS to investigate the physical parameters of adult steers (oxen) and young cattle meat samples in order to evaluate their quality attributes. As a result, samples of *Longissimus thoracis* muscle obtained from young animals reared under extensive conditions were analyzed for pH, colour (L*, a*, b*), water holding capacity (WHC) and Warner–Braztler shear force (WBSF). The meat colour values are usually highly important since they are closely related to factors such as freshness, ripeness, desirability and food safety and often affect dramatically the consumers' decision. The effective prediction of these parameters in young cattle meat samples was highly correlated with the intramuscular fat content and was characterised by higher accuracy (L*—redness; $R^2 = 0.869$; SECV = 1.56 and b*—yellowness; $R^2 = 0.901$; SECV = 1.08) in comparison to oxen samples.

Al-Jowder et al. (1999) used mid-infrared (MIR) spectroscopy in an effort to authenticate beef and ox kidney and liver. At any spectra recorded (800 to 4000 cm⁻¹), sixty-four interferograms have been co-added and a triangular apodization was used prior to Fourier transformation. All absorbance spectra were truncated to 470 data points in the range of 900–1800 cm⁻¹ (the "fingerprint" region), which is frequently the most useful part of the MIR spectrum. High variation between the sets of spectra was observed for the group of features in the region 1000–1200 cm^{-1}, present in the spectra of liver and attributable to the glycogen content of the specimens. The fat concentration was determined by the slight shoulder at 1744 cm^{-1} and was highest in the brisket and silverside specimens, lower for the cuts of neck, while the lowest values or even absence reported for offal.

Rannou and Downey (1997) discriminated between raw pork, turkey and chicken meat (n = 74) using recorded spectra in the visible, NIR and MIR regions. Discriminant models were developed leading to positive results in all ranges. Specifically, the best results from NIR spectra were obtained using the 400–1100 nm range, with seven principal components working optimally for all calibration samples and giving successful results on 91.9% of the cases. The development of the optimal factorial discriminant model was based on MIR spectra using five principal components for all calibration samples and led to 86.5% correct identification in the validation step. It was proved that the optimum results were obtained by combining NIR and MIR spectra and by using the data in the wavelength ranges 400–750 plus 5000–12500 nm. Six principal components were used to efficiently and effectively classify the calibration sample set in comparison with the 94.6% of samples in the validation set from the poultry while some overlap of chicken and turkey samples was observed. The individual spectral regions were effectively separated.

Xia et al. (2007) employed a fibre optical probe for counting spatially resolved diffuse reflectance from beef samples in Vis-NIR bandwidth of 450–950 nm. It is widely accepted that beef absorption coefficients are highly connected with the chemical synthesis of the samples, such as the myoglobin content; while scattering coefficients are mainly related to meat structural characteristics such as sarcomere length and collagen content. Due to the fact that the evaluation of the beef tenderness is also affected by the above-mentioned parameters, the hypothetical existence of a correlation between optical scattering and cooked WBSF was justifiable. Therefore, the higher the scattering coefficient, the higher the cooked WBSF association, as was proved from the linear regression analysis (p < 0.0001 vs. R^2 of 0.59, respectively).

Visible and near infrared reflectance spectroscopy (VIS–NIRS) was employed for the discrimination of meat and meat juices from three livestock species. In a first trial, samples of Longissimus lumborum muscle, corresponding to beef (31), llamas (21) and horses (27), were subjected to homogenization and their spectra were collected in reflectance (NIR Systems 6500 scanning monochromator, in the range of 400–2500 nm). In the second trial, the samples (20 beef, 19 llamas and 19 horses) were scanned in folded transmission (transflectance). The development of discriminating models (PLS regression) was performed against "dummy" variables, testing different mathematical treatments of the spectra. The best models achieved identification of almost all species by their meat (reflectance) or meat juice (transflectance) spectra. A few (three of beef and one of llama, for meat samples; one of beef and one of horse, for juice samples) were classified as uncertain. It was finally proved that NIRS is an effective method of discrimination of meat and meat juice from beef, llamas and horses (Mamani-Linares et al., 2012).

Sultan et al. (2004) developed an innovative assay, relied on Western blotting, for detecting brain tissue in heat-treated meat products through developing central nervous system (CNS)-specific antigens. The visualisation of the bands of antigen-bound primary antibodies were performed using secondary anti-antibodies labelled with peroxidase, which was required for the production of chemiluminescence for the photographic film documentation. Ponceau-S staining was applied before antibody incubation, providing an intermittent control for the efficacy of the extraction and blotting procedures by judging the intensity of the protein profile on the blotted paper. The molecular mass information on detected antigens after immunoreactions gave further information in order to correctly identify brain tissue in the meat products. Thus, the B50/growth-associated protein (B50), glial fibrillary acidic protein (GFAP), myelin basic protein (MBP), neurofilament (NF), neuron-specific enolase (NSE) and synaptophysin proteins were found in raw luncheon meat and a liver product that contained brain tissue at a level of 5% (m/m). However, only MBP and NSE were proved to be suitable biomarkers for detecting 1% (m/m) brain tissue in meat products after pasteurisation at 70°C or sterilisation at 115°C. On the other hand, the use of an anti-monkey MBP instead of anti-human MBP allowed speciation of the CNS material from bovine, ovine or porcine brain tissue.

One of the most common adulterants in meat foodstuffs, the addition of non-declared and lower-quality protein, was examined by Flores-Munguia et al. (2000) using an immunodiffusion assay in agar gel. The added quantities of protein from bovine, porcine, equine and avian species, in uncooked commercial hamburger and Mexican sausage (chorizo) were effectively identified. Thus, in a total of 40 samples from local food stores, undeclared equine species was found in nine of the 23 of the hamburger meat samples, and in five of the 17 chorizo samples.

Skarpeid et al. (1998) used intensity profiles from isoelectric focusing of water-soluble proteins in conjunction with a multivariate calibration model to identify animal species in ground meat mixtures. The isoelectric focusing images in immobilized pH gradients from samples were enriched with different quantities of beef and pork meat and were analysed and converted to a digital format. The images were further analysed to reduce the background and optimize the signal strength, thereby providing profiles, which were used to determine the composition of the samples with prediction errors of about 10%.

The high significance of the development of a clear, trustworthy, quick, sensitive and highly specific identification method for species in meat samples, led to the implementation of enzyme-linked immunosorbent assay (ELISA) (Asensio et al., 2008). Martin et al. (1998) effectively used ELISA to quantitatively evaluate the adulteration of raw ground beef with pork. Chen and Hsieh (2000) reported the development of an ELISA characterised by a monoclonal antibody to a porcine thermal-stable muscle, for detecting pork in cooked meat foodstuffs. The method can be used for the detection of porcine skeletal muscle, but not for detecting cardiac and smooth muscles or blood and non-muscle organs. There was no cross-reactivity with common proteins found in foods. The detection limit was about 0.5% (w/w) pork in meat combinations. The intra- and inter-assay coefficients of variation were defined at 5.8 and 7.9%, respectively. The analysis of market samples indicated that the technique was 100% accurate. The verification of the results was performed through product labelling and a confirmation was obtained using a commercial polyclonal antibody test kit. Furthermore, in the study of Liu et al. (2006), a monoclonal antibody-based sandwich ELISA was used to detect porcine skeletal muscle in raw and thermally treated meat and feed products. It was proved that heat processing of samples up to 132°C for 2 h had no effect on the efficiency of the assay. Two monoclonal antibodies (MAbs 8F10 and 5H9) characterised by specificity to skeletal muscle troponin I (TnI) were used.

The assay was effectively used for the detection of 0.05% (w/w) of pork in chicken, 0.1% (w/w/) pork in beef and 1% commercial meat and MBM enriched with an unknown portion of pork, in soy-based feed. The method was proved to be both rapid and accurate. According to Cawthorn et al. (2013), various molecular techniques were applied for the evaluation of the extent of meat product in South Africa. A total of 139 processed meat products (minced meats, burger patties, deli meats, sausages and dried meats) were collected from retail outlets and butcheries in South Africa. The enzyme-linked immunosorbent assay (ELISA) was used to detect undeclared plant proteins (soya and gluten) in the samples. A commercial DNA-based LCD array was employed for screening the samples and detecting the possible presence of 14 animal species. The results were confirmed using a species-specific polymerase chain reaction (PCR) and in some cases also DNA sequencing. It was proved that 95 of 139 (68%) samples contained species, not declared on labelling, while the highest percentage of mislabelling was related to sausages, burger patties and deli meats. Soya and gluten were found as undeclared plant proteins in > 28% of samples, while undeclared pork was found in 37% of samples. Species such as donkey, goat and water buffalo were also detected in some products. It was finally proved that the mislabelling of processed meats is quite common in South Africa.

The adulteration of meat products through addition of low cost and/or quality meat is quite common in cheap fast foods, such as hamburgers, where meat of various animal species can be used. Regarding this product, Macedo-Silva et al. (2000) used the dot-ELISA and anti-sera to bovine, swine and horse albumin detecting homologous species at portions as low as 0.6%. The developed anti-albumin anti-sera was used for the detection of the above-mentioned species with satisfactory specificity and sensitivity. The commercial samples examined were not found to be adulterated.

The qualitative identification of hemoglobins from cow, lamb or pork meat samples (extracted with Milli-Q water and filtered on a cellulose acetate filter) was performed using cation exchange chromatographic separation and diode array detection from the different peak patterns at 416 nm. The common characteristic in the chromatograms of the examined meat species was an intense peak at its beginning: 1.6 min for the pork meat, and 1.7 min and 1.9 min for cow meat and lamb meat partially separated, which could correspond to myoglobin or denatured hemoglobin, both being strongly retained on the analytical column during the analysis. The characteristic hemoglobin peaks for pork, cow and lamb were eluted at 27.5 min and at 30.5 min, at 8.7 min and 13.2 min, and at 8.6 min, 11.3 min and 13.3 min, respectively. Thus, the detection of approximately 10% of a meat species in another can effectively be performed, while lamb and cow meat cannot be easily differentiated because of the co-elution of characteristic peaks (Wissiack et al., 2003).

The high-performance liquid chromatography method with electrochemical detection (HPLC-EC) was used by Chou et al. (2007) for the differentiation between fresh or cooked meat products from fifteen food animal species, such as cattle, pigs, deer, horses, etc. Satisfactory coefficients of variation (< 6%) were obtained for species-specific markers displaying reproducible peak retention times, irregardless different runs, body regions and subjects. It is important to note that either the incubation of fresh beef and pork at room temperature for 24 h or the repeated freezing and thawing, minimally affected the intensity and by no means the pattern of species-specific peaks. In the study of Giaretta et al. (2013), myoglobin was used as a molecular marker for evaluation of the non-declared meat content in raw beef burgers. Ultra-performance liquid chromatography (UPLC) was used for separating and identifying edible animal species (beef, horse, pig and water buffalo). Sodium nitrite was used for the pre-treatment of samples in order to transform oxymyoglobin and deoxymyoglobin to the more stable metmyoglobin. The validation of the method was performed through the preparation of mixtures with different percentages of pork and beef minced meat. It was proved that using Mb as marker, the detection of 5% (25 mg/500 mg) of pork or beef meat can be achieved in premixed minced meat samples.

A comprehensive summary of methods used for determination of meat authenticity is given in Table 3.1.

3.3 Geographic Origin Determination

According to Verbeke and Viane (1999), one of the consumers' strongest demands as regards food and food products over at least the last two centuries is reliability for guaranteeing quality, safety and animal welfare. However, this is not possible without reliable scientific data obtained with independent technological means that can protect regional designations and ensure fair competition (Ilbery et al., 2000). The need for supply of such information becomes even more intense when the consumers' awareness is related to health concerns and animal diseases such as BSE.

Both globalisation phenomenon and the ease of transport of goods between countries and continents are a constant source of awareness. The origin of various food materials, directly associated with specific geographic regions, is protected by prescribed techniques such as the European adapted name schemes PDO (protected designation of origin), PGI (protected geographical indication) and CSC (certificate of specific character) (Regulation (EEC) No. 2081/92). Nevertheless, this is a highly complicated subject since geographic origin can easily be affected by various parameters such as water, feeding, feed supplements, breed-strain, geology, pollution and housing system (Franke et al., 2005).

Towards that, numerous regulations and methods have been adopted or applied (Regulation (ECC) No. 2081/1992; Regulation (EC) No. 2772/1999; Renou et al., 2004; Schmidt et al., 2005; Hintze et al., 2001; Kelly et al., 2005; Piasentier et al., 2003) and a recent review by Luykx and van Ruth (2008). Isotope Ratio Mass Spectrometry (IRMS) analysis is one of the most important techniques for determining the geographical origin of foods, since their isotopic compositions reflect many factors in natural environment. This results from the high-precision analysis of numerous naturally present stable isotope pairs of light elements in materials of biological origin such as honeys, juices, spirits, wines, oils and various foods of plant origin (Kelly, 2003; Renou et al., 2004; Boner and Förstel, 2004; Rossmann, 2001). Recently, the IRMS technique has been implemented to authenticate organically farmed Atlantic salmon (Molkentin et al.,

Table 3.1. Meat authenticity: meat species/type, treatment, quality control (QC) method, correct classification, and advantages and disadvantages of the applied method.

No. of samples	Type of meat/ product(s)	Treatment/ Package	Type of meat(s) to be detected/ excluded	QC method	Special remarks	LOD/% correct classification	Advantages and disadvantages	References
-	Beef	Cooked at 150°C and 200°C for 30 min	Seagull meat	TaqMan probe-based RT-PCR	Mitochondrial NADH dehydrogenase subunit 2 gene	Seagull meat detected at the level of 0.1% in raw and heat-treated test mixtures	Rapid and highly specific	Kesmen et al., 2013
180	Complex food products	Minced, homogenised, boiled for 10 min	Beef, pork and sheep	RT-PCR	Cyt-b target genes	95–100%	Fast, efficient	Cammà et al., 2012
10	Red meat animal species	No	Wild boar, cat, dog, horse, venison, rabbit, kangaroo and pig	RAPD-PCR	29 10-nucleotide primers	-	Rapid Qualitative Reproducible Discriminative	Koh et al, 1998
-	Beef meatballs	Mixture of beef with tapioca starch, salt, spices. Emulsification, cooking for 15 min	Porcine	TaqMan probe RT-PCR	109 bp fragment of porcine cytochrome b gene	≤ 6% relative errors	High precision and accuracy	Ali et al., 2012
-	Sausages	Cooked	Swine, beef and donkey	ss-PCR	ss-primers of (bp): 227 153 145	1% (w/w)	Rapid Cost effective	Kesmen et al., 2007

No.	Product	Treatment	Species	Method	Target / Primers		Detection limit	Characteristics	Reference
103	Beef	Raw Cooked Autoclaved	Equine, swine, ovine, goat, deer, camel, donkey, mule, mouse, buffalo and yak	PCR	bovine 1.709 satellite DNA		33.6 fg 0.324 pg 0.322 pg	Sensitive Specific Rapid Convenient	Guoli et al., 1999
18	Various products	Raw Cooked at 100°C and 120°C	Cattle, pig, sheep, goat and horse	PCR	mt *cyt b* gene 6 primers		0.25 ng	Rapid Sensitive Qualitative	Matsunaga et al., 1999
13	Foodstuffs and SRMs	Severe heat-processing under overpressure	Cattle, sheep, goat, pig, wild boar, roe deer, ovine, rabbit and horse	PCR	CYTbos1(fw) CYTbos2(rv)	115 bp specific amplif. of mt *cyt b*	2.5 pg bovine DNA	Direct Specific	Pascoal et al., 2005
60	Several (very commercial)	Wide variety of raw and cooked methods	Beef, pig, wild boar, deer, roe deer, rabbit, sheep, goat and horse	PCR-RFLP	CYTb1/CYTb2 universal primers		-	Rapid Easy-to-perform Qualitative	Pascoal et al., 2004
16	Commercial	Raw	Cattle, beef, sheep and goat	PCR-RFLP	mt 12S rRNA		-	Qualitative Good efficiency	Girish et al., 2005
150	Dry salt-cured Heated	Raw muscle	Fallow deer, cattle, sheep, goat, roe deer	PCR-RFLP	mt 12S rRNA 12S-FW and 12S-REV primers		-	Simple Rapid Cost effective	Fajardo et al., 2006
	Muscle of: Camel, pirenean ibex, mouflon	Raw	Red deer, fallow deer, roe deer, cattle, sheep, goat, and swine	PCR	mt D-loop region		0.1%	Rapid Straight-forward	Fajardo et al., 2007a
27	Beef Sheep Goat	Raw	Heef, dog, cat, and pig	PCR	ss-mtDNA primers 35 PCR cycles		0.1%/-	Rapid Easy Reliable	Ilhak and Arslam, 2007

Table 3.1. contd....

Table 3.1. contd.

No. of samples	Type of meat/ product(s)	Treatment/ Package	Type of meat(s) to be detected/ excluded	QC method	Special remarks		LOD/% correct classification		Advantages and disadvantages	References
35	Commercial	Raw Autoclaved	Ruminants } Poultry } Pork	Multiplex-PCR	Primers based on: 12S Rrna tRNA Val } 16S rRNA		0.004%/- 0.002%/-		Advanced Wide use	Dalmasso et al., 2004
11	Meat mixtures	Heated Dry-cured	Pork	PCR-RFLP	A 531 bp of the D-loop mt DNA region		5%/-		High ss primers Reliable Rapid	Montiel-Sosa et al., 2000
300	Venison (mixt of roe deer, red deer and fallow deer)	Raw Pasteurized Dry-cured	Mouflon, piyrenean ibex, cattle, sheep, goat, swine, horse and rabbit	PCR	~170–175 bp fragments mt12S rRNA gene		-/-		Reliable High specific Rapid, Simple	Fajardo et al., 2007b
25	Meat mixtures	Minced Lean	Pork, buffalo and lamb	ssRT-PCR	A <150 bp fragment of mt cyt b gene		0.5%/- 0.1%/-		Reliable Accurate	Dooley et al., 2004
30	Horse sausages	Fresh	Pork	Duplex-PCR	Two 398 and 439 bp fragments of mt cyt b gene		-		Qualitative Rapid Accurate	Di Pinto et al., 2005
21	Complex meat products/samples	Freeze-dried	Pork, cattle and lamb	RT-PCR	TaqMan^R MGB probe		>1% 0.03–0.80 pg >5%		Cost effective Quantitative Good accuracy	López-Andreo et al., 2005
69	Numerous commercial products	Heated Canned Sausages	Pork	PCR	SINE motif 161bp ampl. fragment	Cycles 30 20	0.005%/- 1%/-		Rapid Simple Specific Sensitive Cost effective	Calvo et al., 2001; 2002
10	Beef	Pâtés Raw				20	1%/-			
-	Commercial products	Raw meats Sausages Salami, etc.	Horse and donkey	RT-PCR	mt cyt b gene	119bp 69bp	-/25 pg -/1 pg		Specific High sensitive Wide applicability	Chisholm et al., 2005

No.	Sample	State	Species	Method	Primers: Species specific	Universal		2%/-	Quantitative / Ease of handle of unknown samples	Reference
8e	Beef and lamb admixtures	Raw	Buffalo	RT-PCR	COW1 COW2	UNIV P UNIV Q	104 bp / 237 bp	-	Rapid; Low cost; Large N°samples/run	Sawyer et al., 2003
96	Beef-meat mixtures	Minced	Pork, lamb, venison, mutton	PAGE	280–300 V 1½–2¼ h				Rapid; Low cost; Large N°samples/run	Skrökk et al., 1994
18	Meat and meat products	Raw	Pork, buffalo and lamb	RAPD-PCR AP-PCR	Primers: OPL-04 and OPL-05 M-13			Differentiation ss-detection	Reproducible; Rapid; Simple	Saez et al., 2004
15	Meat mixtures in swine	Raw	Red deer, Fallow deer and roe deer	RT-PCR	Primers: 134 bp 169 bp 120 bp	of 12SrRNA		0.1–0.8%/-	Accurate; Simple; Relatively Rapid	Fajardo et al., 2008
80	Beef mixtures	Raw	Pork	RT-PCR	Primer: 411 bp	of mt 12SrRNA		0.5–5.0%	Quantitative; Rapid; High-sensitive	Rodriguez et al., 2005
72	Food products and mixtures	Raw	Wallaroo, pork, horse and cattle	RT-PCR uniplex duplex	205 & 100 bp 250 & 92 bp	SYBR Green I TaqMan probe		-/0.04 pg -/0.4 pg	Simple/cost effective	López-Andreo et al., 2006
8	Meat and meat mixtures	Raw	Pig, beef, cow, frog and snake	PCR-: RAPD actin barcoding	ss "fingerprinting"	mt 16SrDNA & ND4 actin gene		-	Discriminative; Accurate & efficient/-	Rastogi et al., 2007
7	Meat and meat mixtures	Raw	Pork, lamb and beef	HATR FT-IR Raman spect/py	942, 988, 1382, 1413, 1444, 1575, 1606 and 1729 nm	GA-D-MLR PC-DFA		-	Rapid; Cost effective	Ellis et al., 2005
332	Muscles	Raw and Homogenized	Beef, lamb, chicken and pork	Vis-NIR	400–2500 nm	PLS PCA DA		> 85% correct classification	Reliable; Rapid; Objective	Cozzolino and Murray, 2004

Table 3.1. contd....

Table 3.1. contd.

No. of samples	Type of meat/ product(s)	Treatment/ Package	Type of meat(s) to be detected/ excluded	QC method	Special remarks		LOD/% correct classification	Advantages and disadvantages	References
100	Spanish sausages	Dry-cured (sliced)	Homogenized	DAD-NIRS 400–1700 nm	Fat Moisture Protein	PCA	-	Rapid Cost effective Non-distractive	Gaitán-Jurado et al., 2007
150	Iberian pork	Homogenized: Dry-cured Fresh	Standard pork and admixtures	NIRS 400–1700 nm	critical wavelengths: 975 nm ~1210 nm ~1455 nm	DA PCA	> 80% correct classification	Rapid Cost effective Non-distractive	Ortiz-Somovilla et al., 2005
127	Several types of bovine meat	Minced Frozen Thawed	Handle Type	NIRS 400–2500 nm		PLS ANOVA	Good accuracy	Rapid Effective	Alomar et al., 2003
194	Beef hamburgers	Raw Cooked Minced	5–25% of mutton, pork, skim milk powder, wheat flour	NIRS 400–2500 nm	966, 1212, 1396, 1732, 1748, 1870, 1900, 2310 and 330 nm.	CDA KNN PCA	Up to 92.7% accuracy	Rapid	Ding and Xu, 2000
41	Beef	Frozen then: Minced or Cut	Kangaroo	VIS/NIRS 400–2500 nm	926, 1204, 1246, 1308, 1680, 1742, 2096, 2210, 2236, 2278 and 2384 nm	CDA MLR	Up to 100% accuracy	Rapid Effective	Ding and Xu, 1999
120	Cattle	Muscle (raw)	Ox	NIRS Physical properties	1100–2500 nm pH, L*, a*, b* colour, WHC, WBSF	Colour parameters	Successful	-	Prieto et al., 2008
185	Beef Ox kidney Ox liver	Minced	Fresh frozen-thawed mixtures	MIRS	900–1800 nm	PLS/CVA, SIMCA, and PLS regression	< 10%	-	Al-Jowder et al., 1999
32	Beef muscles	Raw	Tenderness	VIS-NIRS	450-950 nm	LGA	-	Non-destructive Practical Fundamental Sensitive	Xia et al., 2007

	Products of: Meat Liver	Heated	Brain tissue	Western blotting	Biomarkers: MBP and NSE SDS-PAGE	1% (m/m)	70°C pasteurization 115°C sterilization	Advanced Sensitive High throughput	Sultan et al., 2004
2	Products of: Meat Liver	Heated	Brain tissue	Western blotting	Biomarkers: MBP and NSE SDS-PAGE			Advanced Sensitive High throughput	Sultan et al., 2004
139	Commercial meats	Processed	Soya and gluten 14 animal species	ELISA LCD array PVR					Cawthorn et al., 2013
88	Minced beef meat	Raw	Turkey meat	UV-vis NIR MIR				UV-vis not satisfactory	Alamprese et al., 2013
137	Meat products	Raw	Cattle, llama and horse meat and juices	VIS-NIRS	NIRSystems 6500 scanning monochromator, in the range of 400–2500 nm		-	Accurate	Mamani-Linares et al., 2012
30	Buffalo meat products	Raw	Buffalo meat	PCR HRM	Specific 12S and universal 18S rRNA primers		0.1% HMR	Rapid Accurate	Sakaridis et al., 2013
40	Minced lamb meat	Raw	Minced pork	NIR	-		0.01%	Rapid Low-cost Non-destructive	Kamruzzaman et al., 2012
-	Pork meat	Frozen	Hare meat	PCR	Species-specific primers for hare meat detection		0.01%	Accuracy	Santos et al., 2012
336	Beef	Fresh, frozen, thawed	Pork, beef offal, beef fat and connective tissue	VIS-NIR	400–2500 nm wavelength		-	Precision Accuracy	Morsy and Sun, 2013
75	Pork meat (PSE, RFN and DFD quality grades)	Raw	Grading and classification	Near-infrared hyperspectra imaging, PCA	900 to 1700 nm wavelength		Over 96% accuracy	Fast Accurate	Barbin et al., 2012
22	Animal feeds	Processed	Bovine, porcine, ovine, chicken	PCR	90 bp – 12SrRNA bovine, 85 bp – COX1 porcine, 67 bp – COX1 ovine, 66 bp – 16SrRNA chicken	0.09% for bovine, porcine, ovine	0.08% chicken	High sensitivity, Broad range of animal products	Natonek-Wisniewska et al., 2013

Table 3.1. contd....

Table 3.1. contd.

No. of samples	Type of meat/ product(s)	Treatment/ Package	Type of meat(s) to be detected/ excluded	QC method	Special remarks		LOD/% correct classification	Advantages and disadvantages	References
-	Beef	Fresh/ processed/ autoclaved	Authentication	PCR	513 bp DNA fragments	Level of detection: <1%	-	Accurate, no adverse effects of heat treatment	Mane et al., 2012
40	Commercial hamburger Mexican sausage	Slightly processed	Poultry, bovine, equine, and pork	Immunodiffusion	Agarose gel assay	3% 1% 1% 10%	-	Rapid High specific Reproducible	Flores-Maguia et al., 2000

Notes: SRMs: Specific risk materials, MGB: Minor Groove Binding. a: Pork in beef, b: Pork in beef, c: 0.1%, 1%, 2%, 5%, 10%, 25%, 50% and 100% beef in lamb, SINE: Short Interspersed Nuclear Element, PAGE: Polyacrylamide Gel Electrophoresis, RAPD-PCR: Random Amplified Polymorphic DNA Polymerase Chain Reaction, AP-PCR: Arbitrarily Primed Polymerase Chain Reaction, ND4: NADH dehydrogenase subunit 4, ss: Species specific, HATR: Horizontal Attenuated Total Reflectance, GA-D-MLR: Genetic Algorithms Discriminant Multiple Linear Regression, PLS: Partial Least Squares, DA: Discriminant Analysis, PCA: Principal Component Analysis, PC-DFA: Principal Components Discriminant Function Analysis, CDA: Canonical Discriminant Analysis, KNN: K-nearest-neighbour method, WBSF: Warner–Bratzler Shear Force, 2D-CA: 2D Correlation Analysis, CVA: Canonical Variate Analysis Modeling, SIMCA: Soft Independent Modeling of Class Analogy, PLOT: Porous Layer Open Tubular.

2007). The H and O isotope analysis has been used particularly for authenticating regional origins connected with regional climatic conditions (Boner and Förstel, 2004; Renou et al., 2004). Furthermore, the analysis of the same isotopes has been widely used for the detection of dietary components, such as maize or concentrates (Boner and Förstel, 2004; Gebbing et al., 2004; Quilter, 2002), whereas S has been of limited use.

Schmidt et al. (2005) used a successful combination of the above-mentioned types of the IRMS technique for the analysis of natural stable isotope compositions of carbon, nitrogen and sulphur in order to verify the geographical origin and feeding history of beef cattle. As a result, beef reared in the USA (23 samples) and Brazil (10 samples) was found to be isotopically different from beef bred in northern Europe (35 samples), mainly because of contrasting proportion of plants with C_3 and C_4 photosynthetic pathways in the cattle diets. The measurement of the stable-isotope ratios of carbon ($^{13}C/^{12}C$), nitrogen ($^{15}N/^{14}N$) and sulphur ($^{34}S/^{32}S$) was performed on de-fatted muscle residue by continuous flow IRMS (Scrimgeour and Robinson, 2004). The significant difference from the mean values of terrestrial C_3 and C_4 plants of $\delta^{13}C$ values (–27‰ and –13‰, respectively; Kelly, 2000) and the determined $\delta^{13}C$ values of the conventional Irish (–24.5‰ ± 0.7‰) and other European (–21.6‰ ± 1.0‰) samples and those from US (12. 3‰ ± 0.1‰) and Brazilian (–10.0‰ ± 0.6‰) beef were used as a reflexion of the different types of foods.

The determination of the isotopic ratios of D/H, $^{18}O/^{16}O$, $^{15}N/^{14}N$, $^{34}S/^{32}S$ and $^{13}C/^{12}C$ of two hundred and twenty three (223) beef samples from Germany and Chile was performed in an effort to authenticate them and detect possible fraud (Boner and Förstel, 2004). The isotopic composition was successfully used for the identification of the geographical origin, the specific location of the breeding and the kind of feed supplied. Data from the same seasonal trend and D/H-$^{18}O/^{16}O$ have earlier been used to correlate samples from northern and southern Germany (D = 4.5 × ^{18}O–18.6 {‰}, r = 0.88; and D = 5.5 × ^{18}O–22.2 {‰}, r = 0.89, respectively), with the first factor to be the most determinative. As a result, the seasonal variation stands out the geographical pattern through the isotope ratios' decrease from north to south for D/H from about –45 to –85‰ and for $^{18}O/^{16}O$ from –7 to –12‰ (IAEA, 1983). The geographic origin within Germany was plotted and revealed mean and median values of the D/H ratios in north German samples of –36‰; south German samples lower between mean –56‰ and median –55‰, being in agreement with the expected north-south difference resulted from the isotopic fractionation called the "continental effect". The $^{18}O/^{16}O$ ratios were almost the same with the mean and the median of northern samples to present a slight difference (northern –3.7‰ and –4.1‰; southern –5.9‰ and –6.3‰). Conventional and organic farming could be differentiated through the evaluation of the $^{13}C/^{12}C$ ratios of samples (21) from conventional farming compared to samples (223) from organic farming.

By combining the $^{15}N/^{14}N$ and (between 5–6‰ and 5–7‰, respectively) isotopes from northern German farms, three farms were directly differentiated from eleven by their low $^{34}S/^{32}S$ and high $^{15}N/^{14}N$ values. Despite the relatively small overlapping, the correlation of $^{34}S/^{32}S$ ratios above 8.5‰ with Argentinean origin beef-samples was effectively performed, while $^{34}S/^{32}S$ ratios below 4.8‰ were definitely of German origin. The samples were further differentiated according to the $^{15}N/^{14}N$ ratios data (Argentinean and Chilean sample-values ranging above 5.9‰ and 15.2‰, respectively, and 75% of the German ones below 6.1‰) and the $^{18}O/^{16}O$ data. As a result, a multi-elemental analysis seems to be necessary for enhancing the reliability of authentication (Hegerding et al., 2002).

In the study of Nakashita et al. (2008) isotopic signatures (C, N and O) of beef samples from Australia, Japan, and USA were analysed in order to declare the geographical origin of products commercially distributed in Japan. The carbon isotopic composition of beef defatted dry matter for USA, Japan and Australia was −13.6‰ to −11.1‰, −19.6‰ to −17.0‰ and −23.6‰ to −18.7‰, respectively. On the other hand, the oxygen isotopic composition was according to the following ranges: +15.0‰ to +19.4‰, +7.3‰ to +13.6‰ and +9.5‰ to +11.7‰ for Australia, Japan and USA, respectively. Finally, the nitrogen isotopic composition values from Japan were a bit higher (+7.2‰ to +8.1‰) than that from USA (+5.1‰ to +7.8‰) and Australia (+5.7‰ to +9.3‰); not allowing a direct use apart from the similarity reflectance in the diet $\delta^{15}N$ values among these countries, resulting mainly from the cattle diet (deNiro and Epstein, 1978; 1981). Thus, the first two parameters were shown to be more powerful as regards both provenance and region discrimination possibility.

In the study of Heaton et al. (2008), a high number (> 200) of beef samples of worldwide origin (Europe, USA, South America, Australia and New Zealand) were examined using both IRMS and ICP-MS. The determination of the C and N and H and O isotopic composition of the specimens was performed in defatted dry mass and the corresponding lipid fractions. It was observed that the type of pasture feeding affected the ^{13}C content of beef allowing the discrimination between Brazil-USA and British cattles. The mean δ^2H‰ and $\delta^{18}O$‰ values of beef lipid was satisfactory correlated with the latitude of production regions and went with the Meteoric Water Line supporting the hypothesis that the systematic global variations in the ^2H and ^{18}O content of precipitation are transferred through drinking water and feed into beef lipid. Canonical discriminant analysis was used to obtain the optimal result as regards the broad geographical areas (Europe, South America and Australasia) through six key variables [$\delta^{13}C$‰ (defatted dry mass) and Sr (function 1) and Fe, δ^2H‰ (lipid), Rb and Se (function 2)].

In the study of Bong et al. (2010), an attempt was made to geographically discriminate between American, Mexican, Australian, New Zealand and Korean beefs sold in the Korean markets. The carbon, nitrogen and oxygen isotopic compositions of defatted dry samples were used for the analysis. The isotopic compositions presented statistical differences, particularly in $\delta^{18}O$‰ and $\delta^{13}C$‰ values due to the different isotopic percentages in their water and feed. The $\delta^{13}C$ values of Korean beef ranged from −19.10 to −14.21%, while those of New Zealand beef presented the lowest and most distinctive values having some common values with only those of Australian beefs which covered the entire range of the analysis. The highest $\delta^{13}C$ values were obtained from the analysis of the USA and Mexican samples. Finally, the ^{18}O isotopes were particularly high in Mexican and Australian beef, while low ^{18}O values were obtained from samples originated from Korean, USA and New Zealand.

Franke et al. (2007) employed inductively coupled plasma high resolution mass spectrometry (ICP-HRMS) to measure the concentrations of 72 elements in dried beef, targeting at the determination of its geographic origin. The samples were obtained from *M. biceps femoris* and *M. semitendinosus* produced in Switzerland, Austria, Australia,

United States, and Canada out of raw meat originating either from these or from other countries. Sixty-six out of 72 elements/isotopes (all except ^{9}Be, ^{149}Sm, ^{165}Ho, ^{172}Yb, ^{202}Hg and ^{209}Bi) were detected in the samples, while significant statistical differences between countries of origin and site of processing were found in 15 and 16 samples, respectively. The following eighteen elements were analysed: B, Ca, Cd, Cu, Dy, Eu, Ga, Li, Ni, Pd, Rb, Sr, Te, Tl, Tm, V, Yb and Zn. The highest concentrations of elements were found in samples originated from Australia (Ca, Cd, Ga, Pd and Tl), Switzerland (Ca, Cu and Ni), USA (Ca, Cd, Ga and Pd), Austria (Cu and Sr), Canada and Brazil (Ni and Rb). Lower Zinc (Zn) concentrations were detected in Australian and US samples in comparison to samples from other countries.

The stable isotope ratios (^{13}C/^{12}C and ^{15}N/^{14}N) of lamb meat were also analysed using IRMS in order to authenticate the feeding and geographical origin of the product. Meat from twelve lamb types, produced in couples in six European countries (country of origin) was used and divided into three groups in accordance with the feeding regime during their finishing period: suckled milk only, pasture without any solid supplementation and supplementation containing maize grain (feeding regime). The analysis of the samples was performed in duplicate, displaying significant differences in δ^{13}C values. Higher values were found for protein than for fat (average difference 5.0‰). However, high correlation ($r = 0.976$) was found between the pairs δ^{13}C values of crude fat and protein. The δ^{15}N values of meat protein fraction were significantly different between lamb types, but no correlation with the feeding regime was found. In fact, lambs fed on similar diets, but in different countries, gave meat with different ^{15}N relative abundances and allowed the discrimination of lamb types within the same feeding regime (Piasentier et al., 2003).

The concentrations of 72 different trace elements and the oxygen isotope ratio of 74 dried beef samples were analysed for determining whether the accuracy of the prediction of the geographic origin can be enhanced by using a combination of promising techniques. The method was validated through determination of the origin of a smaller sub-group using a statistical model developed from the data of the second, larger, sub-group. It was shown that the geographic origin of the samples could be effectively predicted. Nevertheless, the combined method did not limit the percentage of incorrect classifications. The cross-validation and validation led to 73% and 43% correct classifications, respectively. It was finally proved that the combination of the methods led to the same level of classification errors appeared when each method was applied separately (Franke et al., 2008).

The significant advantages of nuclear magnetic resonance (NMR) as a non-destructive and quick method, which requires short preparation time of samples, were used by Shintu et al. (2007) who employed ^{1}D ^{1}H HR-MAS NMR spectroscopy in order to select potential molecular markers of one specific geographic origin. A confirmation of this trial had been earlier obtained for beef through the analysis of the acid contents of the animals' diets (Renou et al., 2004). Therefore, in the study of Shintu et al. (2007) dried beef samples of certified origin were examined providing positive results, even with a low number of samples, thus confirming that the method can be effectively used for rapid food analysis. Fat content as well as specific metabolites, connected with the feeding system, were used as markers of origin. It was proved that American and Australian samples were richer in fatty acids, while the Canadian and Brazilian samples were richer in numerous small metabolites such as proline phenylalanine, glutamic acid and/or glutamine. The Swiss sample was characterised by high phenylalanine and lower alanine and/or methionine contents. The discrimination between the North American and Australian-Brazilian samples was performed through the levels of carnitine and succinate.

Sasazaki et al. (2007) employed six DNA markers in order to discriminate between Japanese and Australian beef. Two Bos indicus-specific markers SRY (803 bp) and ND5 (527 bp); and the MC1R (219/218 bp) marker were selected as possible candidate markers, while the subsequent development of more markers was performed with AFLP (BIMA100—465 bp; BIMA118—153/148 bp; and BIMA119—78/76 bp). The 1564 primer combinations provided three markers in the form of single nucleotide polymorphisms markers for high-throughput genotyping. The allele frequencies of these markers were examined in cattles from both countries using PCR-RFLP in an effort to evaluate their discrimination potential. The probability of Australian beef identification was 0.933 while the probability of misjudgement was 0.017 using the above-mentioned markers. It is believed that these markers could also be used for the discrimination of samples from other countries. In the study of Haider et al. (2012) PCR-RFLP of a part of the mitochondrial cytochrome c oxidase subunit 1 (COI) gene was used to identify the species origin of raw meat samples of cow, sheep, pig, buffalo, camel and donkey. PCR yielded a 710-bp fragment for all meat species. The digestion of the amplicons was performed using seven restriction endonucleases (Hind II, Ava II, Rsa I, Taq I, Hpa II, Tru 1I and Xba I) the selection of which was relied on the preliminary in silico analysis. Several levels of polymorphism were found among samples. The level of COI variation revealed only with the use of Hpa II was sufficient for the generation of easily analyzable species-specific restriction profiles that could be used for distinguishing all species examined. The method was cheaper and faster in comparison to previously reported methods developed for determining meat origin at the molecular level, and thus could be applied to routinely identify meats.

In the study of Sacco et al. (2005), twenty-five lamb meat samples from three areas located in Apulia (Southern Italy) were analyzed for moisture, ash, fat and protein content, stable isotope ratios ($^{15}N/^{14}N$ and $^{13}C/^{12}C$), major elements (Ca, Mg, Na and K) and trace metals (Zn, Cu, Fe and Cr). The ^1H high resolution magic angle spinning (HR-MAS) NMR spectra was successfully used. The main benefits of the method are the short time required for preparing the samples and the potential of analysing different compounds such as amino acids, fatty acids, sugars, etc., in a single experiment. The same signals were seen in all meat spectra, but the intensities differed. Thus, 100% classification capacity and 96% prediction ability of origin were obtained for lamb from stable isotope ratios and NMR data.

Camin et al. (2007) employed a multi-element (H, C, N and S) stable isotope ratio analysis in order to geographically differentiate between lamb meats from several European regions. The defatted dry matter (crude protein fraction) from lamb meat was characterised as a "suitable probe for light element stable isotope ratio analysis" because the isotope ratios differed significantly between the different geographical areas. The mean hydrogen isotopic ratios of the defatted dry matter from lamb presented good correlation with the mean hydrogen isotopic ratios of precipitation and groundwater in the production regions. The highest deuterium content for lamb was found in Greek lamb (Chalkidiki), with a mean value of −80‰, which is in accordance with the high deuterium content found in eastern Mediterranean precipitation and ground waters. The above trends were in agreement with the results obtained from the other Greek region examined (Lakonia). The mean values ratios from UK-Sicily, Central Europe and Alpine Mountains were: −80 − (−90)‰, −90 − (−100)‰ and −100 − (−110)‰, respectively. Both feeding practices and climate had a significant effect on carbon and nitrogen isotopic ratios. Thus, the nitrogen isotopic values of the lamb defatted dry matter did not vary significantly between different European regions (range is from +3.8 to +9.2‰). The north-western European regions of the UK and Ireland had relatively high $\delta^{15}N$‰ values from +7.4 to +9.2‰, while other regions provided more homogeneous results for $\delta^{15}N$‰ (from +5 to +6‰). It should be mentioned that the results obtained for the mountain area of Trentino were significantly ($p = 0.001$) lower (+3.8‰), while the results obtained for Greece (Chalkidiki) were quite high (+6.5‰). Sulphur isotopic ratios varied significantly between different geographical regions, being affected by the surface geology of the production region, with $\delta^{34}S$‰ values ranging from +1.6 to +12.8‰.

Selenium (Se) is one of the most important elements for the human organism and one of the highest sources of Se is beef. The Se content of beef products is highly variable and is potentially a unique supplemental source of dietary Se (Hintze et al., 2002). In the study of Hintze et al. (2001), the determination of the Se content of 138 cull cows from 21 ranches was determined in five distinct geographic regions of North Dakota, and analysed on the basis of soil parent material, reports of Se deficiency, and previous soil and forage Se surveys. The determination of the Se content of all samples was achieved using hydride generation atomic absorption spectroscopy. It was found that the Se content of all samples varied significantly ($p < 0.05$) in different geographic areas, whereas any additional Se intake from mineral supplementation had insignificant effects on the Se content of the samples ($p > 0.05$). As a result, the geographic origin of the animals could be more effectively used for the determination of the Se content of beef in comparison to the presence or absence of Se supplements. The results from the different geographical regions ranged from 0.27 to 0.67 μg of Se/g, whereas the use of supplemental Se increased the mean Se content from 0.41 μg/g (no supplemental Se) to only 0.46 μg of Se/g ($p > 0.07$).

The geographical origin determination of beef and lamb products with isotopic ratio methods is shown in Table 3.2, while Fig. 3.1 diagrammatically depicts the seasonal variation in ^{13}C composition of beef samples obtained from retailers in Dublin and lamb muscle samples obtained during a diet switch.

3.4 Multivariate Analysis

The foods are consisted of a huge variety of different chemical substances and physical structures. As a result, those chemical substances that can most easily be detected affect significantly the perception of the quality of a food product. The need for analysis, understanding and improvement of a foodstuff is highly dependent on the determination of a vast variety of properties and the use of analytical techniques that can handle the large and complex data matrices produced (Næs et al., 1996).

Table 3.2. Geographical origin determination of beef and lamb products with isotopic ratio methods.

Country	No of samples	Measured parameter(s)	QC method	Special remarks on samples	Statistical method(s)	δ (‰) Means/Ranges					References
						^{13}C or $^{13}C/^{12}C$ (*)	^{18}O or $^{18}O/^{16}O$ (*)	^{14}N or $^{15}N/^{14}N$ (*)	^{2}H or $D/^{2}H$ (*)	^{32}S or $^{34}S/^{32}S$ (*)	
EU vs Ireland USA Brazil	68	C, N and S	SIRA[a]	Conventional and organic farming	MANOVA	−21.6 −24.5 −12.3 −10.0	—	7.1	—	7.6	Schmidt et al., 2005
Germany vs Argentina Chile	244	H, O, C, N and S	SIRA[a]	Freeze-dried Extracted water Defatted raw protein Organic farming	PCA DA	(*)	North: −7.9 South: −11.2 - -	<6.1 >5.9 5.0	−36 −55	<4.8 >8.5 15.2	Boner and Förstel, 2004
Japan vs Australia USA	22	C, N and O	EA/ IRMS[b]	Defatted dry	–	(−)19.6−(−)17.0 (−)23.6−(−)18.7 (−)13.6−(−)11.1	7.3–13.6 15.0–19.4 9.5–11.7	7.8–8.1 5.7–9.3 5.1–7.8	–	–	Nakashita et al., 2008
EU GB, Irl, Sctl USA Australia S. America Brazil N. Zealand S. Africa	>200	C, N O, H	IRMS ICP-MS	Defatted dry mass Lipid fractions	CDA	(−)22.6−(−)21.2 (−)25.8−(−)24.3 −11.1 −16.9 −20.6 −11.4 −26.32 (−)15.3−(−)12.5	17.7–20.3 17.9–18.3 17.9 23.4 22.3–22.8 21.4 19.4 26.1–32.3	4.7–6.6 6.6–7.1 6.0 6.1 6.6–6.8 6.7 5.8 7.3	−220.0−−176.4 (−)208.0− (−)189.8 −214.2 −157.7 −180.8 −183.0 −214.6 (−)151.0− (−)140.5	–	Heaton et al., 2008

Table 3.2. contd....

Countries	n	Analyte	Method	Sample prep	Statistics	Data / values	Reference
USA, Mexico, Australia, N. Zealand, Korea	85	C, N and O	EA/IRMS[b]	Freeze-dried	ANOVA	$\delta^{13}C$ (‰): ~−12.0, ~−12.0, ~−17.0, ~−23.0, ~−16.0; $\delta^{15}N$ (‰): ~9.0, ~15.0, ~14.0, ~10.0, ~10.0; $\delta^{18}O$: 6.3, 7.0, ~7.0, 5.6, 6.15; (–, –)	Bong et al., 2010
Switzerland, Australia, USA, Brazil and Canada	72	18O/O, 87Sr/86Sr	ICP-MS IRMS	Dried sample	One-way ANOVA	−1.84, 1.84, −3.35, −1.03, −5.59; (–, –, –)	Franke et al., 2008
Australia, Brazil, Canada, Switzerland, USA	23	1D 1H	HR-MAS NMR	Dried	One-way ANOVA, PCA, Stepwise DA	See Compounds used (ppm) below	Shintu et al., 2007
Switzerland, Austria, Australia, USA, Brazil and Canada	23	72 elements	ICP-HRMS	Dried sample	ANOVA, LDA, PCA	See element table below	Franke et al., 2007
Japan, Australia	782	DNA markers	PCR-AFLP	Tissue and blood	-	See marker table below	Sasazaki et al., 2007

Compounds used (ppm):
[c]carns/tyr: 3.20–3.25
[c]carns: 2.70–2.75
[d]glut/pro: 2.35–2.40
carnt/suc/[d]UB: 2.40–2.45
[c]carnt: 2.45–2.50
[d]ala: 3.80–3.85
[c]phe: 3.30–3.35
[e]UA: 3.65–3.70
[e]pro/carnt: 3.40–3.45

Element table (Franke et al., 2007):

	Ca	Cd	Cu	Ga	Ni	Pd	Rb	Sr	Tl	Zn
Switzerland	√[c]		√		√			√		√
Austria					√				√	
Australia		√		√		√				
USA		√			√	√	√			√
Brazil and Canada					√		√			

Marker table (Sasazaki et al., 2007):

Marker	SRY	ND5	MC1R	BIMA100	BIMA118	BIMA119
Product size (bp)	803	527	219/218	465	153/148	78/76

Table 3.2. contd.

The columns **13C or 13C/12C (*)**, **18O or 18O/16O (*)**, **14N or 15N/14N (*)**, **2H or D/2H (*)**, and **33S or 34S/32S (*)** are grouped under the heading **δ (‰) Means/Ranges**. For the USA row these carry the regional sub-headers *Region / Muscle (mg/kg)* — NW, S Central, SW, Central, SE.

Country	No of samples	Measured parameter(s)	QC method	Special remarks on samples	Statistical method(s)	13C or 13C/12C (*)	18O or 18O/16O (*)	14N or 15N/14N (*)	2H or D/2H (*)	33S or 34S/32S (*)	References
						Region / Muscle (mg/kg)	*NW*	*S Central*	*SW*	*Central*	*SE*	
USA (N Dakota)	138	Se	HG-AAS	Several tissues		0.67	0.47	0.40	0.38	0.27	Hintze et al., 2001
GB, Spain, France, Greece, Iceland, Italy	360 lamb	$^{13}C/^{12}C$, $^{15}N/^{14}N$	IRMS	Meat fat and protein fractions	ANOVA, CDA	−26.75 d, −26.25, −29.30, −24.80, −31.50, −28.45 √◦	√ √ √ √ √	6.75 d, 6.15, 4.55, 5.75, 2.50, 5.7 √	√	√	Piasentier et al., 2003
Greece, UK (+Sicily), Centr. Europe, Alpine Mnts	203 lamb	H, C, N, S	SIRA	Dried	ANOVA, PCA, LDA	−21.2–−23.0, −26.2–−28.5, −24.6–−26.1	–	+5.0–+6.0, +5.0–+6.0, +7.4–+9.2, +5.0–+6.0	−80, −80–−90, −90–−100, −100–−115	+1.6–+12.8	Camin et al., 2007
Italy (Apulia) — N, C, S	25 lamb	$^{13}C/^{12}C$ $^{15}N/^{14}N$; Li, Ca, Mg, Na, K; Zn, Cu, Fe, Cr, Mn; ^{1}H (500MHz)	IRMS; HPIC; ICP-AES; HR-MAS NMR	Dried	PCA, DA, One-way ANOVA	−25.65, −25.89, −26.02	–	6.9, 5.53, 7.05	–	–	Sacco et al., 2005

Notes: [a] SIRA: Stable Isotope Ratio Analysis, [b] EA/IRMS: Elemental Analyzer/isotope ratio mass spectrometry, [c] Measured higher levels, [d] Mean values of different lamb types and feeding regimes based on protein, [e] carns: Carnosine; tyr: Tyrosine; glut: Glutamine; pro: Proline; ala: Alanine; phe: Phenylalanine, HR-MAS NMR: High Resolution-Magic Angle Spinning Nuclear Magnetic Resonance, HPIC: High Performance Ion Chromatography, ICP-AES: Inductively Coupled Plasma Atomic Emission Spectrometer, HG-AAS: Hydride Generation Atomic Absorption Spectroscopy

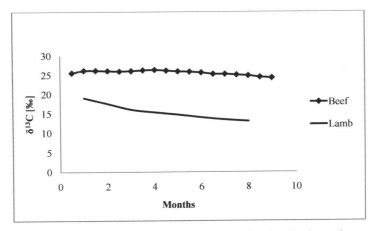

Figure 3.1. Seasonal variation of ^{13}C composition of beef and lamb samples.

Piasentier et al. (2003) employed canonical discriminant analysis to examine whether lamb meat from different CO (country of origin) or FR (feeding regime) or different CO:FR combinations could be mathematically differentiated using its stable isotope ratios. As regards the CO, the corrected empirical allocation of 79.2% of the initial observations and the corrected cross-validation of two thirds of the individual meat samples were obtained. The data obtained from FR showed that 91.7% of all meat samples presented correct allocation and cross-validation, indicating that the stable isotope ratio analysis could be successfully used as a tool for lamb diet characterisation. The most efficient classification was obtained by employing K-means clustering technique and canonical discriminant analysis, thus permitting a high-quality resolution of six CO:FR groups of lamb types: Icelandic fed on pasture; British and French grazing; Italian; suckled and Karagouniko concentrates finished; French Lacaune; and Ternasco de Aragon. Therefore, a multi-element stable isotope analysis can be effectively used for authenticating lamb meat as was also proved in the case of wine, fruit juice, honey and dairy foodstuffs.

According to Camin et al. (2007), the use of stable isotope data on the hydrogen, carbon, nitrogen and sulphur content of lamb meat from different European regions in combination with PCA and linear discriminant analysis (LDA) has been successfully used for the origin differentiation of the samples. In this study, PCA was used to analyse the European lamb protein isotope data (δ^2H‰, δ^{13}C‰, δ^{15}N‰ and δ^{34}S‰) leading to interesting differentiations among the different geographical areas. The relative plot of the first two principal components for samples of the year 2005 depicted the differences between the north-western regions of Orkney, Ireland and Cornwall from Greece, from the Mediterranean areas of Sicily, Toscana, Carpentras, and from the Alpine sites of Trentino and Allgäu with Mühlviertel. Furthermore, the LDA for the same data was used to effectively differentiate between lamb products from Bavaria (Germany), France, Greece, Ireland, and the UK. After double-checking the results, it was concluded that correct classification was achieved for 78.7% of the original grouped cases, 77.6% of the cross-validated grouped cases, and 100% of the Carpentras, Cornish and Mühlviertel lamb samples. The classification errors detected (Limousin vs

Carpentras, 50% France; Orkney vs. Cornish, 35% UK) were considered to be due to barn and feeding practices. Incorrect classification was obtained for 40% of Franconian samples (Bavaria), which were characterised as being from Limousin (10%), Allgäu (5%), Trentino (10%) and Sicily (15%). The classification of the remaining regions was successful in 80% of samples, or more, with the exception of Ireland where only 69.2% of samples were classified correctly. In another study, the reliability and precision of hyperspectral imaging technique combined with multivariate analyses were examined in an effort to identify and authenticate different red meat species. Hyperspectral images were obtained from *longissimus dorsi* muscle of pork, beef and lamb and the extraction and analysis of their spectral data were performed using principal component analysis (PCA) and partial least squares discriminant analysis (PLS-DA). The pre-treatment of the spectra was performed by second derivative and six wavelengths (957, 1071, 1121, 1144, 1368 and 1394 nm) were identified as important wavelengths from the 2nd derivative spectra. These wavelengths were applied in a pattern recognition algorithms for classifying meat samples with PLS-DA yielding 98.67% overall classification accuracy in the validation sets. The obtained classification algorithms were subsequently used in the independent testing set for authenticating minced meat. It was found that by combining hyperspectral imaging, multivariate analysis and image processing a both objective and rapid method was developed for identifying and authenticating red meat species (Kamruzzaman et al., 2012).

Al-Jowder et al. (2002) employed chemometric methods (PCA, PLS and linear discriminant analysis) to MID-IR spectra. The potential of discriminating between the pure and adulterated sample types, was highly connected with the cooking regime. It was found that on pure beef and beef containing 20% w/w of a range of potential adulterants (heart, tripe, kidney and liver), the obtained spectra of the two distinct cooking regimes (8 min at 240 W or 400 W-'level 1' and 'level 2', respectively) resulted in the successful classification of ~97% of samples. The use of PCA indicated that at least two dimensions were required to reveal a distinct group for a particular sample type. The spectra of the cooked samples differed significantly from the corresponding spectra of the raw samples, as a result of the compositional changes taking place during cooking. Thus, since dehydration is the most easily detected effect of cooking, the water peak was reduced at 1650 cm^{-1}, and relatively more prominent fat (1725 cm^{-1}) and protein (1650 and 1550 cm^{-1}) peaks were obtained. However, the spectra quality was significantly reduced depending on the level of cooking.

An effort to classify meat samples as either fresh or frozen-then-thawed was made using NIRS and two chemometric techniques: factorial discriminant analysis and SIMCA. A fibre optic probe for spectral acquisition was used. Samples of meat (*M. longissimus dorsi*) were collected from 32 animals of a single species (female Hereford-cross) from a commercial abattoir. By applying a standard freeze-thaw regime, it was shown that fresh and frozen-then-thawed beef could be effectively discriminated using the factorial discriminant procedure, while the results obtained using SIMCA were less satisfactory (Downey and Beauchêne, 1997).

Samples of various minced beef (32), lamb (33) and lamb-in-beef mixtures (5%, 10% and 20% (w/w), 33 each) were evaluated for their potential of reflecting visible, near and MIR spectral regions. PLS models, using each spectral region separately and combinations of all three, were used to predict the portion of lamb in samples. It was proved that the most satisfactory results were obtained from those models applied to MIR (800–2000 cm⁻¹) and near IR (1100–2498 nm) spectral data following 2nd derivatization (McElhinney et al., 1999).

The spectra results obtained from beef and ox kidney and liver with MIR spectroscopy were used by Al-Jowder et al. (1999) for authentication of specimens. The authentication was achieved through developing a combined technique of PLS/ CVA, predictive models. Although the differences between beef and offal were easily detected, the distinction among the three cuts of beef was not so clear. The neck specimens consisted a separate group, but the separation between silverside and brisket samples was not correctly implemented due to an inherent overlap between these two specimen types. The specimen condition (fresh or frozen-thawed) did not affect the effectiveness of the method, as earlier reported by Downey and Beauchêne (1997) who employed FDA (factorial discriminant analysis) on Vis/NIR date of beef samples. The use of SIMCA class modelling led to the development of a single class model for the pure beef specimens characterised by a type I error rate of ~8%. By challenging this model with beef specimens adulterated with portions in the range of 10–100% w/w kidney or liver, no specimens were accepted by the model indicating that the detection limit for these adulterants was < 10% w/w. The results obtained when PLS regression was used to quantify the amount of added kidney and liver, were in agreement with the results obtained using SIMCA analysis. PLS in combination with canonical variate analysis, led to satisfactory spectra identification of non-adulterated beef, with an acceptable error rate, while spectra of specimens containing 10–100% w/w kidney or liver were rejected. In comparison, good results were obtained when PLS regressions were used to quantify the added offal.

Ding and Xu (1999) differentiated beef from kangaroo meat using a canonical discriminant analysis (CDA)-PCA performance and stepwise multiple linear regression (MLR) based on visible/near-infrared (NIR) spectroscopic data. Standard normal variance and detrend (SNVD) and/or second derivative operation, with gap and smooth both set at 8 nm were used to initially analyse the spectral data. An improved classification of minced beef was only possible after applying scatter correction and derivative treatment of reflectance spectra. On the other hand, the classification of cut meat could be easily performed from the original reflectance spectra. It was proved that 83–100% of overall samples were accurately classified while all kangaroo samples were classified correctly.

A neural network multilayer perceptron (MLP) was used to correlate FT-IR spectral data with beef spoilage over aerobic storage at chill and maltreatment temperatures. After packaging fresh beef fillets under aerobic conditions and spoiling them at several temperatures—from 0 to 20°C—for almost two weeks, FT-IR scattering was employed directly to the surface of meat samples while total viable counts (TVC) were measured with plating techniques. The initial sensorial categorisation of the fillets included three quality classes: fresh, semi-fresh and spoiled. After analysing the FTIR spectra data of the biochemical profile and the microbial load (as TVC) of meat surface, it was proved that the method was very accurate and correctly classified 91.7% of fresh samples, 94.1% of spoiled samples, and 81.2% semi-fresh samples. The microbial load on beef surface and the spectral data were highly correlated, making this method a possible valuable tool for quickly assessing meat spoilage (Argyri et al., 2010).

Arvanitoyannis et al. (2000) analysed the physico-chemical and sensory characteristics of 48 samples of Greek Cavourmas, supplied from 16 producers. The analysis included the determination of moisture, crude protein, ether-extractable fat, ash, NaCl, $NaNO_2$, $NaNO_3$, pH, 2-thiobarbituric acid (TBA) value, lightness (L*), redness (a*), and yellowness (b*). Principal component analysis based on three main axes (PC1, PC2 and PC3) accounting for 56.4% of overall variation indicated that the consumer acceptability of the product is connected with taste and odour as well as to the presence of meat in pieces with red colour and white fat. Fatty appearance, high portions of melted fat and rancid products reduced the acceptability, as confirmed through determining the TBA values, high fat content, high percentage moisture and lightness. Arvanitoyannis and van Houwelingen-Koukaliaroglou (2003) employed MVA to classify and group individual products. PCA, canonical analysis, cluster and PLS were used for the classification of food products according to variety and/ or geographical origin. The classification of meat and meat products was effectively carried out for authentication purposes, following instrumental and/or sensory analyses.

In the study of Papadima et al. (1999), the conduction of chemical, physical, microbiological and sensory analyses was carried out on 31 samples of Greek traditional sausages. The results were as follows: fat 15.49–56.86%, moisture 21.92–65.40%, protein 14.73–26.74%, sodium chloride 2.36–4.13%, nitrites 0.0–3.26 ppm, mean nitrates 38.19 ppm, TBA value 0.42–5.33 mg malonaldehyde/kg, pH 4.74–6.74, water activity (a_w) 0.88–0.97, firmness 0–64 Zwick units, lightness (L*) 25.03–35.37, redness (a*) 2.55–11.42, yellowness 4.42–12.96, aerobic plate count 5.48–9.32 cfu/g, lactic acid bacteria (LAB) 5.26–9.08 cfu/g, micrococci/staphylococci 4.11–6.91 cfu/g and Gram (-) bacteria 1.78–6.15 cfu/g. The data were statistically analysed using Praxitele and SPSS in an effort to characterise and evaluate the properties of the products. The first two principal components (PC1-2) derived by SPSS (50.5% variance) led to a more satisfactory description of the variance in comparison to PC1-2 and PC1-3 obtained by Praxitele (40.4% variance).

Pappa et al. (2000) applied a response surface methodology for the determination of the optimum salt level (1.3–2.1%) and pectin level (0.25–1.0%) of low-fat frankfurters pork backfat (9% fat, 13% protein), when olive oil partially or totally replaced pork backfat. The test ingredients had significant ($P < 0.05$) effects on jelly separation of the batter, skin strength, hardness, saltiness, odour and taste and the overall level of acceptability of the products. Low-fat frankfurters with high salt levels were characterised by very hard skins and increased ($P < 0.05$) saltiness while products with high pectin level were characterised by softness and creamy texture, while they also had the lowest ($P < 0.05$) score for odour and taste. The low-fat frankfurters with 1.8–2.1% salt, 0–35% olive oil and 0.25–0.45% pectin were the most acceptable by consumers. However, low-fat frankfurters with 1.3% salt, 0.25–0.30% pectin and 80–100% olive oil were also characterised by satisfactory acceptability. These low-fat frankfurters have 48% less salt in comparison to commercial frankfurters and 66.6% less fat (from 30 to 10%), of which 80–100% of the added fat is olive oil. More studies are required to enhance the acceptability of these frankfurters.

The main points of MAV are summarized in Table 3.3.

Table 3.3. Application of multivariate analysis to meat and meat products; adulterant, QC method, (pre)treatment, correct classification and detection limit.

No of samples	Type of meat	Adulterant(s)/ mixed with	QC Method	(Pre)Treatment	MV method	Mean Correct Classification (%)	Detection limit (%)	References
12	Beef	Heart, tripe, kidney, liver	MIR	Two levels microwave	PCA PLS LDA	~97%	≥20	Al-Jowder et al., 2002
185	Beef	Ox kidney and liver	FTIR	Fresh Frozen-thawed	PLS-CVA SIMCA	> 95%	> 10 < 10	Al-Jowder et al., 1999
41	Beef	Kangaroo	VIS-NIR	Freezed	CDA Stepwise MLR	80–100	–	Ding and Xu, 1999
32	Beef	Frozen-then-thawed	NIR	None	FDA SIMCA	89.1–95.3	–	Downey and Beauchêne, 1997
154	Beef	Lamb	VIS-NIR/MIR	None	PLSR	~100	<20	McElhinney et al., 1999
74	Beef	Microbial load	FTIR/ATR	Incubation at 0, 5, 10, 15 and 20°C for 350 h	MLP NN	80–90	–	Argyri et al., 2010
203	Lamb	Origin	IRMS (H, C, O, N)	Dried	ANOVA PCA LDA	~78	–	Camin et al., 2007
360	Lamb	Origin	IRMS (C, N)	Vacuum-packed Freezed-dried	ANOVA CDA	~66.7	–	Piasentier et al., 2003
194	Beef (hamburger)	Mutton Pork Skim milk powder Wheat flour	NIRS	None	CDA KNN PCA Mpls	92.7	·	Ding et al., 2000
205	Pig fat	Different subcutaneous layers	Tristimulus colorimetry	Refrigerated	Signal processing and feature selection techniques.	78.1	·	Foca et al., 2013
			FT-NIR	FT-NIR NIR hyperspectral imaging		> 95%		
			NIR hyperspectral imaging			> 95%		
225	Pork Beef Lamb	Authentication	Hyperspectral imaging	VP, refrigerated	Hyperspectral imaging, multivariate analysis and image processing	98.67% overall classification accuracy	-	Kamruzzaman et al., 2012

3.5 Sensory Analysis

One of the most significant sensory properties of food and food products is the external appearance and specifically the colour, which, in the case of meat, is highly dependent on myoglobin and pigment redox stability. The appearance of meat is a subject of study for many fields of science such as animal genetics, *ante-* and *post-mortem* conditions, fundamental muscle chemistry, and fields of meat processing, packaging, distribution, storage, display, and final preparation for consumption (Mancini and Hunt, 2005). Rousset-Akrim et al. (1997) examined how sheep age and diet could affect the odour and flavour of meat. Various feed and growth patterns were examined. It was found that meat flavour was highly connected with the aroma volatiles formed during cooking, since raw meat has little or no aroma and only a blood-like taste. These volatiles can be categorised into water-soluble substances and lipids. During cooking, the main reactions taking place are Maillard reaction and the thermal degradation of lipid (Mottram, 1998). Various techniques have been used for separating, isolating and identifying volatiles such as NPSD (nitrogen purge and steam distillation), SPME (solid phase micro extraction), SE (solvent extraction), SDE (simultaneous distillation extraction) and DHS (dynamic headspace) combined with GC/MS (Ai-Nong and Bao-Guo, 2005; Ai-Nong et al., 2008; Sun et al., 2010; Jerković et al., 2007; Jerković et al., 2010; Bianchi et al., 2007). Moreover, various analytical techniques, such as VIS-NIR spectroscopy, electronic noses, image processing and RT-PCR (Eklöv et al., 1998; Balamatsia et al., 2007; Otero et al., 2003; Bernard et al., 2007; Gerrard et al., 1996; Berzaghi and Riovanto, 2009) and sensory panels (Bernard et al., 2007; Kriese et al., 2007; O'Sullivan et al., 2003b; Balamatsia et al., 2007) have been used for analysing the sensorial characteristics of meat and meat products.

Ahn et al. (2000) examined the formation of volatile compounds in irradiated and non-irradiated pig samples (*Longissimus dorsi* muscle strips) stored under air or vacuum packaging conditions at 4°C for five days. The fluorescence TBARS (2-thiobarbituric acid reactive substances) technique was used for analysing the lipid oxidation, while a purge-and-trap/GC/MS method was used to assess and quantify the produced volatile compounds. Although, irradiation had minimal effects on the formation of volatile compounds relative to lipid oxidation, it had significant effects on the formation of few sulphur-containing compounds not found in non-irradiated samples. Lipid oxidation was not the main source of off-odours in irradiated meat. Radiolytic breakdown of sulphur-containing amino acids led to the formation of thio-bismethane, 3-methoxy-1-propene, thioacetic acid methyl ester, 2,3-dimethyl trisulphide, toluene, and 2,3-dimethyl disulphide; with the latter constituting 75% of the new volatiles produced. A significant decrease of several irradiation-dependent volatiles was also observed during storage in air. Finally, ~70% of sensory panels described the odour of irradiated samples as barbecued-corn-like odour while irradiation did not negatively affect the acceptance of the meat.

In the study of Bernard et al. (2007), new molecular markers connected with the sensorial characteristics of beef, were produced from the transcriptomes of *Longissimus thoracis* muscle from Charolais bull calves. The samples were analysed with microarrays and differentiation was achieved between high- and low-quality meat. A total of 215 genes were thereby differentially expressed according to tenderness, juiciness, and/or flavour. A group of 23 were up-regulated to the tenderest, juiciest, and tastiest meats, and eighteen presented high correlation with both flavour and juiciness (e.g., PRKAG1), thus giving an interpretation to 60% of their variability. Nine were down-regulated in the same samples, but only DNAJA1, which is characterised by the encoding of a heat shock protein, presented a highly negative correlation with tenderness, that alone explained 63% of its variability, thus making DNAJA1 a possible new marker of beef sensory quality.

Eklöv et al. (1998) employed a sensor array and pattern recognition routines (electronic nose) for recording the variability of emitted volatile substances during critical and common fermentation of sausages. Ten metal oxide semi-conductor field-effect transistors (MOFSET) of different gate metal (Pd, Pt, Ir), metal structure and operating temperature were used in combination with principal component regression and an artificial neural network (ANN). It was found that the method was effective in detecting various substances. The results were compared with those extracted from a seventeen-member sensory panel analysis, performed both at the early stage of the process and on the final product (sausages). Both the sensitivity and the discrimination potential of the two methods were of the same grade, allowing any difference to be detected between the batches just after 4 h and by the end of the whole process. This is comparable to predicting the final sensory quality of the product at the beginning of the process.

Garcia et al. (1991) used GC-MS to analyse the taste and flavour of aged Iberian hams. The isolation of the distilled volatiles was performed under vacuum conditions in cool traps and the collection of the samples was accomplished with dichloromethane. Seventy-seven volatile substances were detected, including alkanes (12), branched alkanes (14), aldehydes (13), and aliphatic alcohols (9). Furthermore, small concentrations of lactones (5), esters (9) and ketones (7) and other miscellaneous compounds were also present. The large amount of olfactory volatiles detected in products was attributed to an intense proteolytic and lipolytic breakdown during maturation.

Gerrard et al. (1996) used a sensory panel of ten members in combination with image processing to evaluate steaks of different levels of marbling and colour. These factors were assigned to each steak with USDA marbling score cards and a lean colour guide. The evaluation of the colour was performed by a trained panel according to an eight-point scale from 1 (bleached red) to 8 (very dark red), and in a nine-point scale from 1 (devoid) to 9 (moderately abundant) for marbling. The conduction of the tests was performed under the same conditions. The predictions of the lean colour ($R^2 = 0.86$) and marbling scores ($R^2 = 0.84$), were effectively performed using image processing thus indicating that the method is effective in determining USDA quality attributes of fresh meat.

While Ellekjær and Isaksson (1992) evaluated the use of NIR for quantitatively assessing the major beef components (i.e., protein, fat and water) and the effects of heat treatment, Hildrum et al. (1994) used the same technique to evaluate the potential changes of beef muscles during conditioning and ageing and examined how were these treatments correlated to sensory characteristics. The use of reflection (NIRR) and transmission (NIRT) analysis in a period of two weeks of aging (at 20°C) gave important information about the sensory hardness, tenderness and juiciness of

M. Longissimus dorsi muscles. It was proved that NIRR combined with principal component regression effectively predicted only the first two sensory variables and led to coefficients of 0.80 to 0.90. Moreover, NIRR derived data were useful in obtaining good predictions for frozen and thawed samples. In contrast, no variables were satisfactorily predicted using the NIRT mode.

Liu et al. (2003b) used VIS/NIRS in the 400–1080 nm region to predict the colour, instrumental texture, and sensory characteristics of beef steaks in a certain day post-mortem. By randomly taking measurements of the Hunter L (lightness), α (redness), b (yellowness) and the indirect fraction E* (redness relative to those of yellowness and lightness) obtained from the modified equation of Liu et al. (2003b): $E^* = \alpha/b + \alpha/L$, the prediction coefficients of determination (R_2) in calibration between 0.78 and 0.90 (Hunter α, b and E*), and 0.49 and 0.55 (tenderness, Hunter L, sensory chewiness and juiciness) were obtained. The prediction R2 for tenderness ranged between 0.22 and 0.72 after segregation of samples according to the aging days. The use of the PLS and soft independent modelling of class analogy of PCA (SIMCA/PCA) models led to effective classification of samples into tender and tough (83% and 96%, respectively). It was shown that Bratzler shear force, sensory chewiness and E* (possible loss of meat redness) were reduced with time (Hildrum et al., 1995; Liu et al., 2003a).

Mor-Mur and Yuste (2003) examined the differences in colour, texture and yield between sausages cooked at 500 MPa for 5 or at 65°C for 15 min and stored under VP, and sausages treated under pasteurization conditions (80–85°C for 40 min). It was shown that pressurized sausages were cohesive and less firm, and gave higher yield in comparison to heat-treated products. The work of the sensory panel was quite challenging due to the difficulties in detecting the differences between the two types of products. Also, despite the different characteristics of the sausages, the heat-treated product was more frequently chosen due to its better appearance (lesser gelatin on the surface), taste (stronger and more pleasant, especially in 15 min pressurised sausages) and texture (juicier, less grainy and of more uniform consistency).

Ai-Nong et al. (2005) studied the volatile compounds obtained from Chinese traditional smoke-cured bacon (CSCB). The examination was performed after condensing and dissolving the volatiles in organic solvent (ether and n-pentane) using the NPSD technique. This extract was subsequently analysed using GC-MS and gas chromatography (GC-FID) as well as a flame ionization detector. Twenty-seven new compounds, including four phenols (o-tertbutylphenol, butyl hydroxy toluene, 2,3,5-trimethoxytoluene and 2,2'-methylenebis[6-(1,1-dimethylethyl)-4-methyl]-phenol), were detected, thus proving that the NPSD method was highly effective in carrying the aroma ingredients and contemporarily inhibiting the oxidation of the cellular components. It was proved that most volatiles obtained from the Chinese smoked bacon were phenolic derivatives, mainly formed during smoking, and were significantly different from the volatiles obtained from fresh pork and Jinhua ham. In the study of Ai-Nong et al. (2008), the volatile compounds extracted from the same type of bacon (CSCB) were analysed using SPMS-GC/MS. Forty eight volatile compounds were found and quantified, including several chemical classes: 1 alkane, 16 aldehydes, 5 ketones, 9 alcohols, 4 thioethers and thiols, 3 furans and 10 phenols compounds. Most of the volatiles, with the exception of alkane, contributed to the aroma of the product and were formed through processes such as smoking, oxidation and Maillard reaction, etc. An increased number of volatiles was detected mainly due to the advanced isolation methodology, which included the use of fibres coating such as CAR/PDMS (carboxen/polydimethylsiloxane; with the highest area counts for most volatile compounds) and DVB/CAR/PDMS (divinylbenzene/carboxen/polydimethylsiloxane), and led to better extraction of compounds characterised by low and high linear retention indices (LRI), respectively.

In the study of Jerković et al. (2007), the isolation of volatile compounds from Dalmatian prosciutto was performed through solvent extraction (SE), simultaneous distillation extraction (SDE) and nitrogen purge and steam distillation (NPSD) and the samples were analysed using GC and GC-MS. Forty six substances were detected using SDE and SE (including fatty acids, aldehydes, phenols, esters, ketones and others), while 81 substances were found using NPSD (headspace volatiles including phenols, aldehydes, hydrocarbons, ketones, alcohols, esters and heterocyclic compounds). It was proved that ripening of prosciutto led to an increase in the levels of aldehydes and esters. The levels of volatiles differed significantly between fried and raw samples especially as regards the aldehydes (SDE and SE). The NPSD method was used to obtain additional information for the volatiles from fried ham, since the isolation of the pyrazines and most of the lower aldehydes, that are significant thermally-derived flavour compounds, was only achieved using NPSD.

The characterisation of two kinds of typical Italian dry-sausages, namely "Salame Mantovano" and "Salame Cremonese" was performed through analysis of their volatile composition. Seven samples of "Salame Mantovano" and five samples of "Salame Cremonese" were examined. The analysis was performed using dynamic headspace extraction technique (DHS) coupled with gas chromatography-mass spectrometry (GC-MS). Among the 104 volatiles detected, terpenes, aldehydes, ketones and alcohols were the most abundant. Peak area data for all the compounds were used to statistically analyse the results. Firstly, principal component analysis (PCA) was used towards the visualisation of the data trends and the detection of possible clusters within samples. Afterwards, linear discriminant analysis (LDA) was carried out towards the detection of volatiles that could be used for the differentiation of the two kinds of sausages. The data obtained by GC-MS indicated that the most important substances that could be used to differentiate the two products were seven volatile compounds, i.e., 3-methylbutanal, 6-camphenol, dimethyl disulphide, 1-propene-3,3'-thiobis, ethyl propanoate, 1,4-p-menthadiene and 2,6-dimethyl-1,3,5,7-octatetraene. The prediction potential of the model was estimated to be 100% by the "leave-one-out" cross-validation (Bianchi et al., 2007).

In the study of Jerković et al. (2010), 119 organic compounds were isolated from the Savorian salami "kulen". The isolation was performed using nitrogen purge and steam distillation (NPSD) while the analysis was performed using GC and GC-MS. The volatiles were formed during oxidation of lipids, amino acid degradation, smoking and addition of spices. The NPSD method permitted comprehensive profiling with almost exclusive distribution of numerous major substances in particular trap, with little or no interference from abundant lipid constituent in the products. The most significant substances found were methylphenols, methoxyphenols, organosulphur compounds (diallyl sulphide, diallyl disulphide, methylallyl disulphide, diallyl trisulphide and methional), while some of the less abundant substances included high-molecular fatty acids, alcohols and aldehydes. Possible discrimination between "kulen" volatile profile and other European salami volatiles could be based on the lower amount of terpenes and higher percentages of diallyl sulphide, methoxyphenols, methylphenols and 2-cyclopenten-1-one derivatives detected.

O'Sullivan et al. (2003a) collected pork muscle samples (*M. longissimus dorsi* and *M. psoas major*) from pigs fed with one of four dietary treatments—(1) control diet, (2) supplemental iron (7 g iron (II) sulphate/kg feed), (3) supplemental vitamin E (200 mg dl-a-tocopheryl acetate/kg of feed) and (4) supplemental vitamin E+supplemental iron. Supplements of vitamin C were provided to all dietary treatments to enhance iron uptake. Determination of vitamin E and iron tissue levels were performed for each treatment. The evaluation of the warmed-over flavour (WOF) was performed by a trained sensory panel (n = 8) for the four treatments which were cooked and stored under refrigeration conditions at 4°C for up to five days. Thawing loss, driploss and thiobarbituric acid reactive substances (TBARS) were also evaluated. For *M. longissimus dorsi*, the highest vitamin E muscle tissue levels were found in the iron/vitamin E-treated group followed by the vitamin E group, control and iron treated groups. As regards *M. psoas major*, the vitamin E tissue levels were in the following order of magnitude: vitamin E > iron/vitamin E > iron > control group. Iron tissue levels followed the following order of magnitude: vitamin E > iron/vitamin E > control > iron for *M. longissimus dorsi* and iron > vitamin E > control > iron/vitamin E for *M. psoas major*. Thus, vitamin E and vitamin C enhanced non-supplemental iron absorption in the vitamin E-treated group for *M. longissimus dorsi* and to a lesser extent for *M. psoas major*. It was proved that *M. psoas major* was more susceptible to warmed-over flavour formation in comparison to *M. longissimus dorsi* for all treatments as detected by sensory profiling, due to higher tissue iron levels. WOF development in *M. longissimus dorsi* and *M. psoas major* was highest in the iron-supplemented groups followed by the control and vitamin E-supplemented groups.

In the study of O'Sullivan et al. (2003b), pork muscle samples (*M. longissimus dorsi* and *M. psoas major*) obtained from pigs fed with the same four dietary treatments [(i) control diet, (ii) supplemental iron (300 mg iron (II) sulphate/kg feed), (iii) supplemental vitamin E (200 mg dl-a-tocopheryl acetate/kg of feed) and (iv) supplemental vitamin E+ supplemental iron] were analysed using gas chromatography mass spectrometry (GC/MS) and electronic nose analysis. The analysis was performed on a subset of the full design which included samples of *M. longissimus dorsi*, treatments (ii) and (iii) and *M. psoas major* with treatment (i) for 0 days of WOF development. Day 5 of WOF development was included in the subset and was represented by samples of *M. longissimus dorsi*, treatment (iv) and *M. psoas major*, treatments (ii) and (iii). Bi-linear modelling was employed for the determination of the correlation of GC/MS and electronic nose data to sensory data. The evaluation of the reproducibility and reliability of electronic nose data was performed by repeating the analysis of samples in a different laboratory and with a time difference of approximately 11 months. The data from these two different electronic noise data sets were normalized using mean centring. GC/MS data correlated to sensory data with specific compounds (e.g., pentanal, 2-pentylfuran, octanal, nonanal, 1-octen-3-ol and hexanal), and were proved to be promising indices of oxidation in cooked samples of *M. longissimus dorsi* and *M. psoas major*. Electronic nose data correlated to sensory data and separated the sensory variation. The reproducibility of this data was high with the second set of samples being predictive of the first set.

A semiconductor multisensorial system constructed with tin oxide layers—pure or enriched with Pt, Pd and TiO_2—was used for the qualitative assessment of dry-cured Iberian hams with satisfactory responses obtained from the 12 elements forming the multi-sensor at different operating temperatures (Otero et al., 2003). The electrical resistance of each sensor was measured in air (R_a) and in the sample (R_s) to assess the sample sensitivity (S) of each sensor and was defined as: $R(\%) = \dfrac{(R_a - R_S)}{R_a} \times 100$.

Two types of ham were effectively discriminated using PCA, while the sensitivity to ham aroma gave optimum results at elevated temperature, with 250°C detected as the temperature of maximum response. Furthermore, the thinnest sensor revealed maximal response at 150 and 200°C, with totally minor differences and best sensitivities and discriminations when functioning at 250°C. The dopant incorporation rarely enhanced the sensitivity to ham aroma.

Salchichón is the second most widely consumed dry-cured sausage in Spain after chorizo. This product is a mixture of chopped meat (pork, beef/pork or beef), lard, salt, additives (nitrate, nitrite, antioxidants), starter cultures (optional) and spices. A well-trained sensory panel was employed to record and describe the main sensory characteristics of this foodstuff (Ruiz Pérez-Cacho et al., 2005). The initial 108-term vocabulary was limited down to 15 attributes relative to appearance, odour, texture and flavour. The use of Kruskal-Wallis test revealed that the samples differed significantly ($p < 0.001$) allowing categorisation into three groups: group I (strong other spices smell and aroma and high juiciness), group II (high mould smell and aroma), and group III (acid and salty taste).

A summary of sensory analysis surveys is presented in Table 3.4.

3.6 Genetically Modified Organisms (GMO)

During the last decades, an impressive and stable increase of GM crops occurred. The most common GM crops are corn, soybean and oilseed rape (canola). Nowadays, soybean proteins are commonly added in the processed meat-based products (Belloque et al., 2002; Cardarelli et al., 2005). Processed meat-based foodstuffs are characterised by elevated fat content and low protein content. Meat proteins acting as food emulsifiers inhibit the coalescence of fat during thermal processing. Soybean protein is added to meat products with low protein content to prevent coalescence of fat (Belloque et al., 2002). In the study of Olsman (1979), soybean proteins were determined by two different methods, a direct and an indirect method. The direct method included detection of soybean protein, while the indirect method was mainly based on determining the substances or compounds related to soybean proteins (Olsman, 1979).

Rencova and Tremlova (2009) employed indirect competitive ELISA method towards the detection of soybean protein in meat products. The detection limit of ELISA was 0.5% of the added soybean protein weight. A total of 131 meat products were collected and examined for the presence of soybean protein. It was found that 84% of samples were positive to the presence of soybean protein.

Table 3.4. Sensory analysis of meat and meat products; type of meat, no of panelists (trained or untrained or semi-trained) and attributes investigated.

No of samples	Type	Part/Treatment	QC method	No of panelists	Trained	Attributes	Range	Results	Remarks	References
	Meat/Product									
25	Beef	Muscle	RT-PCR	10–12	Y	Tenderness Juiciness Flavour	1–10	No of specific expressed genes/ markers: 29 21 15 DNAJA1 new!	Expensive Time consuming	Bernard et al., 2007
48	Pork	Muscle	NIR, PLS-DA	–	–	Freshness	Refrigerated or VP and frozen	100% classification accuracy	Precise, non-destructive	Barbin et al., 2013
72	Sausage	Fermented	Electronic nose of 14 sensors (2 × 5 MOSFET[a])	17	Y	19	1–9	Good agreement	Rapid Simple Inexpensive	Eklöv et al., 1998
10	Ham	Dry cured	CG-LC GC-MS	–	–	Alkanes Alkanes branched Aldehydes Alcohols aliphatic	–	12 (6.82%) 14 (12.31%) 13 (30.69%) 9 (22.27%)	Intense proteolytic and lipolytic breakdown Questions on the volatiles real origin	Garcia et al., 1991
60	Beef	Steak	Image processing	10	Y	Colour Marbling	1–8 1–9	$R^2 = 0.86$ $R^2 = 0.84$	Inexpensive Rapid Difficulty on visual evaluation	Gerrard et al., 1996
								No effect		
								More chickeny and tender		

30	Beef	Fresh Raw Raw frozen-then-thawed	NIR-Reflection NIR-Transmission	9	Y	Hardness Tenderness Juiciness	1–9	2.9–7.1 (4.7) 2.9–7.4 (5.6) 4.6–6.1 (5.4)	Great effect of the age, weight, fat content and shear-press Rapid Satisfactory results	Hildrum et al., 1994
177	Beef	Steaks	Frozen	VIS-NIR Sensory Analysis	7	Y	Colour values[d] Shear force Chewiness Juiciness	1–5	5.46 3.19 2.37	> 90% correct classification Good accuracy [d]	Liu et al., 2003b
28	Pork	Muscle	Cooked Refrige-rated	GC-MS Electronic nose Sensory analysis	8	N	Odour, flavour, taste and after taste sensory	0–150	Samples with high [E]WOF values: - 5d refrigerated - From iron enriched feeding - *M. psoas major muscle* vs *M. longissimus dorsi*	Indestructible sensors High reproducibility	O'Sullivan et al., 2003b
-	Ham	Dry-cured 20°C 30°C	Electronic nose	-	-	-	-	Flavour: Sweet Intense and pleasant	12 TiO$_2$ sensors Good responses Best results at 250°C working temperature Non-destructive Non-contact	Otero et al., 2003

Table 3.4. contd....

Table 3.4. contd.

No of samples	Meat/Product		QC method	No of panelists	Trained	Attributes	Range	Results	Remarks	References
	Type	Part/Treatment								
4	Pork	Muscle Ir NIr	GC-MSD Sensory analysis	13	Y	Lipid oxidation Volatiles Sensory analysis	0–15	Irradiation had no (negative) effect on the acceptance of the meat	TBARS are: -responsible for the off-odour in Ir samples - non-dose dependent at < 10 kGy - related to radiolytic degradation of AAs	Ahn et al., 2000
117	Ham	Dry-cured	VIS/NIR Sensory analysis	2	Y	Pastiness, colour, crusting, marbling and ring colour	1–10	$Y(\%)^h$: 80.0 91.1 42.8 70.1 69.9	Use of remote reflectance fiber optic probe Online classification possibility	Ortiz-Somovilla et al., 2007

Notes: [a] MOSFET: Metal Oxide Semiconductor Field-Effect Transistors, [b] Fresh chickeny, Bloody, Oxidized, Sweet aromatic, Colour, Moistness & Glossiness, [c] Colour, Juiciness, Sweet, Salt, Sour, Bitter, Chickeny, Warmed-over, Brothy, Sweet-aromatic, Cardboard, Tenderness & Graininess. Reflection Transmittance, [d] Hunter L (lightness), a (redness), and b (yellowness), [E] WOF: Warm-Over Flavour, [f] I: Irradiated, NI: Non irradiated, [g] A: Air; VP: Vacuum Package; M1: Modified Atmosphere 1, 30%/65%/5% ($CO_2/N_2/O_2$); and M2: Modified Atmosphere, 65%/30%/5% ($CO_2/N_2/O_2$), [h] is the percent variance of the binary response explained by PLS, GC-MSD: Gas Chromatography-Mass Selective Detector, MFI: Myofibrillar Fragmentation Index.

DNA-based techniques are commonly applied to determine small amounts of DNA in foods. The PCR technique has been used for various foodstuffs, such as meat and blended products (Meyer et al., 1996). The PCR assay was used for the detection of trace amounts of soybean in meat sausages by Soares et al. (2010). Soybean proteins are commonly used in the food industry. Since they are included amongst the basic allergenic materials, they should always be declared to avoid allergic reactions among individuals with allergies. In this study, an optimised PCR was used, targeting the soybean lecithin gene and thus enabling the addition of 0.1% and 0.5% hydrated texture protein, which corresponds to 0.01% and 0.06% (w/w) of soybean protein in raw and thermally processed pork meat, respectively. By applying the technique to 18 samples of meat sausages, it was proved that the presence of soybean could be effectively detected. Nine samples, obtained from products with no labelled information about the protein, were found positive.

Treml and Maisonnave-Arisi (2008) analysed 47 samples of processed meat products using a DNA-based technique, in an effort to detect possible contamination with soybean proteins. The soy-specific primers LEC1/LEC2 intensified a 164 bp fragment of the lectin gene. Additional tests were performed to positive samples using nested PCR and the primers GMO5/GMO9 and GMO7/GMO8 (for 447 bp and 169 bp fragments, respectively). The lectin fragment was not detected in 5 of 47 samples. Forty-two samples were found positive to soybean protein. It was concluded that six meat samples contained roundup ready soybean.

Immunochemical methods are employed to detect soybean proteins in processed meat-based foods and are both highly specific and sensitive. The AOAC International granted ELISA as the AOAC official method for determining soybean proteins in raw and heat processed meat (AOAC, 1998). Castro-Rubio et al. (2005) used the above-mentioned technique to develop a simpler and cheaper method for determining soybean proteins in meat products. The method includes defattening of samples with acetone and solubilisation of the soybean proteins in a 30 mM Tris-HCl buffer containing 0.5% (v/v) 2-mercaptoethanol. The chromatograph indicated two peaks from soybean proteins. The method was characterised by high specificity and sensitivity, allowing the detection and quantification of additions of 0.07% (w/w) and 0.25% (w/w), respectively of soybean proteins in meat products (related to 1 g of initial product). Criado et al. (2005) employed perfusion liquid chromatography for detecting soybean proteins in cured meat products. As reported above, the samples were defatted with acetone and soybean proteins were solubilized with a buffer solution at basic pH. The use of water-aceto-nitrile-trifluoroacetic acid and water-tetrahydrofuran-trifluoroacetic acid linear binary gradients at a flow rate of 3 mL/min at a temperature of 50°C and the use of UV at 280 nm, allowed the chromatographic analysis of soybean proteins in less than 3 min.

3.7 Conclusions

Over the last years, legislation was established around labelling of foodstuffs in an effort to protect consumers against food adulteration. Various studies have been performed on authenticity and quality control and assurance. However, all analytical methods have certain limitations, which limit their applicability.

Nevertheless, the continuous research towards the development and improvement of new and existing analytical and chemometric techniques, respectively, allows the development of more accurate and sensitive methods of assessing food authenticity. Most new applications include the use of new technologies such as DNA-based techniques, SNIF-NMR and IRMS. The use of DNA-based techniques on the assessment of authenticity of animal-origin samples present many advantages but these methods are very expensive. Technologies such as SNIF-NMR and IRMS could not currently be widely used because of their high instrumentation cost. However, these techniques are both highly sensitive and specific since they are based on the analysis of atoms found in molecules within the food sample.

IR spectroscopy is an expeditious technique characterized by high sensitivity and can effectively be used for detecting adulteration in a vast variety of foodstuffs. It can be easily applied in conjunction with chemometric analysis, but it is gas chromatography, which significantly reduced the analysis times (chromatographic runs holding only a few minutes), making the use of technique efficient for application in the industry. Another technology, the electric nose, offers the advantages of low cost, rapidity and ease of operation. However, the problems connected with the application of this technique are related with the sensors used as part of the electronic nose instrumentation.

Future research should focus on the development of fast, accurate and non-complicated techniques, capable of detecting food adulteration.

References

Ahn, D.U., C. Jo and D.G. Olson. 2000. Analysis of volatile components and the sensory characteristics of irradiated raw pork. Meat Science. 54: 209–215.

Ai-Nong, Y. and S. Bao-Guo. 2005. Flavour substances of Chinese traditional smoke-cured bacon. Food Chemistry. 89: 227–233.

Ai-Nong, Y., S. Bao-Guo, T. Da-Ting and Q. Wan-Yun. 2008. Analysis of volatile compounds in traditional smoke-cured bacon (CSCB) with different fiber coatings using SPME. Food Chemistry. 110: 233–238.

Al-Jowder, O., M. Defernez, E.K. Kemsley and R.H. Wilson. 1999. Mid-Infrared Spectroscopy and Chemometrics for the Authentication of Meat Products. Journal of Agricultural and Food Chemistry. 47: 3210–3218.

Al-Jowder, O., E.K. Kemsley and R.H. Wilson. 2002. Detection of Adulteration in Cooked Meat Products by Mid-Infrared Spectroscopy. Journal of Agricultural and Food Chemistry. 50: 1325–1329.

Alamprese, C., M. Casale, N. Sinelli, S. Lanteri and E. Casiraghi. 2013. Detection of minced beef adulteration with turkey meat by UV-vis, NIR and MIR spectroscopy. LWT—Food Science and Technology. 53: 225–232.

Ali, M.E., U. Hashim, S. Mustafa, Y.B. Che Man, Th.S. Dhahi, M. Kashif, Md.K. Uddin and S.B.A. Hamid. 2012. Analysis of pork adulteration in commercial meatballs targeting porcine-specific mitochondrial cytochrome b gene by TaqMan probe real-time polymerase chain reaction. Meat Science. 91: 454–459.

Alomar, D., C. Gallo, M. Castañeda and R. Fuchslocher. 2003. Chemical and discriminant analysis of bovine meat by near infrared reflectance spectroscopy (NIRS). Meat Science. 63: 441–450.

AOAC Official Method 998.10. 1998. Soy protein in raw and heat processed meat products, enzyme-linked immunosorbent assay: Official methods of Analysis. 16th ed., Rev. 1996, 1997, 1998. AOAC International.

Argyri, A.A., E.Z. Panagou, P.A. Tarantilis, M. Polysiou and G.J.E. Nychas. 2010. Rapid qualitative and quantitative detection of beef fillets spoilage based on Fourier transform infrared spectroscopy data and artificial neural networks. Sensors and Actuators. B.145: 146–154.

Arvanitoyannis, I.S., J.G. Bloukas, I. Pappa and E. Psomiadou. 2000. Multivariate analysis of Cavourmas—a Greek cooked meat product. Meat Science. 54(1): 71–75.

Arvanitoyannis, I.S. and M. van Houwelingen-Koukaliaroglou. 2003. Implementation of chemometrics for quality control and authentication of meat and meat products. Critical Reviews in Food Science and Nutrition. 43: 173–218.

Asensio, L., I. González, T. García and R. Martín. 2008. Determination of food authenticity by enzyme-linked immunosorbent assay (ELISA)—Review. Food Control. 19: 1–8.

Aslan, O., R.M. Hamill, T. Sweeney, W. Reardon and A.M. Mullen. 2009. Integrity of nuclear genomic deoxyribonucleic acid in cooked meat: Implications for food traceability. Journal of Animal Science. 87: 57–61.

Balamatsia, C.C., A. Patsias, M.G. Kontominas and I.N. Savvaidis. 2007. Possible role of volatile amines as quality-indicating metabolites in modified atmosphere-packaged chicken fillets: Correlation with microbiological and sensory attributes. Food Chemistry. 104: 1622–1628.

Ballin, N.Z. and R. Lametsch. 2008. Analytical methods for authentication of fresh vs. thawed meat—A review. Meat Science. 80: 151–158.

Ballin, N.Z., F.K. Vogensen and A.H. Karlsson. 2009. Review Species determination—Can we detect and quantify meat adulteration? Meat Science. 83: 165–174.

Barbin, D., G. Elmasry, D.-W. Sun and P. Allen. 2012. Near-infrared hyperspectral imaging for grading and classification of pork. Meat science. 90: 259–268.

Barbin, D., D.-W. Sun and C. Su. 2013. NIR hyperspectral imaging as non-destructive evaluation tool for the recognition of fresh and frozen–thawed porcine *longissimus dorsi* muscles. Innovative Food Science and Emerging Technologies. 18: 226–236.

Belloque, J., M.C. Garcia, M. Torre and M.L. Marina. 2002. Analysis of Soyabean Proteins in Meat Products: A Review. Critical Reviews in Food Science and Nutrition 42: 507–532.

Bernard, C., I. Cassar-Malek, M. Le Cunff, H. Dubroeuco, G. Renand and J.F. Hocquette. 2007. New Indicators of Beef Sensory Quality Revealed by Expression of Specific Genes. Journal of Agricultural and Food Chemistry. 55: 5229–5237.

Berzaghi, P. and R. Riovanto. 2009. Near infrared spectroscopy in animal science production: principles and applications. Italian Journal of Animal Science. 8: 39–62.

Bianchi, F., C. Cantoni, M. Careri, L. Chiesa, M. Musci and A. Pinna. 2007. Characterization of the aromatic profile for the authentication and differentiation of typical Italian dry-sausages. Talanta. 72: 1552–1563.

Boner, M. and H. Förstel. 2004. Stable isotope variation as a tool to trace the authenticity of beef. Analytical and Bioanalytical Chemistry. 378: 301–310.

Bong, Y.S., W.J. Shin, A.R. Lee, Y.S Kim, K. Kim and K.S. Lee. 2010. Tracing the geographical origin of beefs being circulated in Korean markets based on stable isotopes. Rapid Communications in Mass Spectrometry. 24: 155–159.

Bottero, M.T., I.A. Dalmasso, D. Nucera, R.M Turi, S. Rosati, S. Squadrone et al. 2003. Development of a PCR assay for the detection of animal tissues in ruminant feeds. Journal of Food Protection. 66: 2307–2312.

Brodmann, P.D. and D. Moor. 2003. Sensitive and semi-quantitative TaqMan™ real-time polymerase chain reaction systems for the detection of beef (*Bos taurus*) and the detection of the family Mammalia in food and feed. Meat Science. 65(1): 599–607.

Calvo, J.H., R. Osta and P. Zaragoza. 2002. Quantitative PCR Detection of Pork in Raw and Heated Ground Beef and Pâté. Journal of Agricultural and Food Chemistry. 50: 5265–5267.

Calvo, J.H., P. Zaragoza and R. Osta. 2001. Technical note: A quick and more sensitive method to identify pork in processed and unprocessed food by PCR amplification of a new specific DNA fragment. Journal of the Animal Science. 79: 2108–2112.

Cammà, C., M.D. Domenico and F. Monaco. 2012. Development and validation of fast Real-Time PCR assays for species identification in raw and cooked meat mixtures. Food Control. 23: 400–404.

Camin, F., L. Bontempo, K. Heinrich, M. Horacek, S.D. Kelly, C. Schlicht, F. Thomas, F.J. Monahan, J. Hoogewerff and A. Rossmann. 2007. Multi-element (H, C, N, S) stable isotope characteristics of lamb meat from different European regions. Analytical and Bioanalytical Chemistry. 389: 309–320.

Cardarelli, P., M.R. Branquinho, R.T.B. Ferreira, F.P. da Cruz and A.L. Gemal. 2005. Detection of GMO in food products in Brazil: the INCQs experience. Food Control. 16: 859–866.

Castro-Rubio, F., M. Concepcion Garcia, R. Rodriguez and M. Luisa Marina. 2005. Simple and inexpensive method for the reliable determination of additions of soybean proteins in heat-processed meat products: An alternative to the AOAC Official Method. Journal of Agricultural and Food Chemistry 53: 220–226.

Cawthorn, D.-M., H.A. Steinman and L.C. Hoffman. 2013. A high incidence of species substitution and mislabelling detected in meat products sold in South Africa. Food Control. 32: 440–449.

Chan, D.E., P.N. Walker and E.W. Mills. 2002. Prediction of pork quality characteristics using visible and near-infrared spectroscopy. Transactions of the ASAE. 45: 1519–1527.

Chen, F.C. and Y.H. Hsieh. 2000. Detection of pork in heat-processed meat products by monoclonal antibody-based ELISA. Journal of AOAC International. 83: 79–85.

Chisholm, J., C. Conyers, C. Booth, W. Lawley and H. Hird. 2005. The detection of horse and donkey using real-time PCR. Meat Science. 70: 727–732.

Chou, C.C., S.P. Lin, K.M. Lee, C.T. Hsub, T.W. Vickroy and J.M. Zenb. 2007. Fast differentiation of meats from fifteen animal species by liquid chromatography with electrochemical detection using copper nanoparticle plated electrodes. Journal of Chromatography B. 846: 230–239.

Cozzolino, D. and I. Murray. 2004. Identification of animal meat muscles by visible and near infrared reflectance spectroscopy. Lebensmittel-Wissenschaft & Technologie. 37: 447–452.

Criado, M., F. Castro-Rubio, C. Garcia-Ruiz, M. Concepcion Garcia and M.L. Marina. 2005. Detection and quantification of additions of soybean proteins in cured meat products by perfusion reversed-phase high-performance liquid chromatography. Journal of Separation Science. 28: 987–995.

Dalmasso, A., E. Fontanella, P. Piatti, T. Civera, S. Rosati and M.T. Bottero. 2004. A multiplex PCR assay for the identification of animal species in feedstuffs. Molecular and Cellular Probes. 18: 81–87.

DeNiro, M.J. and S. Epstein. 1978. Influence of diet on the distribution of carbon isotopes in animals. Geochimica et Cosmochimica Acta. 42: 495–506.

DeNiro, M.J. and S. Epstein. 1981. Influence of diet on the distribution of nitrogen isotopes in animals. Geochimica et Cosmochimica Acta. 45: 341–351.

Di Pinto, A., V.T. Forte, M.C. Conversano and G.M. Tantillo. 2005. Duplex polymerase chain reaction for detection of pork meat in horse meat fresh sausages from Italian retail sources. Food Control. 16: 391–394.

Ding, H.B. and R.J. Xu. 1999. Differentiation of beef and kangaroo meat by visible/near infrared reflectance spectroscopy. Journal of Food Science. 64: 814–817.

Ding, H.B. and R.J. Xu. 2000. Near-Infrared Spectroscopic Technique for Detection of Beef Hamburger Adulteration. Journal of Agricultural and Food Chemistry. 48: 2193–2198.

Dooley, J.J., K.E. Paine, S.D. Garrett and H.M. Brown. 2004. Detection of meat species using TaqMan real-time PCR assays. Meat Science. 68: 431–438.

Downey, G. and D. Beauchêne. 1997. Discrimination between Fresh and Frozen-then-thawed Beef m. *longissimus dorsi* by Combined Visible-near Infrared Reflectance Spectroscopy: A Feasibility Study. Meat Science. 45: 353–363.

Drummond, M.G., B.S.A.F. Brasil, L.S. Dalsecco, R.S.A.F. Brasil, L.V. Teixeira and D.A.A. Oliveira. 2013. A versatile real-time PCR method to quantify bovine contamination in buffalo products. Food Control. 29: 131–137.

Ebbehoj, K.F. and P.D. Thomsen. 1991a. Species Differentiation of Heated Meat Products by DNA Hybridization. Meat Science. 30: 221–234.

Ebbehoj, K.F. and P.D. Thomsen. 1991b. Differentiation of Closely Related Species by DNA Hybridization. Meat Science. 30: 359–366.

Eklöv, T., G. Johansson, F. Winquist and I. Lundsröm. 1998. Monitoring Sausage Fermentation Using an Electronic Nose. Journal of the Science of Food and Agriculture. 76: 525–532.

Ellekjær, M.R. and T. Isaksson. 1992. Assessment of maximum cooking temperature in previous heat treated beef. Part 1: Near Infrared Spectroscopy. Journal of the Science of Food and Agriculture. 59: 335–343.

Ellis, D.I., D. Broadhurst, S.J. Clarke and R. Goodacre. 2005. Rapid identification of closely related muscle foods by vibrational spectroscopy and machine learning. Analyst. 130: 1648–1654.

Fajardo, V., I. González, I. López-Calleja, I. Martín, E.P. Hernádez, T. García and R. Martín. 2006. PCR-RFLP Authentication of Meats from Red Deer (*Cervus elaphus*), Fallow Deer (*Dama dama*), Roe Deer (*Capreolus capreolus*), Cattle (*Bos taurus*), Sheep (*Ovis aries*), and Goat (*Capra hircus*). Journal of Agricultural and Food Chemistry. 54: 1144–1150.

Fajardo, V., I. González, I. López-Calleja, I. Martín, M. Rojas, T. García, P.E. Hernández and R. Martín. 2007a. PCR identification of meats from chamois (*Rupicapra rupicapra*), pyrenean ibex (*Capra pyrenaica*), and mouflon (*Ovis ammon*) targeting specific sequences from the mitochondrial D-loop region. Meat Science. 76: 644–652.

Fajardo, V., I. González, I. López-Calleja, M. Martín, M. Rojas, P.E. Hernández, T. García and R. Martín. 2007b. Identification of meats from red deer (*Cervus elaphus*), fallow deer (*Dama dama*), and roe deer (*Capreolus capreolus*) using polymerase chain reaction targeting specific sequences from the mitochondrial 12S rRNA gene. Meat Science. 76: 234–240.

Fajardo, V., I. González, I. Martín, M. Rojas, P.E. Hernández, T. García and R. Martín. 2008. Real-time PCR for detection and quantification of red deer (*Cervus elaphus*), fallow deer (*Dama dama*), and roe deer (*Capreolus capreolus*) in meat mixtures. Meat Science. 79: 289–298.

Fajardo, V., I. González, J. Dooley, S. Garret, H.M. Brown, T. García and R. Martín. 2009. Application of polymerase chain reaction-restriction fragment length polymorphism analysis and lab-on-a-chip capillary electrophoresis for the specific identification of game and domestic meats. Journal of the Science of Food and Agriculture. 89: 843–847.

Fei, S., T. Okayama, M. Yamanoue, I. Nishikawa, H. Mannen and S. Tsuji. 1996. Species identification of meats and meat products by PCR. Animal Science and Technology. 67: 900–905.

Flores-Munguia, M.E., M.C. Bermudez-Alamada and L. Vaquez-Moreno. 2000. A research note: Detection of adulteration in processed traditional meat products. Journal of Muscle Foods. 11: 319–325.

Foca, G., D. Salvo, A. Cino, C. Ferrari, D.P.L. Fiego, G. Minelli and A. Ulrici. 2013. Classification of pig fat samples from different subcutaneous layers by means of fast and non-destructive analytical techniques. Food Research International. 52: 185–197.

Franke, B.M., G. Gremaud, R. Hadorn and M. Kreuzer. 2005. Geographic origin of meat-elements of an analytical approach to its authentication. European Food and Research Technology. 221: 493–503.

Franke, B.M., R. Hadorn, J.O. Bosset, G. Gremaud and M. Kreuzer. 2008. Is authentication of the geographic origin of poultry meat and dried beef improved by combining multiple trace element and oxygen isotope analysis? Meat Science. 80: 944–947.

Franke, B.M., M. Haldimann, J. Reimann, B. Baumer, G. Gremaud, R. Hadorn, J.O. Bosset and M. Kreuzer. 2007. Indications for the applicability of element signature analysis for the determination of the geographic origin of dried beef and poultry meat. European Food and Research Technology. 221: 501–509.

Gaitán-Jurado, A.J., V. Ortiz-Somovilla, F. España-España, J. Pérez-Aparicio and E.J. De Pedro-Sanz. 2008. Quantitative analysis of pork dry-cured sausages to quality control by NIR spectroscopy. Meat Science. 78: 391–399.

Gao, H.W., C.Z. Liang, Y.B. Zhang and L.H. Zhu. 2004. Polymerase chain reaction method to detect canis materials by amplification of species-specific DNA fragment. Journal of AOAC International. 87: 1195–1199.

Garcia, C., J.J. Berdagué, T. Antequera, C. López-Bote, Córdoba and J. Ventanas. 1991. Volatile Components of Dry Cured Iberian Ham. Food Chemistry. 41: 23–32.

Gebbing, T., J. Schellberg and W. Kühbauch. 2004. Switching from grass to maize diet changes the C isotope signature of meat and fat during fattening of steers. pp. 1130–1132. *In*: Proceedings of the 20th general meeting of the European grassland federation. Grassland Science in Europe (Vol. 9). Lucern: European Grassland Federation.

Gerrard, D.E., X. Gao and J. Tan. 1996. Beef Marbling and Color Score Determination by Image Processing. Journal of Food Science. 61: 145–148.

Ghivizzani, S.C., S.L.D. Mackay, C.S. Madsen, P.J. Laipis and W.W. Hauswirth. 1993. Transcribed heteroplasmic repeated sequences in the porcine mitochondrial DNA D-loop region. Journal of Molecular Evolution. 37: 36–47.

Giaretta, N., A.M.A. Di Giuseppe, M. Lippert, A. Parente and A. Di Maro. 2013. Myoglobin as marker in meat adulteration: a UPLC method for determining the presence of pork meat in raw beef burger. Food Chemistry. doi: http://dx.doi.org/10.1016/j.foodchem.2013.04.124.

Gilg, A. and M. Battershill. 1998. Quality farm food in Europe: A possible alternative to the industrialised food market and to current agrienvironmental policies: Lessons from France. Food Policy. 23: 25–40.

Girish, P.S., A.S.R. Anjaneyulu, K.N. Viswas, M. Anand, N. Rajkumar, B.M. Shivakumar et al. 2004. Sequence analysis of mitochondrial 12S rRNA gene can identify meat species. Meat Science. 66: 551–556.

Girish, P.S., A.S.R. Anjaneyulu, K.N. Viswas, B.M. Shivakumar, M. Anand, M. Patel et al. 2005. Meat species identification by polymerase chain reaction-restriction fragment length polymorphism (PCR-RFLP) of mitochondrial 12S rRNA gene. Meat Science. 70: 107–112.

Guoli, Z., Z. Mingguang, Z. Zhijiang, O. Hongsheng and L. Qiang. 1999. Establishment and application of a polymerase chain reaction for the identification of beef. Meat Science. 51: 233–236.

Haider, N., I. Nabulsi and B. Al-Safadi. 2012. Identification of meat species by PCR-RFLP of the mitochondrial *COI* gene. Meat Science. 90: 490–493.

Harrison, S.M., O. Schmidt, A.P. Moloney, S.D. Kelly, A. Rossmann, A. Schellenberg, F. Camin, M. Perini, J. Hoogewerff and F.J. Monahan. 2011. Tissue turnover in ovine muscles and lipids as recorded by multiple (H, C, O, S) stable isotope ratios. Food Chemistry. 124: 291–297.

Heaton, K., S.D. Kelly, J. Hoogewerff and M. Woolfe. 2008. Verifying the geographical origin of beef: The application of multi-element isotope and trace element analysis. Food Chemistry. 107: 506–515.

Hegerding, L., D. Seidler, H.J. Danneel, A. Gessler and B. Nowak. 2002. Oxygenisotope-ratio-analysis for the determination of the origin of beef. Fleischwirtschaft. 82(4): 95–100.

Hildrum, K.I., T. Isaksson, T. Naes, B.N. Nilsen, R. Rodbotten and P. Lea. 1995. Near infrared reflectance spectroscopy on the prediction of sensory properties of beef. Journal of Near Infrared Spectroscopy. 3: 81–87.

Hildrum, K.I., B.N. Nilsen, M. Mielnik and T. Næs. 1994. Prediction of Sensory Characteristics of Beef by Near-infrared Spectroscopy. Meat Science. 38: 67–80.

Hintze, K.J., G.P. Lardy, M.J. Marchello and J.W. Finley. 2001. Areas with High Concentrations of Selenium in the Soil and Forage Produce Beef with Enhanced Concentrations of Selenium. Journal of Agricultural and Food Chemistry. 49: 1062–1067.

Hintze, K.J., G.P. Lardy, M.J. Marchello and J.W. Finley. 2002. Selenium accumulation in beef: Effect of dietary selenium and geographical area of animal origin. Journal of Agricultural and Food Chemistry. 50(14): 3938–3942.

Hopwood, A.J., K.S. Fairbrother, A.K. Lockley and R.G. Bardsley. 1999. An actin gene-related polymerase chain reaction (PCR) test for identification of chicken in meat mixtures. Meat Science. 53: 227–231.

IAEA. 1983. Technical Rep. Ser. No. 226.

Ilbery, B. and M. Kneafsey. 1998. Product and place: Promoting quality products and services in the lagging rural regions of the European Union. European Urban and Regional Studies. 5: 329–341.

Ilbery, B., M. Kneafsey and M. Bamford. 2000. Protecting and promoting regional speciality food and drink products in the European Union. Outlook on Agriculture. 29: 31–37.

Ilhak, I. and A. Arslan. 2007. Identification of Meat Species by Polymerase Chain Reaction (PCR) Technique. Turkish Journal of Veteranian and Animal Science. 31(3): 159–163.

Jerković, I., D. Kovačević, D. Šubarić, Z. Marijanović, K. Mastanjević and K. Suman. 2010. Authentication study of volatile flavour compounds composition in Slavonian traditional dry fermented salami "kulen". Food Chemistry. 119: 813–822.

Jerković, I., J. Mastelić and S. Tartaglia. 2007. A study of volatile flavour substances in Dalmatian traditional smoked ham: Impact of dry-curing and frying. Food Chemistry. 104: 1030–1039.

Kamruzzaman, M., D. Barbin, G. ElMasry, D.-W. Sun and P. Allen. 2012. Potential of hyperspectral imaging and pattern recognition for categorization and authentication of red meat. Innovative Food Science and Emerging Technologies. 16: 316–325.

Kang, J.O., J.Y. Park and H. Choy. 2001. Effect of grinding on color and chemical composition of pork sausages by near infrared spectrophotometric analyses. Asian-Australian Journal of Animal Science. 6: 858–861.

Kelly, J.F. 2000. Stable isotopes of carbon and nitrogen in the study of avian and mammalian trophic ecology. Canadian Journal of Zoology. 78: 1–27.

Kelly, S.D. 2003. Using stable isotope ratio mass spectrometry (IRMS) in food authentication and traceability. pp. 156–183. In: M. Lees (ed.), Food Authenticity and Traceability. Cambridge: Woodhead Publishing.

Kelly, S., K. Heaton and J. Hoogewerff. 2005. Tracing the geographical origin of food: The application of multi-element and multi-isotope analysis. Trends in Food Science & Technology. 16: 555–567.

Kesmen, Z., F. Sahin and H. Yetim. 2007. PCR assay for the identification of animal species in cooked sausages. Meat Science. 77: 649–653.

Kesmen, Z., Y. Celebi, A. Güllüce and H. Yetim. 2013. Detection of seagull meat in meat mixtures using real-time PCR analysis. Food Control. 34: 47–49.

Koh, M.C., C.H. Lim, S.B. Chu, S.T. Chew and S.T.W. Phang. 1998. Random Amplified Polymorphic DNA (RAPD) Fingerprints for Identification of Red Meat Animal Species. Meat Science. 48: 215–285.

Kriese, P.R., A.L. Soares, P.D. Guarnieri, S.H. Prudencio, E.I. Ida and M. Shimokomaki. 2007. Biochemical and sensorial evaluation of intact and boned broiler breast meat tenderness during ageing. Food Chemistry. 104: 1618–1621.

La Neve, F., T. Civera, N. Mucci and M.T. Bottero. 2008. Authentication of meat from game and domestic species by SNaPshot minisequencing analysis. Meat Science. 80: 210–215.

Laube, I., A. Spiegelberg, A. Butschke, J. Zagon, M. Schauzu, L. Kroh and H. Broll. 2003. Methods for the detection of beef and pork in foods using real-time polymerase chain reaction. International Journal of Food Science and Technology. 38: 111–118.

Liu, L.H., F.C. Chen, J.L. Dorsey and Y.H.P. Hsieh. 2006. Sensitive monoclonal antibody-based sandwich ELISA for the detection of porcine skeletal muscle in meat and feed products. Journal of Food Science. 71: M1–M6.

Liu, Y., X. Fan, Y.R. Chen and D.W. Thayer. 2003a. Changes in structure and color characteristics of irradiated chicken breasts as a function of dosage and storage time. Meat Science. 63: 301–307.

Liu, Y., B.G. Lyon, W.R. Windham, C.E. Realini, T.D. Pringle and S. Duckett. 2003b. Prediction of color, texture, and sensory characteristics of beef steaks by visible and near infrared reflectance spectroscopy. A feasibility study. Meat Science. 65: 1107–1115.

López-Andreo, M., A. Garrodo-Pertierra and A. Puyet. 2006. Evaluation of Post-Polymerase Chain Reaction Melting Temperature Analysis for Meat Species Identification in Mixed DNA Samples. Journal of Agricultural and Food Chemistry. 54: 7973–7978.

López-Andreo, M., L. Lugo, A. Garrido-Pertierra, M.I. Prieto and A. Puyet. 2005. Identification and quantitation of species in complex DNA mixtures by real-time polymerase chain reaction. Analytical Biochemistry. 339: 73–82.

Luykx, D.M.A.M. and S.M. van Ruth. 2008. An overview of analytical methods for determining the geographical origin of food products. Food Chemistry. 107: 897–911.

Macedo-Silva, A., S.F.C. Barbosa, M.G.A. Alkmin, A.J. Vaz, M. Shimokomaki and A. Tenuta-Filho. 2000. Hamburger meat identification by dot-ELISA. Meat Science. 56: 189–192.

Mafra, I., I.M.P.L.V.O. Ferreira and M.B.P.P. Oliveira. 2008. Food authentication by PCR-based methods. European Food Research and Technology. 227: 649–665.

Mamani-Linares, L.W., C. Gallo and D. Alomar. 2012. Identification of cattle, llama and horse meat by near infrared reflectance or transflectance spectroscopy. Meat Science. 90: 378–385.

Mancini, R.A. and M.C. Hunt. 2005. Current research in meat color. Meat Science. 71: 100–121.

Mane, B.G., S.K. Mendiratta and A.K. Tiwari. 2012. Beef specific polymerase chain reaction assay for authentication of meat and meat products. Food Control. 28: 246–249.

Martin, D.R., J. Chan and J.Y. Chiu. 1998. Quantitative evaluation of pork adulteration in raw ground beef by radial immunodiffusion and enzyme-linked immunosorbent assay. Journal of Food Protection. 61: 1686–1690.

Martín, I., T. García, V. Fajardo, I. López-Calleja, M. Rojas, P.E. Hernández, I. González and R. Martín. 2007a. Mitochondrial markers for the detection of four duck species and the specific identification of Muscovy duck in meat mixtures using the polymerase chain reaction. Meat Science. 76: 721–729.

Martín, I., T. García, V. Fajardo, M. Rojas, P.E. Hernández, I. González and R. Martín. 2007b. Technical Note: Detection of cat, dog, and rat or mouse tissues in food and animal feed using species-specific polymerase chain reaction. Journal of the Animal Science. 85: 2734–2739.

Matsunaga, T., K. Chikuni, R. Tanabe, S. Muroya, H. Nakai, K. Shibata, J. Yamada and Y. Shinmura. 1998. Determination of Mitochondrial Cytochrome B Gene Sequence for Red Deer (*Cervus Elaphus*) and the Differentiation of Closely Related Deer Meats. Meat Science. 49: 379–385.

Matsunaga, T., K. Chikuni, R. Tanabe, S. Muroya, K. Shibata, J. Yamada and Y. Shinmura. 1999. A quick and simple method for the identification of meat species and meat products by PCR assay. Meat Science. 51: 143–148.

McElhinney, J., G. Downey and C. O'Donnell. 1999. Quantitation of Lamb Content in Mixtures with Raw Minced Beef Using Visible, Near and Mid-Infrared Spectroscopy. Journal of Food Science. 64: 587–591.

Meer, D.P. and T.J. Eddinger. 1996. Polymerase chain reaction for detection of male tissue in pork products. Meat Science. 44: 285–291.

Meyer, R., U. Candrian and J. Liithy. 1994a. Detection of pork in heated meat products by the polymerase chain reaction. Journal Association of the Official Analytical Chemists. 77: 617–622.

Meyer, R., C. Höfelein, J. Lüthy and U. Candrian. 1994b. Polymerase chain reaction-restriction fragment length polymorphism analysis: a simple method for species identification in food. Journal of the AOAC International. 78: 542–551.

Meyer, R., F. Chardonnens, P. Hubner and J. Luthy. 1996. Polymerase chain reaction (PCR) in the quality and safety assurance of food: detection of soya in processed meat products. Z. Lebensm. Unters. Forsch. 203: 339–344.

Mohamad, N.A., A.F. El Sheikha, S. Mustafa and N.F.K. Mokhtar. 2013. Comparison of gene nature used in real-time PCR for porcine identification and quantification: A review. Food Research International. 50: 330–338.

Molkentin, J., H. Meisel, I. Lehmann and H. Rehbein. 2007. Identification of organically farmed Atlantic salmon by analysis of stable isotopes and fatty acids. European Food Research and Technology. 224(5): 535–543.

Moloney, A.P., B. Bahar, O. Schmidt, C.M. Scrimgeour, I.S. Begley and F.J. Monahan. 2009. Confirmation of the Dietary Background of Beef from its stable isotope signature, Beef Production Series No. 77, Teagasc, Grange Beef Research Centre, Dunsany, Co. Meath, Ireland, pp. 1–39.

Montowska, M. and E. Pospiech. 2011. Authenticity Determination of Meat and Meat Products on the Protein and DNA Basis. Food Reviews International. 27: 84–100.

Montiel-Sosa, J.F., E. Ruiz-Pesini, J. Montoya, P. Rocalés, M.J. López-Pérez and A. Pérez-Martos. 2000. Direct and highly species specific detection of pork meat and fat in meat products by PCR amplification of mitochondrial DNA. Journal of Agricultural and Food Chemistry. 48: 2829–2832.

Mor-Mur, M. and J. Yuste. 2003. High pressure processing applied to cooked sausage manufacture: Physical properties and sensory analysis. Meat Science. 65: 1187–1191.

Morsy, N. and D.-W. Sun. 2013. Robust linear and non-linear models of NIR spectroscopy for detection and quantification of adulterants in fresh and frozen-thawed minced beef. Meat Science. 93: 292–302.

Mottram, D.S. 1998. Flavour formation in meat and meat products: a review. Food Chemistry. 62: 415–424.

Murugaiah, C., Z.M. Noor, M. Mastakim, L.M. Bilung, J. Selamat and S. Radu. 2009. Meat species identification and Halal authentication analysis using mitochondrial DNA. Meat Science. 83: 57–61.

Næs, T., P. Baardseth, H. Helgesen and T. Isaksson. 1996. Multivariate Techniques in the Analysis of Meat Quality. Meat Science. 43: 135–149.

Nakashita, R., Y. Suzuki, F. Akamatsu, Y. Iizumi, T. Korenaga and Y. Chikaraishi. 2008. Stable carbon, nitrogen, and oxygen isotope analysis as a potential tool for verifying geographical origin of beef. Analytica Chimica Acta. 17(1-2): 148–152.

Natonek-Wiśniewska, M., P. Krzyścin and A. Piestrzyńska-Kajtoch. 2013. The species identification of bovine, porcine, ovine and chicken components in animal meals, feeds and their ingredients, based on COX I analysis and ribosomal DNA sequences. Food Control. 34: 69–78.

Olsman, W.J. 1979. Methods for detection and determination of vegetable proteins in meat products. Journal of the American Oil Chemists' Society. 56: 285–287.

O'Sullivan, M.G., D.V. Byrne, J.H. Nielsen, H.J. Andersen and M. Martens. 2003a. Sensory and chemical assessment of pork supplemented with iron and vitamin E. Meat Science. 64: 175–189.

O'Sullivan, M.G., D.V. Byrne, M.T. Jensen, H.J. Andersen and J. Vestergaard. 2003b. A comparison of warmed-over flavour in pork by sensory analysis, GC/MS and the electronic nose. Meat Science. 65: 1125–1138.

Ortiz-Somovilla, V., F. España-España, E.J. De Pedro-Sanz and A.J. Gaitán-Jurado. 2005. Meat mixture detection in Iberian pork sausages. Meat Science. 71: 490–497.

Ortiz-Somovilla, V., F. España-España, A.J. Gaitán-Jurado, J. Pérez-Aparicio and E.J. De Pedro-Sanz. 2007. Proximate analysis of homogenized and minced mass of pork sausages by NIRS. Food Chemistry. 101: 1031–1040.

Otero, L., M.C. Horrillo, M. García, I. Sayago, M. Aleixandre, M.J. Fernández, L. Arés and J. Gutiérrez. 2003. Detection of Iberian ham aroma by a semiconductor multisensorial system. Meat Science. 65: 1175–1185.

Papadima, S.N., I. Arvanitoyannis, J.G. Bloukas and G.C. Fournitzis. 1999. Chemometric model for describing Greek traditional sausages. Meat Science. 51(3): 271–277.

Pappa, I.C., J.G. Bloukas and I.S. Arvanitoyannis. 2000. Optimization of salt, olive oil, pectin level for low-fat frankfurters produced by replacing pork backfat with olive oil. Meat Science. 56(1): 81–88.

Partis, L., D. Croan, Z. Guo, R. Clark, T. Coldham and J. Murby. 2000. Evaluation of a DNA fingerprinting method for determining the species origin of meats. Meat Science. 54: 369–376.

Pascoal, A., M. Prado, J. Castro, A. Cepeda and J. Barros-Velázquez. 2004. Survey of authenticity of meat species in food products subjected to different technological processes, by means of PCR-RFLP analysis. European Food and Research Technology. 218: 306–312.

Pascoal, A., M. Prado, P. Calo, A. Cepeda and J. Barros-Velázquez. 2005. Detection of bovine DNA in raw and heat-processed foodstuffs,commercial foods and specific risk materials by a novel specific polymerase chain reaction method. European Food and Research Technology. 220: 444–450.

Piasentier, E., R. Valusso, F. Camin and G. Versini. 2003. Stable isotope ratio analysis for authentication of lamb meat. Meat Science. 64(3): 239–247.

Prado, M., P. Calo-Mata, T.G. Villa, A. Cepeda and J. Barros-Velázquez. 2007. Co-amplification and sequencing of a cytochrome b fragment affecting the identification of cattle in PCR-RFLP food authentication studies. Food Chemistry. 105: 436–442.

Prieto, N., S. Andrés, F.J. Giráldez, A.R. Mantecón and P. Lavín. 2008. Ability of near infrared reflectance spectroscopy (NIRS) to estimate physical parameters of adult steers (oxen) and young cattle meat samples. Meat Science. 79: 692–699.

Pun, K.M., C. Albrecht, V. Castella and L. Fumagalli. 2009. Species identification in mammals from mixed biological samples based on mitochondrial DNA control region length polymorphism. Electrophoresis. 30: 1008–1014.

Quilter, J.M. 2002. Determination of the dietary history and geographical origin of bovine muscle using stable isotope analysis. M.Agr.Sc. thesis, Dublin: University College.

Rannou, H. and G. Downey. 1997. Discrimination of Raw Pork, Chicken and Turkey Meat by Spectroscopy in the Visible, Near- and Mid-infrared Ranges. Analytical Communications. 34: 401–404.

Rastogi, G., M.S. Dharne, S. Walujkar, A. Kumar, M.S. Patole and Y.S. Shouche. 2007. Species identification and authentication of tissues of animal origin using mitochondrial and nuclear markers. Meat Science. 76: 666–674.

Regulation (EC) No. 2772/1999 providing for the general rules for a compulsory beef labelling system. Official Journal 1999, L334, 0001–0002.

Regulation (EEC) No. 2081/92. 1992. On the protection of geographical indications and designations of origin for agricultural products and foodstuffs. Official Journal, L 208 (Corrigenda: [Official Journal L 27, 30.01.1997 and L 53, 24.02.1998]. Amended by Council Regulation (EC) No 535/97 of 17 March 1997 (Official Journal L 83, 25.03.1997).

Regulation (EEC) No. 2081/92 on the protection of geographical indications and designations of origin for agricultural products and foodstuffs. Official Journal, 1992, L208 (Corrigenda: [Official Journal L 27, 30.01.1997 and L 53, 24.02.1998]. Amended by Council Regulation (EC) No. 535/97 of 17 March 1997 (Official Journal L 83, 25.03.1997).

Reid, L.M., C.P. O'Donnell and G. Downey. 2006. Recent technological advances for the determination of food authenticity. Trends in Food Science & Technology. 17: 344–353.

Rencova, E. and B. Tremlova. 2009. ELISA for detection of soya proteins in meat products. Acta Veterinaria. 78: 667–671.

Renou, J.P., G. Bielicki, C. Deponge, P. Gachon, D. Micol and P. Ritz. 2004. Characterization of animal products according to geographic origin and feeding diet using nuclear magnetic resonance and isotope ratio mass spectrometry. Part II: Beef meat. Food Chemistry. 86: 251–256.

Rodríguez, M.A., T. García, I. González, L. Asensio, P.E. Hernández and R. Martín. 2004. PCR identification of beef, sheep, goat and pork in raw and heat treated meat mixtures. Journal of Food Protection. 67: 172–177.

Rodríguez, M.A., T. García, I. González, L. Asensio, B. Mayoral, I. López-Calleja et al. 2003a. Development of a polymerase chain reaction assay for species identification of goose and mule duck in *foie gras* products. Meat Science. 65: 1257–1263.

Rodríguez, M.A., T. García, I. González, L. Asensio, B. Mayoral, I. López-Calleja et al. 2003b. Identification of goose, mule duck, chicken, turkey, and swine in *foie gras* by species-specific polymerase chain reaction. Journal of Agricultural and Food Chemistry. 51: 1524–1529.

Rodríguez, M.A., T. García, I. González, L. Asensio, P.E. Hernández and R. Martín. 2003c. Qualitative PCR for the detection of chicken and pork adulteration in goose and mule duck *foie gras*. Journal of the Science of Food and Agriculture. 83: 1176–1181.

Rodríguez, M.A., T. García, I. González, P.E. Hernández and R. Martín. 2005. TaqMan real-time PCR for the detection and quantitation of pork in meat mixtures. Meat Science. 70: 113–120.

Rodriguez-Ramirez, R., A.F. Gonzalez-Cordova and B. Vallejo-Cordiba. 2011. Review: Authentication and traceability of foods from animal origin by polymerase chain reaction-based capillary electrophoresis. Analytica Chimica Acta. 685: 120–126.

Rossmann, A. 2001. Determination of stable isotope ratios in food analysis. Food Reviews International. 17: 347–381.

Rousset-Akrim, S., A. Young and J.L. Berdagué. 1997. Diet and Growth Effects in Panel Assessment of Sheepmeat Odour and Flavour. Meat Science. 45: 169–181.

Ruiz Pérez-Cacho, M.P., H. Galán-Soldevilla, F. León Crespo and G. Molina Recio. 2005. Determination of the sensory attributes of a Spanish dry-cured sausage. Meat Science. 71: 620–633.

Sacco, D., M.A. Brescia, A. Buccolieri and A. Caputi Jambrenghi. 2005. Geographical origin and breed discrimination of Apulian lamb meat samples by means of analytical and spectroscopic determinations. Meat Science. 71: 542–548.

Saez, R., Y. Sanz and F. Toldrá. 2004. PCR-based fingerprinting techniques for rapid detection of animal species in meat products. Meat Science. 66: 659–665.

Sakaridis, I., I. Ganopoulos, A. Argiriou and A. Tsaftaris. 2013. A fast and accurate method for controlling the correct labeling of products containing buffalo meat using High Resolution Melting (HRM) analysis. Meat Science. 94: 84–88.

Santos, C.G., V.S. Melo, J.S. Amaral, L. Estevinho, M.B.P.P. Oliveira and I. Mafra. 2012. Identification of hare meat by a species-specific marker of mitochondrial origin. Meat Science. 90: 836–841.

Sasazaki, S., H. Mutoh, K. Tsurifune and H. Mannen. 2007. Development of DNA markers for discrimination between domestic and imported beef. Meat Science. 77: 161–166.

Sawyer, J., C. Wood, D. Shanahan, S. Gout and D. McDowell. 2003. Real-time PCR for quantitative meat species testing. Food Control. 14: 579–583.

Schmidt, O., J.M. Quilter, B. Bahar, A.P. Moloney, C.M. Scrimgeour, I.S. Begley and F.J. Monahan. 2005. Inferring the origin and dietary history of beef from N and S stable isotope ratio analysis. Food Chemistry. 91: 545–549.

Scrimgeour, C.M. and D. Robinson. 2004. Stable isotope analysis and applications. pp. 381–431. *In*: K.A. Smith and M.S. Cresser (eds.). Soil and Environmental Analysis: Modern instrumental techniques (3rd ed.). New York: Marcel Dekker.

Shintu, L., S. Caldarelli and B.M. Franke. 2007. Pre-selection of potential molecular markers for the geographic origin of dried beef by HR-MAS NMR spectroscopy. Meat Science. 76: 700–707.

Sinclair, A.J., W.J. Slattery and K. O'Dea. 1982. The analysis of polyunsaturated fatty acids in meat by capillary gas-liquid chromatography. Journal of the Science of Food and Agriculture. 33: 771–776.

Skarpeid, H.J., K. Kvaal and K.I. Hildrum. 1998. Identification of animal species in ground meat mixtures by multivariate analysis of isoelectric focusing protein profiles. Electrophoresis. 19(18): 3103–3109.

Skrökki, A. and O. Hormi. 1994. Composition of Minced Meat, Part B: A Survey of Commercial Ground Meat. Meat Science. 38: 503–509.

Soares, S., I. Mafra, J.S. Amaral and M.B.P.P. Oliveira. 2010. A PCR assay to detect trace amounts of soybean in meat sausages. International Journal of Food Science and Technology. 45: 2581–2588.

Sultan, K.R., M.H.G. Tersteeg, P.A. Koolmees, J.A. de Baaij, A.A. Bergwerff and H.P. Haagsman. 2004. Western blot detection of brain material in heated meat products using myelin basic protein and neuron-specific enolase as biomarkers. Analytica Chimica Acta. 520: 183–192.

Sun, W., Q. Zhao, H. Zhao, M. Zhao and B. Yang. 2010. Volatile compounds of Cantonese sausage released at different stages of processing and storage. Food Chemistry. 121: 319–325.

Thyholt, K. and T. Isaksson. 1997. Differentiation of frozen and unfrozen beef using near-infrared spectroscopy. Journal of the Science of Food and Agriculture. 73: 525–532.

Treml, D. and A.C. Maisonnave Arisi. 2008. Monitoring of Roundup Ready soybean in processed meat products sold in Brazilian markets. Universidade Federal de Santa Catarina. First Global Conference on GMO Analysis. Villa Erba, Como, Italy, 24–27 June 2008.

Vallejo-Cordoba, B., A.F. González-Córdova, M.A. Mazorra-Manzano and R. Rodríguez-Ramírez. 2005. Capillary electrophoresis for the analysis of meat authenticity. Journal of Seperation Science. 28: 826–836.

Verbeke, W. and J. Viaene. 1999. Consumer attitude to beef quality labelling and associations with beef quality labels. Journal of International Food and Agribusiness Marketing. 10: 45–65.

Verkaar, E.L.C., I.J. Nijman, K. Boutaga and J.A. Lenstra. 2002. Differentiation of cattle species in beef by PCR-RFLP of mitochondrial and satellite DNA. Meat Science. 60: 365–369.

Walker, J.A., D.A. Hughes, B.A. Anders, J. Shewale, S.K. Sinha and M.A. Batzer. 2003. Quantitative intra-short interspersed element PCR for species-specific DNA identification. Analytical Biochemistry. 316(2): 259–269.

Wang, Q., X. Zhang, H.Y. Zhang, J. Zhang, G.Q. Chen, D.H. Zhao, H.P. Ma and W.J. Liao. 2010. Identification of 12 animal species meat by T-RFLP on the 12S rRNA gene. Meat Science. 85: 265–269.

Wissiack, R., B. de la Calle, G. Bordin and A.R. Rodriguez. 2003. Screening test to detect meat adulteration through the determination of hemoglobin by cation exchange chromatography with diode array detection. Meat Science. 64: 427–432.

Wolf, C. and J. Lüthy. 2001. Quantitative competitive (QC) PCR for quantification of porcine DNA. Meat Science. 57: 161–168.

Wolf, C., J. Rentsch and P. Hübner. 1999. PCR-RFLP analysis of mitochondrial DNA: a reliable method for species identification. Journal of Agricultural and Food Chemistry. 47: 1350–1355.

Xia, J.J., E.P. Berg, J.W. Lee and G. Yao. 2007. Characterizing beef muscles with optical scattering and absorption coefficients in VIS-NIR region. Meat Science. 75: 78–83.

Yman, I.M. and K. Sandberg. 1987. Differentiation of Meat from Horse, Donkey and Their Hybrids (Mule/Hinny) by Electrophoretic Separation of Albumin. Meat Science. 21: 15–23.

Poultry and Eggs Authenticity

*Ioannis S. Arvanitoyannis** and *Konstantinos V. Kotsanopoulos*

4.1 Introduction

While different instrumentation has been extensively used at a laboratory level in order to provide a better understanding of the mechanisms underlying poultry quality, use of this knowledge for evaluating poultry quality at an industrial level is still quite limited. The enhanced international trade of poultry meat, as well as the growing segmentation of the poultry meat market, is accompanied by increased consumer interest in the authenticity and organoleptic characteristics of products marketed under quality labels (geographical origin, breed within a species, production system), and increased requirement from the processing industry as regards the status of meat (ageing, previous freezing, irradiation, undesirable residues, trace elements). As a result, it is now of high importance to develop new methods so that the origin of meat and its freshness can be guaranteed (Baéza et al., 2004).

This chapter reviews the main techniques used for the detection of poultry adulteration and poultry authenticity assurance, analyzing the results of numerous studies carried out in this field. The role of specific compounds as meat authenticity markers and the use of instrumentation for their determination are also discussed.

4.2 Authenticity of Poultry

4.2.1 Use of PCR for the Authentication of Poultries

According to Calvo et al. (2001), some fraudulent or unintentional mislabelling can occur without being detected, resulting in lower quality pâté. Since some population groups avoid eating meat from certain animal species due to philosophical or religious

School of Agricultural Sciences, Department of Agriculture, Ichthyology and Aquatic Environment, University of Thessaly, Fytoko St., 38446 Nea Ionia Magnesias, Volos, Hellas, Greece.
* Corresponding author

reasons, a new technique was developed and evaluated for the detection of pâté products based on randomly amplified polymorphic DNA (RAPD). The RAPD method was applied for the generation of fingerprint patterns for pork, chicken, duck, turkey, and goose meats. Ten DNA samples from pork, chicken, turkey, and duck meats were examined to confirm the effectiveness and specificity of the method. Specific results for each species were obtained. The sensitivity of the method was assessed by DNA dilution, detecting as little as 250 pg of DNA. DNA from 30 pâtés (tinned and untinned) was isolated and used as template DNA in a polymerase chain reaction (PCR). The RAPD-PCR pattern effectively determined the composition of the product, identifying pork, duck, duck-pork, goose, and poultry species. The study demonstrated the usefulness of RAPD fingerprinting in distinguishing between different species in pâtés.

In another study, polymerase chain reaction was used for the amplification of a zone of a gene for the cytochrome b of ostrich (*Struthio camelus*) mitochondrial DNA, using a primer pair, which allows the identification of ostrich and its differentiation from emu (*Dromaius novaehollandiae*), giving different size fragments (about 543 bp for ostrich and 229 bp for emu). The PCR products were revealed using horizontal polyacrylamide electrophoresis. Three emu samples were not effectively amplified probably due to very low template concentrations and the absence of an optimization step in the amplification protocol. The technique used in the study utilizes only one primer pair developed to be specific for two species only. This can be performed by initially designing several primer pairs, specific to only one species, and testing them on the analogous target sequence of the other species until a primer pair is found that provides an amplicon, which significantly differs in size among the two species. The above-mentioned method can be used when only two species are to be differentiated. The simultaneous differentiation of several species can be very difficult (Colombo et al., 2000).

Chisholm et al. (2008) developed species-specific real-time PCR assays for detecting pheasant and quail in commercial food products. Real-time PCR primer and probe sets were developed with the aim of detecting the mitochondrial cytochrome b gene of pheasant (*Phasianus colchicus*) and quail (*Coturnix coturnix*) and were optimized to be highly species-specific. The efficiency and sensitivity of the assays were evaluated and their potential in analyzing commercial samples was assessed. The calculation of the efficiency of each assay was performed using the slope of the C_T values plotted against the log of the DNA dilution over the range of dilutions, when the response was linear. Primer and probe sets performing at 100% efficiency can lead to an approximate increase of 3.3 cycles to the C_T per 1 in 10 dilution of template. The efficiencies of the pheasant and quail assays were found to be 107% and 96%, respectively. The determination of the limits of detection of the pheasant and quail assays was performed using DNA extracted from raw meat. This DNA was serially diluted in either water or non-target DNA (maize DNA to simulate a complex food product) allowing for the determination of the quantity of DNA in a reaction at which C_T values were reproducibly produced (but beyond which the assay failed). The limits of detection of the pheasant and quail assays were determined as 10 and 2 pg, respectively, when diluted in water or maize DNA. It is important to mention that the dilution in non-target DNA did not negatively affect the limit of detection of the assays. Successful detection of pheasant and quail was achieved in complex food matrices of raw, oven-cooked, and autoclaved meat, indicating that the assays are suitable for use in enforcement and food control laboratories.

A species-specific duplex polymerase chain reaction (PCR) assay was designed to simultaneously detect pork and poultry meat species using the mitochondrial cytb and 12S rRNA as target genes for pork and poultry, respectively. Through amplifying binary reference meat mixtures, a linear normalized calibration curve was obtained using the fluorescence intensities of PCR products for pork (149 bp) and poultry (183 bp) species. The technique was effectively used to quantify pork addition to poultry meat in the range of 1–75%, and was characterised by a sensitivity of 0.1%. The in-house validation was performed using samples with pre-determined quantities of pork meat (1.0%, 2.5%, 7.5%, 20.0% and 40%) and proved that the technique is highly reproducible (coefficient of variation from 4.1% to 7.6%). The successful application of the duplex PCR was also shown by the high correlation ($R^2 = 0.99$) obtained from regression analysis between the predicted and the actual values of pork meat added in blind meat mixtures. The proposed technique can be used as a cost-effective, fast, easy and reliable alternative for the estimation of the level of poultry meat adulteration by pork (Soares et al., 2010).

Another study highlighting that species identification in meat products is of high importance nowadays, since many foodstuffs can be susceptible targets for fraudulent labelling, is that of Soares et al. (2013). In this study, a real-time PCR approach based on SYBR Green dye was suggested for quantitatively detecting pork meat in processed meat products. The method was developed using binary meat mixtures that contained known levels of pork meat in poultry. The mixtures were used to obtain a normalised calibration model from 0.1 to 25% with high linear correlation and PCR efficiency. It was proved that the technique was highly specific, and successful validation was performed through application to blind meat mixtures. The effectiveness of the method was further demonstrated in commercial meat products, allowing the verification of labelling compliance and the identification of meat species in processed food products.

"Mortara" goose salami is a traditional poultry product of the Lomellina zone (Italy) and is usually home made. A PCR was employed for the examination of Italian goose meat and Mortara. A zone of the cytochrome b of the mitochondrial DNA of goose was sequenced and a specific primer pair was designed for identifying the species *Anser anser* in salami. The development of the specific primer pair for goose species was performed using the software "Clustal W" by aligning Italian goose sequences to unknown and control species sequences (pig, turkey, ducks). A primer pair was designed which, in theory, leads to amplification of only goose species. The primers were named "Goose1" and "Goose2", and their synthesis was performed by a commercial service, while they were tested using a PCR method. After PCR and electrophoresis optimization, only the goose-specific band appeared. It is important to note that the majority of processes throughout this experiment were performed using commercial kits, thus improving significantly the reproducibility of the method (Colombo et al., 2002).

The aim of the study of Andrée et al. (2004) was the optimization of DNA-based methods used for identifying different poultry species of high significance for human nutrition, as well as the development of techniques for quantifying animal species in meat products. An essential prerequisite for animal species identification and especially quantification is the application of reproducible isolation of intact DNA in traditional and real-time PCR. Among numerous DNA isolation kits, a classical CTAB protocol was proved to be the most suitable, after optimization. The use of sequences of the mitochondrial genome leads to limits of detection of 100 to 1000-fold lower than those observed by using nuclear single-copy genes. Nevertheless, single-copy genes of the nuclear genome can only be used for species quantification by real-time PCR. Some species-specific primer systems were adapted from literature [chicken (*G. gallus*), goose (*A. cygnoides*) and duck (*Cairina moschata*)]. The mitochondrial (mtDNA) sequences coding for *cytochrome b* for turkey (*Meleagris gallopavo*), pheasant (*P. colchicus*), quail (*C. coturnix*) and guinea fowl (*Numida meleagris*), were found in numerous genomic databases. Species specific primer systems relied on the mitochondrial *cytochrome b* gene were designed. It was proved that clear differentiation between poultry species could be achieved by using mitochondrial primer systems.

According to Girish et al. (2007), chicken (*G. gallus*), duck (*Anas platyrhynchos*), turkey (*M. gallopavo*), guinea fowl (*N. meleagris*) and quail (*Coturnix japonica*) are the most widely consumed poultries worldwide. Due to that a molecular technique that can be used for identifying and differentiating between meat originating from these species was developed. The method is based on extracting DNA from a given sample, using PCR to amplify the mitochondrial 12S rRNA gene (by employing universal primers), performing restriction analysis with selected restriction enzymes, and finally identifying the meat species using a restriction fragment length polymorphism (RFLP) pattern. In this experiment, HinfI, Mph1103I, MvaI and Eco47I were employed for identifying and differentiating between poultry species. The technique was equally successful when used for identifying processed meat products including those cooked at 120°C for 30 min. The data extracted could be easily interpreted, thus making the technique a convenient tool for identification of poultry meat species.

Meat authenticity is becoming more and more important for reasons related to economical, religious or public health. The use of PCR-RFLP of a part of the mitochondrial cytochrome c oxidase subunit 1 (CO1) gene was examined as a potential tool for identifying the species origin of raw meat samples of chicken and turkey. The PCR resulted in a 710-bp fragment for both species. The digestion of the amplicons was performed using seven restriction endonucleases (Hind II, Ava II, Rsa I, Taq I, Hpa II, Tru 1I and Xba I) that were selected based on the preliminary in silico analysis. Different levels of polymorphism were found among samples. The level of CO1 variation, demonstrated only by employing Hpa II, was sufficient for the generation of easily analyzable species-specific restriction profiles that could be used for distinguishing the targeted species. This technique was proved to be both cost-effective and cheap, and therefore could be effectively used for identifying meat samples in food control laboratories (Haider et al., 2012).

PCR amplification of a conserved region of the α-actin gene was employed by Rodríguez et al. (2003) for the identification of chicken (*G. gallus*) and pork (*Susscrofa domesticus*) meat in goose (*A. anser*) and mule duck (*Anasplatyrhynchos* x *C. moschata*) *foiegras*. The development of species-specific forward primers, in combination with a reverse universal primer, allowed the generation of amplicons of different lengths in each species. The different sizes of the species-specific amplicons, separated by agarose gel electrophoresis, was important for clearly identifying the presence of chicken and pork in goose and mule duck *foiegras* with a detection limit of 0.1% (w/w). By comparing the sequences obtained in this experiment with those previously found from goose and mule duck, it was demonstrated that the differences in length appear due to the presence of an intron in goose, mule duck and chicken PCR products, which is not detected in pork PCR products. As a result, whereas goose, mule duck and chicken sequences comprised three introns and four exons, pork sequences lacked the first intron. The method could be applied in inspection programmes for enforcing labelling regulation of *foiegras* and other meat foodstuffs.

In the study of Haunshi et al. (2009), a PCR based method aiming at identifying chicken, duck and pigeon was generated through the development of species-specific markers. Mitochondrial sequences were used for designing species-specific primers of 256 bp, 292 bp and 401 bp for chicken, duck and pigeon, respectively. Sequencing of the species-specific PCR products was performed for confirming the specificity of the amplified product. The markers were also tested for cross amplification by checking them with beef, mutton, chevon, pork, rabbit, chicken, duck, turkey and pigeon meat. The DNA markers generated in this experiment can be effectively used for identifying fresh, cooked and autoclaved meat of chicken, duck and pigeon. The identification procedure is quite simple, cost-effective and quick in comparison to other methods such as RAPD, PCR-RFLP and sequencing.

Hird et al. (2005) reported the use of a real-time PCR assay for simultaneously detecting Mallard and Muscovy duck. Species-specific primers were generated for the above-mentioned duck species using the mitochondrial cytochrome b gene sequence. These primer sets were multiplexed with a single duck probe aiming at producing a simple, rapid and robust real-time PCR assay. The assay was proved to be specific for duck and was effectively applied for detecting duck meat in complex food matrices. This was the first report of an assay capable of detecting all species of commercially available duck in products, using real-time PCR.

The development of a quick and highly specific assay suitable for routinely detecting turkey and chicken in processed meat products was reported by Hird et al. (2003). By using PCR amplification of species-specific amplicons and rapidly visualising them using vistra green, the assay can be completed within five hours of samples receipt. The isolation of the DNA from meat samples was performed using a Wizard DNA isolation technology and, subsequently, DNA amplification was performed using the following species-specific primers: chicken forward (CF), chicken reverse (CR), turkey forward (TF) and turkey reverse (TR). The production of an amplicon was detected after the end of the PCR in less than 5 min using vistra green and a fluorescence plate reader.

Adulteration of meat products with seagull meat has been reported in coastal cities. The identification and quantification of seagull meat in meat mixtures can be performed using a real-time PCR assay, that employs species-specific primers and a TaqMan probe and is designed on the mitochondrial NADH dehydrogenase subunit 2 gene. The detection of the template DNA of seagull at a level of 100 pg was also possible and no cross-reactivity with non-target species (chicken, turkey, goose, duck) was reported. The same method could be successfully used for the detection of seagull meat at a level of 0.1% in raw and heat-treated test mixtures, after preparing them by mixing seagull meat with beef and chicken at different levels (0.01–10%).

It was therefore concluded that the above-mentioned assay can be used as a rapid and sensitive method for routinely identifying seagull meat in raw or cooked meat foodstuffs (Kesmen et al., 2013).

Lockley and Bardsley (2002) developed a one-step method for differentiating between chicken and turkey DNA. The method is based on the use of PCR and primers that exploit intron variability in a-cardiac actin for the generation of single products of a characteristic size for each species. No cross-reactivity with porcine, ovine or bovine DNA templates was observed, while by analyzing chicken/turkey admixtures, it was proved that 1% turkey could be detected in 99% chicken and vice versa. Due to the fact that the test is based on a nuclear gene target, it can be used as a valuable complement to other techniques that are relied on mitochondrial DNA sequences. The use of the chicken/turkey test with three primers applied simultaneously, led to detection of each of the species in admixtures with a detection limit of 10% for chicken and 25% for turkey, even after increased cycle numbers. However, when the species-specific primers were implemented (with the common reverse primer) in separate tubes, detection of 1% of a species in 99% of another species was reported. It was finally proved that the chicken-specific primer was significantly more effective than the turkey-specific equivalent under the same experimental conditions.

A PCR method relying on oligonucleotide primers that target the mitochondrial 12S rRNA gene was used to specifically identify meats from quail (*C. coturnix*), pheasant (*P. colchicus*), partridge (*Alectoris* spp.), and guinea fowl (*N. meleagris*). By using specific primer pairs for quail, pheasant, partridge and guinea fowl, selective amplification of the desired avian sequences was performed. Following PCR amplification with the 12S conserved primers, the sequencing of amplicons from at least two individuals from each quail, pheasant, partridge and guinea fowl species was performed with the aim of detecting a DNA segment with sufficient species-to-species variation to allow their discrimination. Furthermore, the establishment and evaluation of a wider spectrum of PCR specificity was possible through obtaining 12S rRNA gene sequences from other game and domestic species such as capercaillie, eurasian woodcock, woodpigeon, song thrush, chicken, turkey, Muscovy duck, and goose and including them in the alignment. Through this technique, specific primers pairs for quail (12SCOT-FW/12SCOT-REV), pheasant (12SPHA-FW/12SPHA-REV), partridge (12SALEC FW/12SALEC-REV), and guinea fowl (12SNUM-FW/12SNUM-REV) were designed. It was shown that the assay can be effective for accurately identifying meat from game bird species (Rojas et al., 2009). Also, Rojas et al. (2010) reported the development of a PCR assay with the potential of identifying meats and commercial meat products from quail (*C. coturnix*), pheasant (*P. colchicus*), partridge (*Alectoris* spp.), guinea fowl (*N. meleagris*), pigeon (*Columba* spp.), Eurasian woodcock (*Scolopax rusticola*), and song thrush (*Turdus philomelos*). The assay relied on oligonucleotide primers targeting specific sequences from the mitochondrial D loop region. The primers designed led to the generation of specific fragments of 96, 100, 104, 106, 147, 127 and 154 bp in length for quail, pheasant, partridge, guinea fowl, pigeon, Eurasian woodcock and song thrush, respectively. The specificity of each primer pair was cross-checked against DNA derived from numerous game and domestic species. Finally, the amplification was satisfactorily performed for experimentally pasteurized (72°C for 30 min) and sterilized (121°C for 20 min) meats, as well as in commercial meat products from the target species. The method was also used to analyze raw and sterilised muscular binary mixtures, characterised by a detection limit of 0.1% (w/w) for each of the targeted species. This PCR assay is a rapid and effective method that can be used for detecting possible mislabelling of game bird meat products.

A PCR method was used to quantitatively identify chicken (*G. gallus*), turkey (*Meleagris gallipavo*), duck (*A. platyrhynchos* X *C. muschata*), and goose (*A. anser*) tissues in feedstuffs. This assay was relied on the use of oligonucleotide primers, specific for each avian species, targeting the 12S rRNA mitochondrial gene. The primers led to the generation of amplicons of 95, 122, 64 and 98 bp lengths for chicken, turkey, duck and goose, respectively. The specificity degree of the primers was tested against 29 animal species, as well as eight plant species. It was proved that detection of each target species in the range of 0.1 to 100% was feasible. The effectiveness of the method was not affected by prolonged heat-treatment (up to 133°C for 20 min at 300 kPa), and therefore it could be successfully used for the accurate identification of tissues from these four avian species in feedstuffs processed using denaturing technologies (Martín et al., 2007).

Jonker et al. (2008) reported the generation of a real-time PCR method that could be used for the identification of processed meat products. Test mixtures containing beef, pork, horse, mutton, chicken and turkey, were tested and identification of the above species down to a level of 0.05% was possible. The adjustion of the number of cycles allowed for the detection of levels as low as 0.01%. No cross-reactivity between these species was detected, except for pure horsemeat (250 ng DNA) in the assay for turkey meat. The cross-reactivity of deer, roe, ostrich, kangaroo, goat, domestic duck, mallard, goose, pigeon, guinea fowl, quail and pheasant was also examined and it was proved that levels as high as 250 ng DNA did not lead to (false) positive signals, with the exception of deer and pigeon DNA where amounts higher than 125 ng for deer DNA and higher than 50 ng for pigeon DNA gave false results when used for determining chicken and beef, respectively. More than 150 meat samples were tested using DNA hybridization and real-time PCR. By comparing the results, it was shown that the method was more effective than the DNA hybridization.

The aim of the study of Natonek-Wiśniewska et al. (2013) was the development of a universal method for identifying chicken DNA using PCR. The primers developed were used for the generation of short amplicons of 66 bp for chickens within the gene, encoding COX1 in the case of 16S rRNA. These primers only amplify chicken DNA and do not cross-react with DNA of other species of animals and plants. According to the authors of this study "use of short amplification products for the indicators allows for the highly effective species identification of chicken, both in raw samples and in samples processed at high temperature and pressure". The PCR products were species-specific for levels as low as 0.1% and for 100% of samples. The limit of quantification for both plant feeds and animal feeds from other species was 0.08% for poultry. The method is believed to be highly specific and sensitive as well as universal, while it can also be used with any degree of processing, type and form of the source material.

Dooley et al. (2004) reported the development of species-specific real-time PCR (TaqMan) assays for detecting chicken and turkey. The development of assays around amplicons of less than 150 base pairs of the mitochondrial cytochrome b (cytb) gene was performed. Speciation was allowed using species-specific primers. A TaqMan probe was also generated for these poultry species. The application of the assays to DNA extracts from raw meat admixtures allowed the detection of each species when added in any other species at a level as low as 0.5%. No determination of the absolute level of detection was performed, but experimentally determined limits for turkey were found to be below 0.1%.

DNA testing for the purpose of food authentication and quality control is performed using sensitive species-specific quantification of nuclear DNA from biological samples. A multiplex assay relied on TaqMan® real-time quantitative PCR (qPCR) was developed aiming at detecting and quantifying chicken (*G. gallus*), duck (*A. platyrhynchos*), and turkey (*Meleagris gallopavo*) nuclear DNA. The assay proposed can be used for the accurate detection of very low levels of species-specific DNA derived from single or multispecies sample mixtures, while its minimum effective quantification range is 5 to 50 pg of starting DNA material. The use of potent inhibitors such as hematin and humic acid, as well as the degradation of template DNA using DNase did not significantly affect the effectiveness of the method. By efficiently determining species and accurately quantifying DNA in samples, fraudulent food distribution could be reduced (Ng et al., 2012).

Rojas et al. (2011) reported the development of a rapid and highly species-specific real-time PCR assay that could be used to rapidly and effectively authenticate ostrich meat (*S. camelus*). The method is based on the combined use of ostrich-specific primers that are employed for the amplification of a 155 bp fragment of the mitochondrial 12S rRNA gene, and a positive control primer pair that is used to amplify a 141 bp fragment of the nuclear 18S rRNA gene from eukaryotic DNA. SYBR® Green dye or TaqMan® fluorogenic probes were employed for monitoring the amplification of target genes. The application of TaqMan® probes enhanced the specificity of the real-time PCR assay. By analyzing 100 commercial ostrich meat products obtained from the market, it was proved that the method was suitable for detecting ostrich DNA. It was also shown that this method could be routinely applied for the verification of the correct labelling of ostrich meat products.

Rojas et al. (2012) described a species-specific real-time PCR assay that uses TaqMan® probes for identifying meat and meat products from common pigeon (*Columba livia*), woodpigeon (*Columba palumbus*) and stock pigeon (*Columba oenas*). A combination of species-specific primers and TaqMan® probes are used for the amplification of small fragments (amplicons < 200 base pairs) of the mitochondrial 12S rRNA gene, and an endogenous control primer pair is used for amplifying a 141 bp fragment of the nuclear 18S rRNA gene from eukaryotic DNA. By analyzing raw and heat-treated binary mixtures as well as commercial meat products, it was proved that the assay was suitable for detecting the target DNAs. This PCR assay could be a valuable tool for inspection agencies that deal with the verification of the correct labelling of raw and heat-treated pigeon meat products.

A TaqMan real-time PCR method relied on nucleotide sequence variation in the D-loop and 12S rRNA mitochondrial genes was generated for specifically detecting chicken, turkey, duck, and goose in animal feeds. The assay is based on the use of four primer/probe sets that target short species-specific mitochondrial sequences together with a positive amplification control based on the eukaryotic 18S rRNA gene. The effectiveness of the real-time PCR assay was evaluated through analyzing a batch of industrial feed samples treated under several temperatures in accordance with the European legislation regulations. This chicken-specific real-time PCR system can be used to quantitatively detect chicken-derived processed animal protein, even in samples containing 0.1% chicken and treated at temperatures as high as 133°C. However, turkey, goose and duck real-time PCR systems similarly detected the same level of target material in binary mixtures (muscle/oat) manufactured in accordance with the minimum legal requirements for sterilisation temperatures (133°C). Quantification results, based on calibration standard curves, were highly reproducible, but the quantitative potential of the assay is limited due to the extreme variability of the composition of feeds as well as the processing methods that can be applied. These factors significantly affect both the amount and quality of amplifiable DNA (Pegels et al., 2012).

According to Alonso et al. (2011), enteropathogenic *E. coli* is a food-borne pathogen that can lead to fatal infant diarrhea. Therefore, the detection of enteropathogenic *E. coli* contamination is essential to ensure the safety of consumers. PCR was employed for the analysis of samples from different stages of the chicken slaughtering process. Swabs were collected from chicken cloacae and washed carcasses. Samples from unwashed eviscerated carcasses were also analyzed. Detection of enteropathogenic *E. coli* was achieved in 6 to 28% of cloacal samples, 39 to 56% of unwashed eviscerated carcasses, and 4 to 58% of washed carcasses. All samples were negative for bfpA, suggesting contamination with a typical enteropathogenic *E. coli*. The detection of enteropathogenic *E. coli* at different stages of the chicken slaughtering process indicated that the proportion of contaminated samples was steady or even increased during the processing procedure. It was therefore proved that the high proportion of contaminated carcasses remaining after processing can be a potential risk for the consumers and a good reason for improving the sanitary conditions during processing.

Finally, in another study, an effort was made to develop a method for detecting reticuloendotheliosis virus (REV) in fowl pox vaccines. Examination of 30 fowl pox vaccine samples was carried out for the presence of REV using *in vitro* and *in vivo* methods. The *in vitro* testing included inoculation of the fowl pox vaccine samples into chicken embryo fibroblast cultures prepared from specific-pathogen-free embryonated chicken eggs, and examination of the cultures using PCR. The *in vivo* testing was performed through inoculation of each fowl pox vaccine sample into 5-days-old specific-pathogen-free chicks, and collection of serum samples at 15, 30 and 45 days post-inoculation for detecting REV-specific antibodies using enzyme-linked immunosorbent assays (ELISA). Tissue samples were collected 8 and 12 weeks after the inoculation and examined histopathologically. By using PCR, it was shown that only one vaccine sample was REV positive. Serum samples collected from chicks infected with the PCR-positive vaccine were also found positive for REV-specific antibodies using ELISA. The results obtained by PCR and ELISA were also confirmed using histopathology (Awad et al., 2010).

4.2.2 Use of Enzyme-linked Immunosorbent Assays (ELISA) for the Detection of Poultry Adulteration

ELISA was used by Berger et al. (1988) for detecting poultry in cooked and canned meat foods. The assay is based on species-specific, polyclonal antibodies raised against heat-resistant antigens. Isolation of heat-resistant antigens from raw skeletal muscle tissue of chicken was performed and immunoreactivity was detected even after thermal treatment at 120°C for 15 min. The method detected chicken or turkey even at levels as low as 126 ppm. The preparation of the samples was performed using simple aqueous extractions. However in the study of Hemmen et al. (1993), an effort was made to improve the texture of "foie gras" by using adulterants such as fresh livers from chicken and turkey. It was proved though that duck or goose "foie gras" were not detected by anti-duck lysozyme monoclonal antibodies (mAbs).

The generation of an antibody, that can bind metronidazole (MNZ), a nitroimidazole drug used for treating poultry for coccidiosis and histomoniasis, was reported by Huet et al. (2005). A direct competitive ELISA (cELISA) was employed for the characterisation of the binding of this antibody to numerous nitroimidazole drugs. Cross-reactivity with dimetridazole (DMZ), ronidazole (RNZ), hydroxydimetridazole (DMZOH), and ipronidazole (IPZ) was observed. The extraction of chicken muscle samples was performed using acetonitrile while the de-fattening was performed through washing with hexane. The detection capabilities (CCβ) were as follows: dimetridazole, < 2 ppb; metronidazole, < 10 ppb; ronidazole and hydroxydimetridazole, < 20 ppb; and ipronidazole, < 40 ppb. Furthermore, in the study of Er et al. (2013), the effects of quinolone antibiotics in chicken and beef used in Ankara (Turkey) were investigated. A total of 127 chicken and 104 beef meat samples were analyzed. Following extraction, the quinolones were determined using ELISA. It was proved that 51.1% of the examined chicken meat and beef samples contained quinolone residues. Specifically, 45.7% of chicken and 57.7% of beef meat samples were tested positive to quinolone residues. The mean levels (±SE) of quinolones found in the chicken and beef samples were 30.81 ± 0.45 µg/kg and 6.64 ± 1.11 µg/kg, respectively.

A cELISA was generated to quantitatively detect diethylstilbesterol (DES). Polyclonal rabbit antisera raised against protein conjugate diethylstilbesterol-mono-caroxyl-propyl-ethyl-bovine-serum-albumin (DES-MCPE-BSA), were used in immobilized antibody-based and competitive immunoassays. Optimization of the assays, including concentrations of antisera and horseradish peroxidase, (HRP)-DES, were carried out. The effects of parameters such as incubation time, surfactant concentration, ionic strength and pH of the medium were also evaluated. The typical calibration curve indicated an average IC(50) value of 2.4 ng/mL, while the calibration range ranged from 0.2 to 30.5 ng/mL and the detection limit was 0.07 ng/mL. The specificity of the technique was assessed against DES structurally related compounds and it was demonstrated that the assay was characterised by high sensitivity for DES. The validation of the assay's performance was performed using spiked chicken meat and liver tissue samples. Furthermore, a comparison with liquid chromatography-tandem mass spectrometry was made. The ion pair for quantification of DES was m/z 267.4/251.4, and the linear equation of DES was $y = 0.1033x + 0.0126$ ($r = 0.9960$).

Both methods can be used for monitoring DES and other steroid residues in foods (Xu et al., 2006).

Zhang et al. (2006) reported the development of an indirect cELISA based on chemiluminescent (CL) detection that can be used for detecting chloramphenicol (CAP) in chicken muscle. CAP-specific polyclonal antibody was raised in rabbit with a CAP-succinate derivative conjugated with bovine serum albumin. Luminol solution was used as the substrate of horseradish peroxidase. The detection limit of the method was 6 ng/L. The sensitivity of the CL-ELISA was 10 times higher in comparison to the colorimetric-ELISA. When levels of CAP as low as 0.05–5 mg/kg were spiked in chicken muscle, recoveries ranged from 97 to 118% with coefficients of variation of 6–22%. The CL-ELISA was used to detect residues and the results obtained were in good correlation with those obtained by gas chromatography (GC) with microcell electron capture detector. The residue levels of CAP in treated chicken decreased with time and the levels dropped rapidly after the first 6 h from around 50 to 10 µg/kg. No CAP was found in chicken after three days. Thus, the method can be effectively used for screening CAP in chicken muscle.

Quantification of residues of the tetracycline group of antibiotics was performed in chicken muscle tissue that had previously been subjected to screening with a microbiological inhibition test and an immunological method. Samples of frozen chicken were subjected to screening on a pH 6 culture medium seeded with *Bacillus subtilis*. An aqueous extract of the inhibitor-positive samples was subsequently screened with a group-specific commercial ELISA kit, aiming at detecting small quantities of oxytetracycline, chlortetracycline, tetracycline and doxycycline. The cut-off value of the ELISA was set at a B/B0 value of 75%. The results were confirmed and quantified using a validated HPLC method with fluorescence detection. A portion of the samples was also analyzed with LC-MS-MS. ELISA analysis indicated that residues of tetracyclines were present in 19 out of 21 inhibitor-positive chicken samples. Detection of doxycycline was confirmed with HPLC in 18 out of 19 chicken samples. No tetracyclines were detected in one ELISA positive chicken meat sample. The results obtained with ELISA B/B0 and the HPLC method were not significantly correlated, whereas a better correlation was found between the inhibition zones and the doxycycline levels. It was proved that an inhibition test with a medium at pH 6 and *B. subtilis* can be used for screening chicken muscle tissue for residues of tetracycline antibiotics (De Wasch et al., 1998). Table 4.1 presents some of the methods/techniques that have been developed for detecting adulterants in poultry products.

Beier et al. (1998) developed a method for the determination of levels of Hal in poultry. The method could be used instead of the current HPLC techniques. It includes the use of a cELISA applied as a screening tool for the determination of Hal in chicken liver tissue. The evaluation of the cELISA was performed using standard curves made in both assay buffer and chicken liver extract. It was indicated that standard curves made in assay buffer could be used for the cELISA. A comparison of the HPLC and cELISA results were made during two studies; the first study used spiked chicken liver tissue, while the second study used both spiked chicken liver tissue and incurred levels of Hal in chicken liver tissue. The data obtained by the two methods were well correlated, but in most cases the recovery was higher using the cELISA method than using the HPLC method. Furthermore, the cELISA method can be performed without

Table 4.1. Methods and techniques developed for detecting adulterants in poultry products.

Products	Treatment/storage conditions	Adulterant(s)/pathogens to be detected	QC method	Effectiveness	References
Plasma, feces	Aqueous extraction, purification, and acid partitioning. LC detection at 313 nm	Ipronidazole, ronidazole, dimetridazole	Liquid chromatographic multicomponent	High sensitivity, very effective	Aerts et al., 1991
Chicken Liver Tissue	Very complex, multiple steps	Halofuginone	HPLC vs. cELISA	Higher recovery with ELISA, very effective	Beier et al., 1998
Poultry tissues	Extraction, application to silica gel cartridges, elution of adulterant, evaporation, dissolution, washing	Dimetridazole	Liquid chromatography–thermospray mass spectrometry	Suitable for statutory residue testing	Cannavan and Kennedy, 1997
Chicken muscle	Extraction, centrifugation, exciting supernatant, measurement of emission	Enrofloxacin	Spectrofluorometric screening	Very reliable (error rate of less than 0.26%)	Chen and Schneider, 2003
	Extraction, fat removal, dilution, filtration	Veterinary drugs	Hydrophilic interaction liquid chromatography–tandem mass spectrometry	Very effective, limit of detection 0.1–20 µgkg^{-1}	Chiaochan et al., 2010
Poultry muscle	Extraction, clean-up, separation	Macrolides	Liquid chromatography–electrospray mass spectrometry	Successful for multiresidue determination	Codony et al., 2002
Poultry feed	Extraction, hexane wash, evaporation, reconstitution, ultracentrifugation	Medicinal additives	Liquid chromatography–tandem mass spectrometry	Detection of very low levels (100 µgkg^{-1})	Cronly et al., 2010
Poultry plasma	Pre-drying, addition to buffer	Monesin	Dry reagent dissociation enhanced lanthanide fluoroimmunoassay	Very user-friendly, effective and rapid	Crooks et al., 1998
Chicken meat	Freezing, micro-screening (*B. subtilis*), ELISA	Tetracycline	HPLC with fluorescence detection	Effective for tetracycline antibiotics	De Wasch et al., 1998
Chicken tissues, poultry feed and litter	Extraction, purification, evaporation, methanol-acetonitrile-water	Nicarbazin	Micro high-performance liquid chromatography (HPLC)	Sensitive, selective, rapid, cost-effective	Draisci et al., 1995

	Preparation of reagents				
Poultry muscle		Chloramphenicol, chloramphenicol glucuronide	Surface plasmon resonance biosensor and Qflex® kit chloramphenicol	Poultry spiked at 0.1 µgkg⁻¹	Ferguson et al., 2005
Broiler serum and muscle	Extraction	Flumequine	Biosensor immunoassay	Robust, specific and fast	Haasnoot et al., 2007
Poultry liver	Extraction, incubation, wash-up	Halofuginone	Time-resolved fluorometry	Effective for both qualitative and quantitative measurements	Hagren et al., 2005
Chicken muscle	Extraction, de-fattening, washing	Nitroimidazoles	ELISA	Very sensitive	Huet et al., 2005
Plasma and tissues of chicken	Deproteinisation, degrease, liquid–liquid extraction	Cyadox and its metabolites	HPLC	Highly sensitive and accurate	Huang et al., 2008
Chicken meat	Raw	Quinolone antibiotic residues	ELISA	45.7% of samples contained quinolones (mean levels: 30.81 µg/kg)	Er et al., 2013
Broiler Chickens	Cleaning using gel permeation chromatography	Pentachlorophenol	Gel permeation chromatography (cleaning), gas chromatography with electron capture detection (analysis)	Very sensitive	Stedman et al., 1980
Chicken muscle	Extraction by acetonitrile and de-fattening by washing with hexane	Nitroimidazoles	cELISA	Detection limits: dimetridazole, < 2 ppb; metronidazole, < 10 ppb; ronidazole and hydroxydimetridazole, < 20 ppb; ipronidazole, < 40 ppb	Huet et al., 2005

Table 4.1. contd....

Table 4.1. contd.

Products	Treatment/storage conditions	Adulterant(s)/pathogens to be detected	QC method	Effectiveness	References
Poultry plasma	Dilution of sample and competitive microwell plate immunoassay combined with the measurement of time resolved fluorescence	Narasin residues	Time-resolved fluoroimmunoassay	Detection capability: 1.5 ng ml^{-1}	Peippo et al., 2005
Goose and mule duck *foie gras*	Extraction of genomic DNA, PCR amplification	Chicken and pork meat	Qualitative PCR	Detection limit of 0.1% (w/w)	Rodriguez et al., 2003
Poultry meat	DNA extraction, DNA quantification	Pork meat	Duplex PCR	Quantification of pork meat addition to poultry meat in the range of 1–75%, sensitivity of 0.1%	Soares et al., 2010
Processed poultry meat			SYBR Green real-time PCR	High specificity	Soares et al., 2013
Chicken and liver tissues	Preparation of samples	Diethylstilbesterol residues	ELISA	ELISA detection limit: 0.07 ng/mL	Xu et al., 2006
Chicken muscle	CAP-specific polyclonal antibody raised in rabbit with a CAP-succinate derivative conjugated with bovine serum albumin. Luminol solution used as substrate of horseradish peroxidase	Chloramphenicol	Chemiluminescent ELISA	Detection limit: 6 ng/L	Zhang et al., 2006

the use of organic solvents. It was therefore proved that the cELISA could be effectively used for detecting and analyzing of Hal in chicken liver tissue.

4.2.3 The Use of Specific Compounds as Poultry Authenticity Markers

A high number of quality marks such as: "Label de Qualité Wallon" (in Belgium) or "Label Rouge" (in France), denominations of geographical origin, organic agriculture, etc. can be found in the European chicken market. The majority of these marks are accompanied by specifications requiring the use of slow-growing chicken strains. The amplified fragment length polymorphism (AFLP) technique was employed for searching for molecular markers capable of discriminating slow-growing and fast-growing chicken strains. Two pairs of restriction enzymes (EcoRI/MseI and EcoRI/TaqI) and 121 selective primer combinations were analyzed on individual DNA samples from chicken products essentially in carcass form that were ascribed as belonging to either slow- or fast-growing strains. Within the resulting fingerprints, two fragments were identified as type-strains specific markers. Specifically, one primer combination leads to a band (333 bp) that is connected to slow-growing chickens, while the other primer pair leads to the generation of a band (372 bp), characteristic of fast-growing chickens. Isolation, cloning and sequencing of the two markers was carried out. Their effectiveness and specificity were evaluated on individuals of two well-known strains (ISA 657 and Cobb 500) and on products obtained from different markets. The determination of the sequence of the band of 372 bp (amplification through EcoRI+AAC/TaqI + ATG primer) was effectively performed. On the other hand, the definition of the sequence connected with the AFLP marker (amplification by EcoRI +AAC/MseI + CAA primer) was more complicated. Different sequences were yielded, but none was found in more than one animal. By sequencing more plasmids, it was finally possible to obtain a fragment common for the five slow-growing chickens. The sequences of chicken genomic DNA fragments, without the selective primers, were, 350 and 311 bp, respectively. They were registered at the DDBJ/EMBL/GenBank databases (accession numbers AF525026 and AF525025) but no homology was detected by comparing them with sequence databanks (Fumière et al., 2003).

The EC Poultry meat Marketing Standards Regulation (1906/90), which is incorporated into Council Regulation 1234/20072, demands that the product is only marketed under certain conditions—'fresh', 'frozen' or 'quick frozen' and gives the definitions of the storage temperatures for each one of the above categories. The modern refrigeration technologies, however, may not be in compliance with the specified conditions and could lead to mislabelling. A method that could be used for measuring the activity of an enzyme (HADH) in meat and poultry has previously been employed for distinguishing between chilled and frozen + thawed poultry meat. Now, the HADH method was used for the examination of chilled and frozen chicken and turkey processed with new refrigeration technologies, in an effort to assess the extent to which the method can differentiate between these products and products treated with conventional chilling and refrigeration methods. It was demonstrated that the HADH assay successfully distinguished between poultry that had been frozen, either conventionally or using a new rapid-freezing technique, and fresh or chilled poultry but no clear differentiation could be made between normally chilled poultry and a new superchilling process (Lawrance et al., 2010).

The purpose of the study of Montowska and Pospiech (2013) was to search for proteins that can differentiate between chicken, turkey, duck and goose and are characterised by a relative stability during the meat aging, while do not readily degrade in ready-made products. A two-dimensional electrophoresis was applied for analyzing the protein profiles of raw meat and frankfurters and sausages (15 products). The observed species-specific differences in protein expression in raw meat were recorded. After identifying regulatory proteins, metabolic enzymes, some myofibrillar and blood plasma proteins, the electrophoretic mobility was used for their characterisation. Large differences in the primary structure were recorded for serum albumin, apolipoprotein B, HSP27, H-FABP, ATP synthase, cytochrome bc-1 subunit 1 and alpha-ETF. It was demonstrated that some of these proteins could be applied as markers for authenticating meat products.

The Nagoya breed native to Japan is commonly used for the production of eggs and meat. A method was described for the discrimination of the Nagoya breed and other breeds and commercial stocks of chicken. Four strains of the Nagoya breed were subjected to analysis using 25 microsatellite markers. In these strains, five of the markers (ABR0015, ABR0257, ABR0417, ABR0495 and ADL0262) had a single allele. Other chicken samples (448) of various breeds and hybrids were also analyzed with the same five markers. No sample had the same allele combination as the Nagoya breed strains. These five microsatellite markers is a valuable tool for the accurate discrimination between the Nagoya breed and other chicken breeds (Nakamura et al., 2006).

The Hinai-dori is a native chicken breed of the Akita Prefecture in Japan. A method for the discrimination of the Hinai-jidori and other chickens was reported. 555 samples of the Hinai-dori breed were subjected to analysis using 37 microsatellite markers on the Z chromosome. Fourteen of the marker loci (ABR1003, ADL0250, ABR0241, ABR0311, ABR1004, ABR1013, ABR0633, ABR1005, ABR0089, ABR1007, ABR1001, ABR1009, ABR1010 and ABR1011) were fixed in the Hinai-dori breed. As a result, the Hinai-jidori chicken, F1 of the Hinai-dori breed, must have at least one of the alleles with all fixed loci. The non-detection of these alleles on 14 loci from the Hinai-dori breed in meat samples could mean that the samples were not from the Hinai-jidori chicken. As a result, these 14 microsatellite markers could be effectively used for the accurate discrimination between the Hinai-jidori chicken and other chickens on the market (Rikimaru and Takahashi, 2007).

The need for effective methods for the identification of the origin of meat species led Stamoulis et al. (2010) to the development of a method, according to which two mitochondrial DNA (mtDNA) genes, cytochrome b (cytb) and 12S ribosomal RNA (12S rRNA), are tested as putative discrimination markers in samples of raw and processed poultry meat (chicken, turkey, duck, goose, pheasant, partridge, woodcock, ostrich, quail and song thrush), using the PCR–RFLP technique with universal primers and ten different restriction enzymes. Digestion of 12S rRNA by AciI led to successful differentiation between all avian species, generating species-specific patterns. It was proved that the 12S rRNA gene is more effective than cytb gene for avian species identification. Moderate processing of the samples did not affect the effectiveness of the method and similar patterns with the raw meat were presented. Therefore, this method was capable of detecting mixtures of meat, and could be used for detecting adulterations.

EU regulation dictates that chicken intended to be sold as "corn-fed" poultry must be fed a diet containing at least 50% (w/w) corn for most of the fattening period. Nevertheless, there are currently no reliable methods for authenticating this dietary claim. The following method exploits the differences in the photosynthetic pathways between maize and other cereals such as wheat, rye, barley and oats and their different enrichments of the ^{13}C stable isotope from atmospheric CO_2. These differences in ^{13}C and ^{12}C are estimated using Stable Isotope Ratio Mass Spectrometry. Several controlled feeding experiments were performed on two breeds of chickens, demonstrating that both the fat and protein concentrations of the meat were significantly altered in line with both the quantity and the duration of corn consumption. Blind testing of the method and analysis of the meat of commercially grown corn-fed chicken proved that the ^{13}C content of the protein could be effectively used as a marker of the dietary status of the chickens (Rhodes et al., 2010).

The 2-alkylcyclobutanones (2-ACBs) are generated from triglycerides through irradiation and can be applied as markers for this type of food processing. A detection method for analyzing monounsaturated alkyl side chain 2-ACBs, formed from monounsaturated fatty acids, was described. The estimated radioproduction yields of the cis-2-(dodec-5'-enyl)-cyclobutanones (cis-2-dDeCB) and the cis-2-(tetradec-5'-enyl)-cyclobutanones (cis-2-tDeCB) were 1.0 +/– 0.5 and 0.9 +/– 0.2 nmol·mmol^{-1} precursor fatty acid·kGy^{-1}, respectively, being at the same levels as those of the saturated 2-ACBs. The stability study of the s- and mu-2-ACBs in poultry meat samples irradiated with 10 kGy and stored for 3-4 weeks at 4°C and 25°C, indicated that these compounds are partially transformed, and their amounts are reduced by about 50%. This reduction is not affected by the saturation state of the alkyl side chain. The EI-MS detection limit of 2-tDeCB is three times higher (0.6 pmol) than that of 2-dodecylcyclobutanone (0.2 pmol). Consequently, when the oleic acid content of the analyzed food is three times higher than the content of palmitic acid, the 2-tDeCB could be used as a marker for detecting the irradiation treatment (Horvatovich et al., 2005).

Tewfik (2008) made an effort to evaluate the qualitative performance of a new rapid and straightforward solvent extraction (DSE) method as a potential tool for differentiating between irradiated and non-irradiated food samples. The trial was performed between four European laboratories that agreed to use 2-dodecylcyclobutanone (DCB) as radiolytic marker for detecting irradiated minced chicken. Chicken samples were treated with irradiation with 3 and 7 kGys. Every laboratory performed the experiments using 12-blind coded 'unknown' and 4-known coded samples. All laboratories successfully identified all 12-blind coded samples as either irradiated (at medium and/or high doses) or non-irradiated samples. The examined DSE method is considered to be fast and cheap, and could be effectively used in laboratories that are involved in screening high number of food samples for irradiation.

The role of volatile amines as indicators of poultry meat spoilage was assessed by Balamatsia et al. (2007). Fresh chicken meat (breast fillet) was packed in four different atmospheres: air (A), vacuum (VP) and two modified atmospheres (MAs),

[M1, 30%/65%/5% ($CO_2/N_2/O_2$) and M2, 65%/30%/5% ($CO_2/N_2/O_2$)]. All samples were stored under refrigeration for two weeks. It was demonstrated that the VP and M1 and M2 gas mixtures effectively delayed the growth of aerobic spoilage microbial flora. *Pseudomonas* spp. populations in chicken samples packaged in M2 gas mixture and VP were significantly reduced after the two-week period. Lactic acid bacteria (LAB) and *Brochothrix thermosphacta* were the dominant microorganisms in both aerobically- and MA-packaged chicken, while yeasts was a highly smaller part of the final microbial flora of chicken meat. The shelf life of the samples was extended by 2, 4 and 9–10 days under VP and M1 & M2 gas mixtures, respectively. It was proved that the limit of sensory acceptability (a score of 6) was reached for the aerobically and vacuum-packaged & M1 gas mixture chicken samples after 6–7 and 9–10 days, respectively. According to the sensory (taste) analysis of the spoilage and freshness of the samples, TMA-N and TVB-N limit values of acceptability, specifically 10.0 mg N/100 g and 40 mg N/100 g for chicken samples stored in air, may be proposed as the upper limit values for spoilage initiation of fresh chicken meat stored aerobically. It is important to mention that the M2 gas mixture sample did not reach these limits during the two-week storage. The generation of volatile amines was well correlated with the increase in microbiological count (TVC) and sensory taste score, with the exception of the M2 gas mixture.

4.2.4 Use of Immunoassays for the Detection of Adulterants in Poultries

The development of immunochemical screening assays using surface plasmon resonance was reported for the detection of chloramphenicol and chloramphenicol glucuronide residues in poultry muscle. The assays were based on the use of a sensor chip coated with a chloramphenicol derivative and an antibody. Cross-reaction of the antibody with chloramphenicol glucuronide (73.8%) was observed but no cross-reaction was shown for similar drugs or antibiotics. The extraction of the poultry samples was performed using ethyl acetate and the extracts were analyzed using the biosensor. The decision limit (CCα) was 0.005 µgkg^{-1} while the detection capability (CCβ) was found to be 0.02 µgkg^{-1}. Poultry muscle was spiked at 0.1 µgkg^{-1} and the intra-assay precision ($n = 10$) was found to be 10.5% (Ferguson et al., 2005).

Flumequine (Flu) is a fluoroquinolone commonly used for treating broilers in The Netherlands. Residues of Flu in blood serum of broilers were detected using a biosensor immunoassay (BIA) characterized by both speed (7.5 min per sample) and specificity (no cross-reaction with other (fluoro)quinolones was observed). This inhibition assay was relied on a rabbit polyclonal anti-Flu serum and a CM5 biosensor chip coated with Flu, capable of being detected in the range of 15–800 ngm^{-1}. The Flu was detected in the samples following an easy extraction process in buffer and the measuring range was between 24 and 4000 ng g^{-1}. Average recoveries of 66 to 75% were achieved in muscle samples spiked at 0.5, 1 and 2 times the maximum residue limit (MRL in muscle = 400 ng g^{-1}) and the decision limit (CCα) and detection capability (CCβ) were found to be 500 and 600 ng g^{-1}, respectively. The analysis of the incurred muscle samples was performed by employing the BIA and by LC-MS/MS and the results were well-correlated ($R^2 = 0.998$). The analysis of serum and muscle samples from Flu treated broilers indicated that the concentrations detected in serum were higher than those detected in muscle (average serum/muscle ratio was 3.5) thus proving that the BIA in serum can be effectively used for predicting the Flu concentration in poultries (Haasnoot et al., 2007).

The study of Hagren et al. (2005) gives a description of the development and validation of an immunoassay for screening coccidiostat halofuginone in poultry liver. Time-resolved fluorometry is employed following an innovation all-in-one dry chemistry assay concept, according to which all the reagents required for the competitive immunoassay are incorporated into a single microtiter well in a dry, stable form. This immunoassay is carried out by an automated immunoanalyser and the whole procedure lasts about 18 min. The assay protocol includes extraction of the sample and addition to a well. After incubation for 15 min, the sample is washed and measurement of the fluorescence signal is taken directly from the surface of the dry well in a time-resolved manner. The analytical limit of detection was found to be 0.02 ng ml^{-1} ($n = 12$) and the functional limit of detection was 1.0 ng g^{-1} ($n = 6$). The assay can become more sensitive through adjusting the dilution factor of the samples. The mean recovery was 106.0% at concentrations of 7.5 and 30 ng g^{-1}. The intra-assay variations were usually lower than 10% and the interassay variations ranged between 7.7 and 11.6%. Validation of the immunoassay was performed in accordance to Commission Decision 2002/657/EC.

The development of a simple and rapid time-resolved fluoroimmunoassay (TR-FIA) method was reported for screening narasin in poultry plasma. This method includes the dilution of a sample and the use of a competitive microwell plate immunoassay in combination with the measurement of time-resolved fluorescence. The performance of the assay was validated in accordance with the Commission's Regulation 2002/657/EC. The decision limit and detection capability of the assay were found to be 1.2 and 1.5 ng ml^{-1}, respectively. The recovery of narasin from plasma ranged between 101.0 and 121.3%. In this experiment, broiler chickens were fed with feed containing 0, 3.5 and 70 mg kg^{-1} narasin for three weeks and the final concentration of narasin in the plasma and muscle of chicken was examined using TR-FIA. Feeding with 70 mg kg^{-1} and no withdrawal period led to a very high narasin levels in blood. It was observed that the concentrations of narasin in plasma and breast muscle ($R^2 = 0.83$) and leg muscle ($R^2 = 0.90$) were correlated. It was therefore suggested that the analysis of poultry blood samples could be applied for predicting the narasin residual levels in broilers (Peippo et al., 2005).

Monensin, a carboxylic acid ionophore, is usually incorporated into poultry feed to control coccidiosis. A method for rapidly analyzing unextracted poultry plasma samples was reported. The method is relied on a novel immunoassay format: one-step all-in-one dry reagent time resolved fluorometry. Pre-drying of all specific components was performed onto microtitration plate wells. The addition of the diluted serum sample in assay buffer was only needed to carry out the analysis. The results were obtained only one hour after adding the sample. The limit of detection (14.2 ng ml^{-1} mean +/– 3s) was found by analyzing 23 known negative samples. Determination of intra- and inter-assay RSD was performed and the RSD values were 15.2 and 7.4%, respectively. Eight broiler chickens were given feeds containing monensin at a dose rate of 120 mg kg^{-1} for seven days. After this period, blood samples were taken and the chickens were slaughtered. Plasma monensin concentrations were found using the fluoroimmunoassay and ranged between 101 and 297 ng ml^{-1}. The results were compared with monensin liver concentrations, as found using LC-MS (13–41 ng g^{-1}). This fluoroimmunoassay can be performed very easily and is very rapid. It can be used for effectively detecting and quantifying plasma monensin residues. It was suggested that analysis of plasma could be useful for the prediction of the levels of monensin in liver (Crooks et al., 1998).

4.2.5 Use of Gel Electrophoresis for the Detection of Poultry Adulteration

Electrophoresis is a common technique usually used for species determination in food products and in the study of Bonnefoi et al. (1986), it was used for detecting fraudulent liver mixing in foie gras canned products. An aqueous extract was prepared by homogenising the sample in 8M urea and 1 mM dithiothreitol and was then centrifuged. Triton X-100 urea disc-electrophoresis in polyacrylamide gel subsequently took place in a cationic system by stacking at pH 5.13 and resolution at pH 4.01. The comparison of the gels was performed after densitometric scanning. Foie gras samples characterised by different proportions of liver mixtures were assayed. It was proved that the technique could effectively detect liver species adulteration at concentrations as low as 10%.

Bowker and Zhuang (2013) aimed at determining the relationship between the quality of meat and the protein content and composition of muscle exudate from broiler breast fillets. Deboned breast fillets ($n = 48$) derived from a commercial processing facility were collected and segregated into two groups based on colour (light and dark). Meat pH, colour, moisture content, water-holding capacity (drip loss, salt-induced water uptake, cook loss), protein solubility, and the protein content of muscle exudates were measured. The evaluation of the protein composition of the muscle exudate was performed using SDS-PAGE analysis. Light breast fillets were characterised by lower meat pH (4 and 24 h postmortem) and higher L* (lightness) and b* (yellowness) values than dark fillets. Light breast fillets showed higher drip loss after 2 and 7 days of storage, lower salt-induced water uptake, and higher cook loss in comparison to dark fillets. The sarcoplasmic and total protein solubility presented no differences between light and dark fillets. The protein concentration of muscle exudates was higher in dark fillets and a negative correlation to drip loss ($r = -0.50$) and salt-induced water uptake ($r = 0.42$) was found after two days of storage. Similar electrophoretic protein banding patterns were found between muscle exudates and sarcoplasmic protein extracts. Gel electrophoresis of muscle exudates revealed that the relative abundance of four bands corresponding to 225, 165, 90 and 71 kDa was greater for dark breast fillets, while the relative abundance of three bands corresponding to 47, 43 and 39 kDa was greater in light breast fillets. The muscle pH and water-holding capacity showed significant correlation with the abundance of numerous individual protein bands within the protein profile of muscle exudates. It was finally proved that there is a relation between protein differences in breast muscle exudates and meat pH, colour, and water-holding capacity and, as a result, the muscle exudate could be potentially used as a source of protein markers for fresh meat quality characteristics in broiler fillets.

Microgel electrophoresis was applied to single cells (DNA comet assay) for the detection of DNA comets in quail meat samples treated with irradiation. The evaluation of the DNA comets was performed using photomicrographic and image analysis. Quail meat samples were irradiated with the following doses: 0.52, 1.05, 1.45, 2.00, 2.92 and 4.00 kGy in gamma cell (gamma cell ^{60}Co, dose rate 1.31 kGy/h) covering the permissible limits for enzymatic decay and were subsequently maintained at 2°C. The cells isolated from muscle (chest, thorax) in cold PBS were subjected to analysis using the DNA comet assay on 1, 2, 3, 4, 7, 8 and 11 day post irradiation. The cells were lysed between 2, 5 and 9 min in 2.5% SDS and electrophoresis was performed at a voltage of 2 V/cm for 2 min. After propidium iodide staining the evaluation of the slides was carried out using a fluorescent microscope. In all samples treated with irradiation, fragmented DNA stretched towards the anode and damaged cells were presented as a comet. The analysis of the measurement results was performed using BS 200 ProP with software image analysis (BS 200 ProP, BAB Imaging System, Ankara, Turkey). Increase of radiation dose led to increase of density of DNA in the tails. On the other hand, in non-irradiated samples, the DNA molecules remained relatively intact and there was insignificant or no migration of DNA. The cells were characterized by round shape or very short tails. It was concluded that the DNA Comet Assay EN 13784 standard method can only be applied as a screening method for detecting irradiated quail meat, depending on storage time and condition, as well as for quantifying the dose used (if combined with image analysis). Image analysis can be very useful for evaluating the head and tail of comet intensity related with applied doses (Erel et al., 2009).

According to Kim and Shelef (1986), electrophoretic patterns of fresh chicken, and turkey sarcoplasmic proteins were examined through thin layer agarose gel electrophoresis. Creatine kinase isozyme MM and myoglobin bands were employed for the identification of the species. Binary mixtures of beef, pork, chicken and turkey (5/95, 25/75, 50/50, 75/25, 95/5% by weight of each species) were also evaluated through their electrophoretic and densitometric patterns. Stability of the electrophoretic patterns was observed at high pH values (7.5) and not under low pH (4.7) conditions. The relative ratios of the bands (characteristic of each species) changed in proportion to the species content in each binary mixture. These ratios make the prediction of the approximate fraction of each species present quite possible in these mixtures.

In the study of Montowska and Pospiech (2011), the interspecies differences in two-dimensional electrophoresis patterns of skeletal muscle myosin light chain (MLC) isoforms between *G. gallus*, *M. gallopavo*, *A. platyrhynchos* and *A. anser*, *Bos Taurus* (cattle) and *Sus scrofa* (pig) were analysed. Two-dimensional electrophoretic separations indicated the presence of species-specific differences in the molecular weight and pI of individual MLC isoforms (MLC1f, MLC2f and MLC3f). As regards closely related animal species such as goose and duck or turkey and chicken, significant differences were presented in MLC1f. As regards MLC2f, differences between cattle and turkey and between pig and chicken were found to be around 1 and 0.3 kDa respectively. By comparing the amino acid sequences, it was proved that even MLCs with a difference degree of just 2% in sequences, differ in electrophoretic mobility. Interspecies differences in skeletal MLC isoforms were presented between cattle, pig, chicken, turkey, duck and goose.

4.2.6 Use of Visible Near Infrared Spectroscopy in Poultry Quality Assurance

A system using a visible/near-infrared (NIR) spectroscopic technique and an intensified multispectral imaging technique was evaluated for its potential in the accurate separation of abnormal (unwholesome) and normal poultry carcasses. The spectroscopic subsystem was used to measure reflectance spectra of poultry carcasses at wavelengths from 471 to 965 nm. For the multispectral imaging subsystem, measurements of the gray-level intensity of whole carcasses were taken by employing six different optical filters of 542, 571, 641, 700, 720 and 847 nm wavelengths. It was shown that the method was very effective since all abnormal carcasses were correctly detected. The use of each of the techniques separately, led to an error degree of 2.6% for the spectroscopic technique and 3.9% for the multispectral imaging subsystem. Therefore, the method could be effectively applied for the separation of carcasses into normal and abnormal streams. With perfect selection of normal carcasses in the normal carcass stream, inspection of the abnormal carcass stream would only be required (Park et al., 1996).

The quantification of the cooking route at the centre point in chicken patties was performed taking into account both the time-temperature integrated indices (C and F) and endpoint temperature (T_{max}). Intact cooked patties were subjected to scanning using reflectance and transmittance spectroscopy. Reflectance resulted in a more effective calibration of the indices of thermal history in comparison to transmittance. The evaluation of three reflectance wavelength ranges, visible (400 to 700 nm), near-infrared (1100 to 2500 nm), and visible/near-infrared (400 to 2500 nm), was performed, indicating that the visible/near-infrared was the most accurate. The best calibration resulted in a standard error of prediction (SEP) of 0.11 $\log_{10}$(min) for $\log_{10}$C, 0.25 $\log_{10}$(min) for $\log_{10}$F, and 2.54°C for T_{max} on an independent validation sample set (Chen and Marks, 1997).

The implementation of a nondestructive and noninvasive technique for detecting abnormal poultry carcasses with the use of a diode array spectrophotometer system was reported. Visible I near-infrared reflectance and interactance spectra (500 to 1113 nm) for both sides of the breast of normal, septicemic, and cadaver poultry carcasses were obtained by employing a fibre optic probe with the spectrophotometer. Optimal wavelengths were used for the correlation of the spectral reflectance and interactance and second difference with the condition of the poultry carcasses. Generally, the differentiation between the normal and cadaver carcasses was more accurately performed than the differentiation between normal and septicemic or between septicemic and cadaver. It was proved that the method could be applied for the separation of normal and abnormal carcasses, providing quite accurate results. The accuracy of the model was estimated at 93.3% for normal carcasses and at 96.2% for the abnormal carcasses. Use of different decision criteria could prevent any misclassifications. A reflectance model requiring four wavelength readings yielded classification accuracy of 94.4% for normal carcasses and 93.1% for abnormal carcasses (Chen and Massie, 1993).

Ding et al. (1999) reported the development of a near-infrared spectroscopic method combined with a dummy regression technique for the differentiation between meat originating from broilers and chickens obtained from markets in Hong Kong. Accurate classifications of 100%, 92%, 96% and 92% were obtained for minced thigh meat, minced breast meat, breast cut without skin and breast cut with skin respectively. By comparing regression models of MLR, PCR, PLS and mPLS, it was proved that the classification accuracy did not differ significantly. Scatter correction and derivative treatment of the spectral data before discriminant analysis led to increased classification accuracy for minced meat, while for meat cuts, no pre-treated spectra yielding higher classification accuracies were obtained. Generally, the use of the full spectrum of 400 ± 2500 nm led to good classification. The spectrum in the visible region of 400 ± 750 nm, the short-wavelength NIR region of 750 ± 1100 nm or the long-wavelength NIR region of 1100 ± 2500 nm can also be used to effectively classify the samples. The physicochemical characteristics of the samples were used to support the results of the spectroscopic technique. It was shown that the collagen and fat contents, as well as pH and chromatic values between the two groups of chickens, differed significantly. It was therefore proved that NIR spectroscopy could be applied for the identification of broiler meat or carcass from chickens collected from markets in Hong Kong.

Generalized two-dimensional (2D) correlation analysis of visible/NIR spectra was used for the characterisation of the spectral intensity variations of chicken muscles induced by either storage time/temperature regime or shear force values. It was proved that the intensities at 445 and 560 nm increased by increasing the temperature under identical treatment, probably due to a colour change caused by frozen storage. The 2D NIR correlation spectra showed that all NIR bands reduce their spectral intensities. This can be due to the water loss and changes in the composition of the samples during the freeze-thaw process as well as due to the tenderization development in muscle storage. The heterospectra correlating the spectral bands in both visible and NIR regions were strongly correlated, thus indicating that there is a sequential change between colour and other developments in muscles. Furthermore, the detection of important differences in spectral features between tender and tough muscles was performed using the shear value-induced NIR spectral intensity variations (Liu et al., 2004).

4.2.7 The Use of Near Infrared Spectroscopy in Poultry Quality Assurance

Abeni and Bergoglio (2001) investigated the differences in the chemical composition and colour parameters of fresh breast muscle between different chicken strains, in an effort to assess the results obtained using different illuminants (A, C and F), towards the verification of the prediction of some meat features using NIRS of freeze-dried breast muscle. The analysis was performed on 39 chicken broilers from three strains, characterised by different body size, and slaughtered at the same age (1 month). Differences between fresh breast samples were detected only in the case of the ash content (P < 0.01). Illuminant effect was found to be significant for all colour variables, depending on the major wavelength characteristics of each illuminant. Significant differences amongst strains were found in redness values. The NIRS technique led to very accurate predictions of fat content of muscle (R^2c = 0.98, SEC = 0.20). The above-mentioned results may be very important for improving the instrumental procedures for on-line evaluation of poultry meat quality.

The aim of the study of Berzaghi et al. (2005) was the evaluation of the NIRS as a technique for predicting the physicochemical composition of breast meat samples of laying hens fed four different diets, a control and three diets containing different sources of n-3 polyunsaturated fatty acids: marine origin, extruded linseed, and ground linseed. Moreover, NIRS was employed for the classification of meat samples in accordance to feeding regimen. The chemical analysis of samples included DM, ash, protein, lipids and fatty acid profile. The collection of absorption spectra was performed in diffuse reflectance mode between 1100 and 2498 nm every 2 nm. The accuracy of the prediction obtained by the calibration results regarding the 72 meat samples was very high for DM, protein, lipids and major fatty acids, while less accurate predictions yielded the calibration equations for ash, pH, colour and lipid oxidation. The development of partial least squares discriminant analysis was performed for differentiating between the breast meat samples originating from hens fed different diets. The discriminant models accurately classified 100% of the control and the enriched diets. It was therefore proved that NIRS could be employed for the prediction of the chemical composition of poultry meat and possibly of some dietary treatments applied to the chickens.

The investigation of physical and colour characteristics of meat from 193 chickens was performed through direct application of a fiberoptic probe to the breast muscle and by employing the visible NIR spectral range from 350 to 1800 nm. Records were kept for pH, lightness (L*), redness (a*) and yellowness (b*), 48 h postmortem, while thawing and cooking losses, as well as shear force, were recorded after freezing. Partial least squares regressions were performed using untreated data, raw absorbance data ($\log(1/R)$), and multiplicative scatter correction plus first or second derivative spectra. The validation of the models was performed through full cross-validation, and their prediction potential was assessed by root mean square error of cross-validation ($RMSE_{cv}$) and correlation coefficient of cross-validation (r_{cv}). The best prediction models were obtained using $\log(1/R)$ spectra for b* ($r_{cv} = 0.93$; $RMSE_{cv} = 1.16$) and a* ($r_{cv} = 0.88$; $RMSE_{cv} = 0.29$), while a lower NIR prediction potential was found for pH, L*, and thawing and cooking losses (r_{cv} from 0.69 to 0.76; $RMSE_{cv}$ from 0.01 to 1.73). The shear force, however, was not accurately predicted. It was shown that NIR could be used for assessing numerous quality traits of intact breast muscle (De Marchi et al., 2011).

Riovanto et al. (2012) evaluated the near infrared transmittance (NIT) spectroscopy as a technique that could potentially be used for predicting the FA profile of ground chicken breast (*Pectoralis superficialis*), taking into account the wavelengths between 850 and 1050 nm. The calibration equations were prepared using reference data expressed as (i) percentage of total fatty acids and (ii) absolute concentration, i.e., mg of FA in 100 g of fresh meat. When the fatty acids were expressed in absolute concentration, much higher accuracy was obtained in comparison to the cases where a percentage of total fatty acids expression was used. The maximization of the prediction accuracy was achieved through sample preprocessing (milling) and different spectra pre-treatments. Polyunsaturated fatty acids were the most difficult substances to predict, and it was proved that NIT spectroscopy could not be used for accurate and reliable predictions of these substances.

4.2.8 The use of Chromatographic-Spectrometric Methods for the Detection of Adulterants in Poultries

The development of a confirmatory method was reported by Cronly et al. (2010) that could be used for the quantification of fourteen prohibited medicinal additives in poultry compound feed. Although these substances are prohibited for use as feed additives, a small number can still be used in medicated feed. The extraction of the samples was performed using acetonitrile and sodium sulphate. The extracts were washed with hexane to be better purified and were subsequently evaporated to dryness and reconstituted in initial mobile phase. An ultracentrifugation stage was carried out prior to injecting the samples onto the LC-MS/MS system. The LC–MS/MS system was run in MRM mode with both positive and negative electrospray ionisation. The validation of the method indicated that it could be used for the quantitative analysis of metronidazole, dimetridazole, ronidazole, ipronidazole, chloramphenicol, sulphamethazine, dinitolimide, ethopabate, carbadox and clopidol. The method could also be employed for the quantitative analysis of sulphadiazine, tylosin, virginiamycin and avilamycin. A level of 100 μgkg^{-1} was used for validating the method, proving that analysis to this level is possible.

According to Chiaochan et al. (2010), a highly sensitive method was used for multiresidue analysis of 24 important veterinary drugs (including 3 aminoglycosides, 3 β-lactams, 2 lincosamides, 4 macrolides, 4 quinolones, 4 sulphonamides, 3 tetracyclines, and amprolium) in chicken muscle. The methodology followed included extraction using 2% trichloroacetic acid in water-acetonitrile (1:1, v/v), removal of fat with hexane, dilution of sample extract, filtration and liquid chromatography-tandem mass spectrometric (LC-MS/MS) analysis. Hydrophilic interaction liquid chromatography (HILIC) was capable of effectively separating various polar and hydrophilic compounds while being very sensitive and more effective than reversed phase and ion-pair separation. The validation of the method was successfully performed in accordance with the European Decision 2002/657/EC. Average recoveries were 53–99% at 0.5-MRL, MRL and 1.5-MRL spiking levels, with satisfactory precision ≤15% RSD. The limit of detection was around 0.1–10 μgkg^{-1} for 22 analytes and 20 μgkg^{-1} for aminoglycosides. These values were lower than the maximum residue limits established by the European Union. This method could very reliably screen, quantify and identify 24 veterinary drug residues in foods of animal origin and its effectiveness was successfully evaluated in chicken meat and egg samples.

The determination of tetracycline antibiotics in broiler meat was performed using HPLC with ion-pairing chromatography and diodearray detection at 355 nm. Optimization and validation of the proposed methods were also carried out. The mean recovery values for tetracycline from breast meat were 76%. The within-day precision was 6.1 to 15.5% and the between-day precision was 5.0%. The limit of detection and the limit of quantitation were 10.5 and 20.9 ng/g respectively. The determination of residue values of tetracycline antibiotics in broiler meat was performed after oral administration of medicated feed. Medicated feed with 480 mg/kg tetracycline was given to broilers for one week. Four days after the end of the administration, the mean tetracycline residue value in breast meat was determined at 86 ng/g, which is below the MRL (De Ruyck et al., 1999).

According to Anderson et al. (1981), the detection of the anti-coccidial drug, halofuginone, in chicken tissue was performed at levels as low as 1 ppb (0.001 ppm) and in chicken feed at levels as low as 3 ppm, using high-performance liquid chromatography (HPLC). The treatment of the meat sample includes: enzymatic release of the halofuginone, ethyl acetate extraction under basic conditions, partition into ammonium acetate buffer and concentration using Sep-pakTM C18 cartridge. The feed is analysed following ethyl acetate extraction under basic conditions, partition into hydrochloric acid, and concentration using XAD-2 column chromatography. HPLC with ultraviolet detection can be used in both methods for analysing the treated samples.

Another method was developed for determining levels of the nitroimidazole drug DMZ in poultry tissues using liquid chromatography (LC)-thermospray mass spectrometry (MS). Deuteriated DMZ was applied as an internal standard. The extraction of the samples was performed using dichloromethane (muscle) or toluene (liver) and the extracted samples were subsequently applied to silica gel cartridges. Dimetridazole was eluted with acetone and the eluate was evaporated to dryness at 40°C under nitrogen. Methanol-water (1 + 1, v/v) was used for re-dissolving the residue and hexane was used for washing it before LC-MS analysis. The quantification was performed by the ratios of the positive $[M + H]+$ ions at m/z 142 and 145 for DMZ and the internal standard, respectively. Internal standard corrected recoveries ranged from 93 to 102% with RSDs and from 1.2 to 7.7% for liver spiked at 5, 10 and 20 ng g^{-1} and muscle spiked at 5 ng g^{-1}. The absolute recoveries were as high as 80%. The method could be effectively used for statutory residue testing (Cannavan and Kennedy, 1997).

Cyadox (CYX), an antimicrobial growth-promoter of the quinoxalines can effectively improve the growth and feed conversion of chicken and is characterised by low toxicity. The development of HPLC-UV methods was reported for determining CYX and its major metabolites, 1,4-bisdesoxycyadox (BDCYX) and quinoxaline-2-carboxylic acid (QCA), in plasma, muscle, liver, kidney and fat of chicken. For CYX and BDCYX, the samples were deproteinised and were subsequently subjected to degreasing and liquid-liquid extraction. As regards the QCA samples, an alkali hydrolysis, a liquid-liquid extraction and a cation exchange column (AG MP-50 resin) clean-up were applied. A RP-C18 column was employed for analyzing the samples using UV detection with a gradient program of wavelength. Gradient elution was carried out at a flow of 1 mL/min. The recoveries of the substances of interest in plasma and tissues ranged from 70 to 87%, while the inter-day relative standard deviation was lower than 10%. The applicability of the above-mentioned methods was experimentally confirmed in real animal samples, indicating that the detection of the substances was possible in almost all tissues. The methods were characterised by high sensitivity and accuracy and as a result, they could be valuable tools in pharmacokinetic and residue studies concerning CYX in chicken (Huang et al., 2008).

The development of a spectrofluorometric method for screening enrofloxacin (ENRO) in chicken muscle that is characterised by high simplicity was reported. A single-step extraction with acidic acetonitrile maximized the efficiency of the technique, while no further clean-up was required. After centrifuging the samples, supernatants were excited at 324 nm and measurements of the emission were taken at 442 nm. In total, 18 chicken breast samples from three producers were examined. It was proved that the background signal levels were not as high as those reported for 300 µg/kg ENRO. By statistically analyzing the results, a threshold was found which can be applied in subsequent screening of ENRO at the tolerance level. The calibration curve indicated the presence of a satisfactory linear relationship (R^2 = 0.9991) in a range of 0–700 µg/kg ENRO in fortified chicken breast. The examination of the ENRO-incurred samples was performed using this methodology and the results were in agreement with those obtained from more extensive separation and subsequent HPLC. Since the threshold can be defined at the 3-sigma limit, effective screening can be performed with an error rate as low as 0.26%. As a result of this study, the use of a high-throughput screening method for ENRO in chicken tissue was suggested (Chen and Schneider, 2003).

A methodology based on a LC-MS method was suggested by Codony et al. (2002) for determining seven macrolides authorised in the EU as veterinary drugs for food-producing animals. Specifically, the sample was extracted using a water-methanol mixture containing metaphosphoric acid and was cleaned-up by SPE with a cation-exchange cartridge. Separation was performed in an end-capped silica-based C18 column and the mobile phases were consisted of water/acetonitrile mixtures and trifluoroacetic acid. A gradient elution, from 28 to 40% acetonitrile was applied. The compounds were detected using MS with electrospray ionisation in the positive mode. Various factors that affect the mass spectra were examined. The protonated molecular ion was chosen and quantified using a selected ion monitoring mode. Detection limits ranged from 1 to 20 µg L^{-1}, while the recoveries were between 56 and 93% when RSD was lower than 12%. The method was successfully used for determining residues of the seven macrolides even at quantities below the Maximum Residual Levels established by the European Union.

The development of a micro HPLC method was reported by Draisci et al. (1995) for determining the anti-coccidial drug nicarbazin in chicken tissues, poultry feeds and litter. The extraction of the 4,4'-dinitrocarbanilide (DNC) component of nicarbazin was performed from food, feed and litter samples using acetonitrile. The purification of the extracts was performed using liquid-liquid partitioning, evaporation to dryness and finally by taking them up in methanol-acetonitrile-water (50:30:20, v/v). Micro HPLC of the 4,4'-dinitrocarbanilide (DNC) portion of nicarbazin was carried out by employing a small bore column (1 mm I.D.) packed with reversed-phase and a UV detector set at 340 nm. According to the authors of the study, "The average recoveries of nicarbazin added to muscle and liver were 92.8, and 84.3, respectively, 95.9% in poultry feed and 76.8–95.9% in different litters". The limit of detection was found to be 25 pg. This method is characterised by high sensitivity, while, at the same time, it is cost-effective and thus more efficient than the HPLC.

Finally, in the study of Stedman et al. (1980), Hubbard-Hubbard broiler chickens were fed graded levels (0, 1, 10, 100 and 1000 ppm) of pentachlorophenol (PCP), which contained < 0.0023% octachlorodibenzo-p-dioxin (OCDD) for eight weeks. Analysis of tissue samples for PCP, OCDD and pentachloroanisole (PCA) was performed using GC with electron capture detection. A considerable increase of kidney weights was observed for the animals fed with the 100 ppm and 1000 ppm PCP diet, while the weights of all other organs was reduced by the 1000 ppm PCP diet. The histopathological examination of the liver indicated bile duct proliferation and some fatty changes in all of the six-week-old birds. After examining the brain, liver, gizzard, pancreas, intestine, proventriculus, spleen, kidney, lung and heart, it was shown that there were no histopathological lesions in neither the treated nor the control animals. The PCP accumulation in tissues and the concentration of dietary PCP were linearly related. Kidney accumulated the highest levels of PCP followed by liver, heart, leg, breast, gizzard and fat. A five-week withdrawal of PCP from the diet revealed that PCP residues were still detectable in the adipose tissue of all birds apart from the control ones.

4.2.9 Sensory Analysis and Microbiological Methods as Tools for Poultry Authentication

In the study of Chartrin et al. (2006), an experiment was performed towards the evaluation of the effects of intramuscular fat levels on the sensory characteristics of duck breast meat. By combining duck genotypes (Muscovy, Pekin, and their crossbreed hinny and mule ducks) and feeding levels (overfeeding between 12 and 14 weeks of age vs. ad libitum feeding), various lipid levels were found in the samples. The average values ranged from 2.55 to 6.40 g/100 g of muscle. Breast samples from overfed ducks were characterised by higher lipid content and lower water content in comparison to breast muscle derived from ducks fed ad libitum. Also, the muscle derived from overfed ducks was found to be paler in colour and was characterised by higher yellowness and cooking loss values. Juiciness values were lower, while the flavour was more pronounced in overfed ducks. Muscovy ducks had a higher breast weight and fewer lipids than the other genotypes. However, Pekin ducks were characterised by the highest lipid content and the lowest breast weights. It was shown that the breast muscle of Muscovy ducks was paler, less red, and more yellow in comparison to the muscle of the other genotypes. The breast muscle of Pekin ducks had the lowest degree of lightness, yellowness, and required the lowest energy demanded for shearing meat. It also had the highest cooking loss values, and was found to be tenderer, juicier and less stringy compared to other genotypes. On the other hand, Muscovy ducks were less tender, juicy and flavourful, and they exhibited the highest degree of stringiness. Breast muscle of hinny and mule ducks exhibited the highest degree of redness. It was proved that the genotype affected sensory quality in a higher degree than feeding levels.

Consumers nowadays show a high interest in products not subjected to irradiation. Therefore, methods for detecting irradiated products and differentiating them from non-irradiated could be valuable tools for protecting the due diligence of manufacturers and retailers, as well as for guaranteeing consumer satisfaction. Hashim et al. (1995) examined how irradiation affects the organoleptic characteristics of refrigerated and frozen chicken samples. Specifically, skinless and boneless breasts and legs were used. It was proved that the treatment had no significant effects on moistness and glossiness of the samples. Leg quarters subjected to irradiation while refrigerated obtained a darker colour ($p \leq 0.05$) in comparison to non-irradiated samples. Raw irradiated chicken was characterised by higher "fresh chickeny," bloody, and sweet aromatic intensities in comparison to non-treated samples. Cooked irradiated frozen legs exhibited a more "chicken" flavour, while cooked irradiated refrigerated legs were tenderer than controls. No other effects on organoleptic characteristics were detected.

It was also shown that the state at which chicken is being irradiated (refrigerated or frozen) does not have any effects on the sensory characteristics.

Liu et al. (2003) examined the effects of irradiation on structural changes and colour characteristics of chicken breasts as well as how the subsequent storage conditions affect some key characteristics of the product. For this purpose, visible spectroscopy and HunterLab measurement were used. Ratios of $R_1 = A_{485nm}/A_{560nm}$ and $R_2 = A_{635nm}/A_{560nm}$, connected to absorbances of the visible bands at 485 nm (metmyoglobin), 560 nm (oxymyoglobin), and 635 nm (sulfmyoglobin), indicated that the relative amount of oxymyoglobin is either enhanced under the effect of irradiation, or decreases with the storage process. The plot of R_1 and R_2 versus storage time indicated that the increments of both R_1 and R_2 strongly depend on the dose of the irradiation and also that the relative amount of oxymyoglobin in products treated with irradiation begins to decompose 7–12 days later than non-treated products. Furthermore, R_1 and R_2 values were in good correlation with colour index E^* of chicken breasts.

Descriptive sensory profiling was performed for the evaluation of the effect of oven-cooking temperature (160, 170, 180, 190°C) on warmed-over flavour (WOF) generation in cooked, chill-stored (at 4°C for 0, 1, 2 and 4 days) and reheated chicken patties, derived from *M. pectoralis* major. Furthermore, GC-MS was employed for the examination of samples stored at 4°C for 0, 1 and 4 days. Multivariate ANOVA-Partial Least Squares Regression (APLSR) was used for analyzing the effects of cooking and WOF on the organoleptic and chemical characteristics of the samples. The use of descriptive profiling showed that WOF development was characterised by an increase of 'rancid' and 'sulphur/rubber' sensory notes and a concurrent decrease of chicken 'meaty' characteristics. An increase of cooking temperature made samples more 'roasted', 'toasted' and 'bitter'. Furthermore, the 'roasted' character was connected to WOF development. By analyzing the volatile compounds from the chicken patties it was demonstrated that lipid oxidation derived substances rapidly developed with chill-storage. These substances are more related to the 'rancid' part of WOF development. In addition, changes in sulphur-containing substances were also connected with WOF development and they probably participate in the lipid oxidation reactions. The sensory effects of these substances are related with the 'sulphur/rubber' aspect of the WOF development. The study suggested that cooking temperature enhanced the development of Maillard-derived substances, but the latter did not cause inhibition of WOF development in the chicken patties (Byrne et al., 2002).

According to Chon et al. (2012), a comparison of the effectiveness of two types of Bolton broths and three selective media used for the isolation of *Campylobacter* spp. from naturally contaminated whole-chicken carcass-rinse samples, was successfully performed. One hundred chickens were rinsed with buffered peptone water and the rinses were added to $2 \times$ Bolton broth. Following incubation, the samples were streaked onto Preston agar, modified cefoperazone charcoal deoxycholate agar (mCCDA), and Campy-Cefex agar and then incubated under microaerobic conditions. It was shown that the isolation rates and selectivity between the two types of Bolton broths were not significantly different ($P > 0.05$). Among the three selective agars, Preston agar was characterised by a significantly ($P < 0.05$) higher isolation rate and selectivity. The Campy-Cefex agar, which has been proposed by food authorities for its high quantitative detection ability, presented high contamination levels with competing microorganisms and was finally characterised by the lowest isolation and selectivity degrees.

Poultry is usually contaminated with Campylobacter, one of the major causal agents of foodborne illness in the United States. However, the route of contamination is still unclear. It is believed that broiler litter may be a potential source, but an effective method for the quantification of Campylobacter from litter has not yet been established. As a result, a method was developed for the evaluation of the sensitivity of certain media used for the quantification of Campylobacter for the assessment of the hypothesis that enrichment can facilitate the detection of stressed or viable but nonculturable cells from broiler litter samples. Specifically, five media (campy-Line agar (CLA), campy-cefex agar (CCA), modified CCA, Campylobacter agar plates (CAP), and modified charcoal cefoperazone deoxycholate agar) and two culturing techniques were employed for the quantification of Campylobacter from broiler litter. Each litter sample was equally divided and diluted by a 10-fold into peptone, for direct plating, or by 4-fold into Campylobacter enrichment broth. Dilution of samples in peptone was followed by direct-plating onto each media and incubation under microaerophilic conditions (for 48 hours at 42°C). Dilution of samples in enrichment broth was followed by incubation under the same conditions for 24 hours, and an additional 10-fold dilution before plating. Incubation of plates from enriched samples was performed for a 24 hours period after plating. All plates (direct and enriched) were subsequently counted and presumptive positive colonies were confirmed using a Campylobacter latex agglutination kit. It was proved that all selective media had the same sensitivity. Direct-plated samples had a higher Campylobacter isolation rate in comparison to enriched samples. The CLA and CAP suppressed total bacterial growth more effectively than modified charcoal cefoperazone deoxycholate, modified CCA, and CCA. The CLA and CAP were the only media capable of detecting total bacterial population shifts over time. It is therefore essential to always take into account the medium's potential to suppress total bacterial growth as well as isolate Campylobacter (Kiess et al., 2010).

4.2.10 Authentication of Geographic Origin and Use of Multivariate Analysis

In the study of Franke et al. (2008), data available on contents of up to 72 different trace elements and the oxygen isotope ratio of 78 poultry breast and 74 dried beef samples were subjected to analysis with the aim of determining whether or not the geographic origin can be predicted more accurately when more than one methods are combined. The method was validated through determination of the origin of a smaller sub-group by employing a statistical model generated from the data of the second, larger, sub-group. It was proved that by combining these data, the geographic origin of the samples was more effectively determined. Nevertheless, the combination of the data did not significantly limit the percentage of incorrect classifications of individual samples in comparison to the two methods applied separately. In poultry, cross-validation and validation led to 83% and 50% correct classifications, respectively, while for dried beef, the same values were 73% and 43%, respectively. It was therefore proved that the combination of the methods did not improve the predictions of origin.

Another study was carried out with the aim of investigating the effectiveness of multi-element and multi-isotopic analysis, together with statistical processing of the extracted data, on the determination of the geographical origin of poultry and thus the effective verification of poultry-origin labels. Cross-validation discriminant analysis was used, correctly classifying the geographical origins of 88.3% of the samples. The individual correct classification rates were as follows: China, 100% (n = 36); Brazil, 94.1% (n = 101); Europe 92% (n = 87); Chile, 82.6% (n = 46); Thailand, 70.3% (n = 46) and Argentina 50% (n = 10). The main identification errors were related to miss-classified Argentinean samples (with those originating from Chile and Thailand). Carbon stable isotope ratios of chicken samples offer an indication of the quantity of maize in the diet and is therefore a valuable tool that can be used for discriminating between European poultry and poultries from South America, Thailand and China (where maize is the main feeding ingredient). It was proved that the stable isotopes of hydrogen and oxygen in chicken meat are positively correlated with the surface waters around the globe and was therefore suggested that the global isotopic variation of stable isotopes in drinking water and feed reflects into the animal tissue and can therefore be used for the determination of the geographic origin of the products. Furthermore, isotopically depleted waters are related to cold regions, while enriched waters are found in warmer regions. The patterns can be detected in poultry. These variations can be used to give a first indication of the geographical origin of the product (Kelly, 2007).

According to Cruz et al. (2012), very few studies have been carried out with the aim of detecting animal byproducts in poultry meat. By developing a technique of stable isotopes, a reassurance of a strong traceability system and the certification of broiler diet patterns can be achieved. Thus knowledge of the behaviour of the isotopic signature of different tissues in birds will be used to properly replace (if needed) a diet containing animal-derived ingredients with a strictly vegetable one and vice versa. As a result, this study evaluated meat from the breast, thigh, drumstick and wings with the aim of tracing the existence of poultry offal meat in broiler feed. Stable isotopes of carbon ($^{13}C/^{12}C$) and nitrogen ($^{15}N/^{14}N$) were used and the analysis was performed using MS. The experiment was performed by separating 720 chickens into six groups: chickens fed with vegetable diet from 1 to 42 days of age; 8% poultry offal meal diet from 1 to 42 days of age; vegetable diet from 1 to 21 days and 8% offal meal diet from 22 to 42 days; vegetable diet from 1 to 35 days and 8% offal meal diet from 36 to 42 days; 8% offal meal diet from 1 to 21 days and vegetable diet from 22 to 42 days; and 8% offal meal diet from 1 to 35 days and vegetable diet from 36 to 42 days of age. The results of the carbon and nitrogen isotopic analysis were analyzed using a multivariate ANOVA and GLM of the SAS statistical program. Through analyzing C and N, it was possible to trace the use of offal meal in broiler feeding when part of the feeding or when used as a substitute of a strictly vegetable diet even up to 35 days from the start of feeding. The substitution of an offal meal diet by a vegetable diet, results in the detection of the animal ingredient only if contained in the feed for 21 days or longer prior to the analysis (Cruz et al., 2012).

It is also important to mention the study of Chao et al. (2007), who described the development of a hyperspectral-multispectral line-scan imaging system that can be used for differentiating wholesome and systemically diseased chickens. In-plant testing was performed on a commercial evisceration line moving at a speed of

70 birds per minute. Hyperspectral image data was obtained for a calibration data set of 543 wholesome and 64 systemically diseased birds and for a testing data set of 381 wholesome and 100 systemically diseased birds. The calibration data set was used for the development of the parameters of the imaging system in order to conduct multispectral inspection based on fuzzy logic detection algorithms, using selected key wavelengths. By using a threshold of 0.4 for fuzzy output decision values, multispectral classification achieved 90.6% accuracy for wholesome birds and 93.8% accuracy for systemically diseased birds in the calibration data set and 97.6% accuracy for wholesome birds and 96.0% accuracy for systemically diseased birds in the testing data set. Following adjustment of the classification threshold, 100% accuracy was obtained for systemically diseased birds and 88.7% accuracy for wholesome birds. This line-scan imaging system is effective for the direct implementation of multispectral classification methods developed from hyperspectral image analysis.

Moreover, the development of a method for the differentiation of normal chickens from chickens with septicemia/toxemia (septox) was reported. The method was based on the use of machine inspection (Hazard Analysis and Critical Control Point-Based Inspection Models Project). Spectral measurements of 300 chicken livers (50% normal and 50% condemned due to septox) were obtained and subjected to analysis. Neural network classification of the spectral data after principal component analysis showed that correct differentiation of normal and septox livers were correctly differentiated by spectroscopy (96% correct classification). Analysis of the data indicated 100% correlation between the spectroscopic differentiation and the subset of samples that were histopathologically examined. In an effort to determine the main causal agent of the contamination, isolates from 30 livers were examined indicating that contamination of the poultry carcasses occurred mainly by coliforms present in the environment, hindering the isolation of pathogenic microorganisms. Thus, to establish the cause of diseased livers, strictly aseptic conditions and procedures must be applied for the collection of samples (Dey et al., 2003). Also, Dickens et al. (1986) described a very accurate method of whole bird rinsing for microbiological sampling of processed turkey carcasses. A unit for sampling two turkey carcasses at the same time was developed, fabricated, and microbiological comparison was made with standard manual rinsing procedures. No significant differences in total aerobic plate counts or Enterobacteriaceae were detected, while the Salmonella recoveries of the two procedures were identical.

4.3 Authenticity of Eggs and Egg Products

Several authors have examined the effectiveness of different techniques for the authentication of eggs and egg-derived products. The techniques employed and the results of the most important studies in this field are presented below. Some representative methods/techniques developed for detecting adulterants in egg and egg products are also presented in Table 2.

According to Herman (2004), species-specific primers for duck were deduced from the mitochondrial ATPase8 gene sequence. Species-specific PCR for turkeys and ducks did not cross react with mixtures from chicken and guinea eggs, and were detected

Table 2. Methods and techniques reported for detecting adulterants eggs and egg products.

Products	Treatment/storage conditions	Adulterant(s)/ pathogens to be detected	QC method	Effectiveness	References
Eggs	Aqueous extraction, purification, and acid partitioning. LC detection at 313 nm	Ipronidazole, ronidazole, dimetridazole	Liquid chromatographic multicomponent	High sensitivity, very effective	Aerts et al., 1991
Eggs	Extraction, application to silica gel cartridges, elution of adulterant, evaporation, dissolution, washing	Dimetridazole	Liquid chromatography-thermospray mass spectrometry.	Suitable for statutory residue testing	Cannavan and Kennedy, 1997
Eggs	Extraction, purification, evaporation, methanol-acetonitrile-water	Nicarbazin	Micro high-performance liquid chromatography (HPLC)	Sensitive, selective, rapid, cost-effective	Draisci et al., 1995
Poultry eggs	Extraction, incubation, wash-up	Halofuginone	Time-resolved fluorometry	Effective for both qualitative and quantitative measurements	Hagren et al., 2005
Eggs	Extraction, de-fattening, washing	Nitroimidazoles	ELISA	Very sensitive	Huet et al., 2005
Egg yolks	Raw	Ampicillin, oxytetrazcycline	Agar diffusion microbiological method	Effective for modelling of residue levels in eggs	Donoghue et al., 1996
Egg contents	Initial incubation (72 hours, 25°C)	*Salmonella enteritidis*	Fluorescence polarization and lateral immunodiffusion	Levels of > 10^7 cfu/ml were effectively detected	Gast et al., 2003
Serum and egg yolks	Incubated for 24 h and testing for antigens	*Salmonella enterica* serovar *enteritidis*	Agar gel precipitin test	Sensitivity slightly lower than standard microagglutination assay/specificity slightly higher, cost-effective, simple	Holt et al., 2000
Eggs	Extraction by acetonitrile and de-fattening by washing with hexane	Nitroimidazoles	cELISA	Detection limits: dimetridazole, < 1 ppb	Huet et al., 2005

when concentrations of 0.1% of homogenized duck egg and 5% of homogenized turkey egg were subjected to PCR analysis of 35 cycles. A PCR of 30 cycles led to detection of 10% homogenized duck egg. The same sensitivities were found in dilutions of homogenized egg yolk, but no PCR signals were found for egg white.

In an effort to improve the texture of "foie gras", the hen egg-white adulterant was employed. It was shown that duck or goose "foie gras" could be detected using anti-hen-egg-white lysozyme (HEWL). Similarly, anti-HEWL mAbs detected chicken or turkey fresh liver in duck or goose "foie gras" even after heating the sample at 80°C. Under 110°C, only mAbs HyHEL (hybridoma antihen-egg-white lysozyme) 5 and HyHEL 10 effectively detected fresh hen livers (Hemmen et al., 1993).

Two panels of mAbs raised against duck (Barbary) egg white lysozyme or hen egg white lysozyme, were tested in antigen-coated plate (ACP) and using double antibody sandwich (DAS) ELISA for detecting cross-reactivity with avian lysozymes. Cross-reaction was observed between the antibodies to hen lysozyme and goose lysozyme, but the majority of antibodies to duck lysozyme reacted with it. It was also shown that one antibody to duck lysozyme reacted more strongly with goose lysozyme than with the homologous antigen. Numerous lysozyme epitopes detected by different antibodies were highly resistant to heat denaturation. Such mAbs could be effectively used for the detection of chicken liver adulterants in "foies gras" of goose or duck origin (Saunal et al., 1995).

The use of an antibody that can bind MNZ was reported by Huet et al. (2005). A cELISA was used to characterise the binding of this antibody to several nitroimidazole drugs. Cross-reactivity with dimetridazole (DMZ), ronidazole (RNZ), hydroxydimetridazole (DMZOH), and ipronidazole (IPZ) was observed. The extraction of egg samples was performed using acetonitrile and the samples were de-fattened by washing them with hexane. The detection capabilities (CCβ) were as follows: dimetridazole, < 1 ppb; metronidazole, < 10 ppb; ronidazole and hydroxydimetridazole, < 20 ppb; and ipronidazole, <40 ppb.

Tewfik (2008) examined the qualitative performance of a rapid solvent extraction (DSE) method that could be used for the differentiation between irradiated and non-irradiated food samples. The trial was carried out between four European laboratories that used DCB as radiolytic marker for detecting irradiated liquid whole egg. The egg samples were treated with 3 and 6 kGys. All laboratories performed the experiments using 12-blind coded 'unknown' and 4-known coded samples and all blind coded samples were correctly identified. Therefore, this DSE method was proved to be fast and cheap, and could be effectively used in laboratories involved with screening high number of food samples for detecting irradiation treatment.

Hagren et al. (2005) developed and validated an immunoassay for screening coccidiostat halofuginone in poultry eggs. Time-resolved fluorometry was used and an all-in-one dry chemistry assay concept was followed based on the incorporation of all reagents into a single microtiter well in a dry, stable form. This immunoassay is carried out by an automated immunoanalyser and the whole procedure lasts about 18 min. The assay protocol includes extraction of the sample and addition to a well. After incubation for 15 min, the sample is washed and measurement of the fluorescence signal is taken directly from the surface of the dry well in a time-resolved manner. The analytical limit of detection was 0.02 ng ml^{-1} (n = 12) and the functional limit of detection was 1.7 and 1.0 ng g^{-1} (n = 6). The sensitivity of the assay can increase through adjusting the dilution factor of the samples. The mean recovery was 90.5% at concentration levels of 15 and 60 ng g^{-1}. Also, according to Gast et al. (2003), effective detection of *Salmonella enteritidis* inside eggs is of paramount importance for the protection of the public health. In the majority of the standard bacteriological culturing methods, an initial incubation step is required to allow the small numbers of *S. enteritidis* cells to multiply in order to be readily detectable. Thus two rapid methods were developed and assessed as alternatives to plating on selective media for the detection of *S. enteritidis* in incubated egg pools. Fluorescence polarization and lateral flow immunodiffusion assays were separately employed to detect *S. enteritidis* in egg pools. Although the rapid assays were significantly less sensitive than culturing, they both effectively detected contamination when pools of 10 eggs were inoculated with approximately 10 cfu of *S. enteritidis* (incubation for 72 h at 25°C).

The determination of tetracycline antibiotics in eggs was carried out using HPLC with ion-pairing chromatography and diodearray detection at 355 nm. The mean recovery values for oxytetracycline from eggs was 76%. The within-day precision was about 8.0 to 11.8%. The between-day precision was 4.8%. The limit of detection and the limit of quantitation were determined at 2.2 and 13.0 ng/g respectively. The determination of residue values of tetracycline antibiotics in eggs was performed after oral administration of medicated feed. Feed enriched with 840 mg/kg oxytetracycline was given to laying hens for one week. Two days after the end of the administration, the mean oxytetracycline residue value in the eggs was lower than the Maximum Residue Limit-level reaching at 118 ng/g (De Ruyck et al., 1999).

Aerts et al. (1991) developed a simple, rapid liquid chromatographic method combined with UV/VIS for determining residues of the histomonostats dimetridazole (DMZ), ronidazole (RON), ipronidazole (IPR), and side-chain hydroxylated metabolites of DMZ and RON in eggs. The samples were pre-treated following an aqueous extraction, purification with an Extrelut cartridge, and acid partitioning with isooctane. Injection of an aliquot of the final aqueous extract was performed into a reverse-phase LC system; detecting compounds at 313 nm. The limits of determination were found to be at 5–10 µg/kg range. A UV/VIS spectrum was yielded at the 10 µg/kg level by employing diode-array UV/VIS detection. The recovery ranged from 80 to 98% with a coefficient of variation of approximately 5%. The detection of a side-chain hydroxylated metabolite of IPR was also performed using the same method. Residues of a single oral dose of the drugs to laying hens could be effectively detected in eggs 5–8 days after dosing. The establishment of plasma distribution and excretion in faeces was performed with and without deconjugation. DMZ and IPR were highly metabolized to hydroxylated nitroimidazole metabolites, while RON did not metabolize in a great extent.

Cannavan and Kennedy (1997) developed a method for determining levels of the nitroimidazole drug DMZ in poultry eggs using LC-MS. Deuteriated DMZ was applied as an internal standard. Internal standard corrected recoveries ranged from 1.2 to 7.7% for eggs spiked at 5 ng g^{-1}. The absolute recoveries were as high as 80%. The method could be effectively used for measuring DMZ residues in eggs from chickens fed a diet containing DMZ (Cannavan and Kennedy, 1997). Also, the development of a micro HPLC method was reported by Draisci et al. (1995) for determining the anti-coccidial drug nicarbazin in chicken eggs. The extraction of the 4,4'-dinitrocarbanilide (DNC) component of nicarbazin was performed using acetonitrile. The extracts were purified using liquid-liquid partitioning, evaporation to dryness and finally by taking them up in methanol-acetonitrile-water (50:30:20, v/v). Micro HPLC of the 4,4'-dinitrocarbanilide (DNC) portion of nicarbazin was performed by using a small bore column (1 mm I.D.) packed with reversed-phase and a UV detector set at 340 nm. The average recovery of nicarbazin added to eggs was 85.2%. The limit of detection was 25 pg. This method is very sensitive and cost-effective.

In the study of Brown et al. (1986), pasteurized, liquid egg products were allowed to deteriorate over a period of days. Evaluation of the products was periodically carried out for odour. The volatile components were analyzed using combined purge and trap gas chromatography. Identification of peaks was achieved using MS. It was shown that the first detection of unacceptable odour in whole egg, albumen, or yolk samples was accompanied by the presence of increased levels of dimethyl sulphide. The concentrations of dimethyl disulphide, dimethyl trisulphide, and ethanol further increased during the deterioration period.

In another study, the effectiveness of an automatic black light egg inspection unit was evaluated in terms of its performance, in combination with an automatic candler-type egg inspection unit, using fresh and stored eggs. The combined assessment of the black light unit with the candler-type unit led to the rejection of the 86.2% of heavily contaminated eggs. The sensory analysis performed at the same time led to rejection of 61.5% of the heavily contaminated eggs. It was proved that the rejection of eggs contaminated with Pseudomonas using the sensorial evaluation, frequently occurring when stored at 23 and at 13°C, was not as effective as the evaluation using the combined assessment. The combined assessment, however, did not lead to rejection of eggs contaminated with certain Enterobacteriaceae, commonly detected in eggs stored at 23°C. None of these methods though could detect Alcaligenes, the growth of which did not alter the appearance or odour of the egg contents. In heavily contaminated eggs, the number of Pseudomonas-contaminated eggs (63) was larger than that of Enterobacteriaceae-contaminated eggs (28). It was thus shown that the combined assessment was more accurate than the sensory tests (Imai and Saito, 1985).

Also, in the study of Donoghue et al. (1996), modelling of the pattern of antibiotic drug uptake within yolks of developing follicles was achieved. According to the authors of this study: "Sixteen hens were divided into equal groups (n = 8) and injected only once with either 400 mg/kg ampicillin or 200 mg/kg oxytetracycline (OTC: total hens = 32) approximately 1 h after oviposition. Twenty-four hours following injections, hens were euthanatized and the ovaries were collected. Yolks were dissected free from the individual follicles with a blunt probe. Individual large yellow yolks (≥ 0.2 g) and a pool of five small yellow yolks (< 0.2 g) were collected for determination of ampicillin or OTC content. Samples were prepared and assayed using an agar diffusion microbiological method. Selected parameters were not different ($P > 0.05$) between Experiments 1 and 2 and the data were combined. Results indicate that short-term drug exposure in hens produced incorporation of drug residues in developing yolks in a specific pattern that does not appear to be drug dependent ($P > 0.05$). These incurred residues are contained in developing yolks that are days to weeks from being ovulated. Drug residues were greater (total microgram content) in some of the less mature yolks vs the largest preovulatory yolk. This may lead to a sequential release of eggs with increasing residue content, even after drug withdrawal. These data were used to construct a model to predict the pattern of incurred residues in formed eggs following a hen's exposure to drugs or other contaminants."

Finally, an agar gel precipitin (AGP) test was employed by Holt et al. (2000) for the detection of antibodies to *S. enteritidis* deposited in egg yolks of infected hens. Yolk or sera from infected birds were administered to wells cut into seven-well clusters in an agar gel plate, and addition of antigens to the centre well was performed. The agar gels were incubated for 24 h and then tested for the presence of precipitin lines formed by the interaction of antibody with antigen. Testing was performed for the following antigens: *S. enteritidis flagella*, SEF14 and a sodium deoxycholate extract of whole *S. enteritidis*. Flagella and *S. enteritidis* extract detected antibodies to *S. enteritidis* in the yolk and sera, while SEF14 was not reactive. According to the authors: "Positive reactions were observed in serum 1 week post-challenge, whereas in yolks, this was further delayed by 1 week. The sensitivity of the test was slightly less than the standard microagglutination assay, although specificity was slightly higher, as indicated by results from sera and yolks from birds infected with *Salmonella enterica serovar Typhimurium*. Simplicity and low labour requirements of the assay would allow for the potential testing of several hundred egg samples within a day, which would make up for test shortcomings due to sensitivity. The AGP test could be an important tool for individuals using serological testing to monitor the *S. enteritidis* situation within their flocks or as a rapid screen for vaccine responses. The assay could also be used in tandem with other AGP tests to screen for the presence of multiple avian pathogens."

4.4 Conclusions

Poultry authenticity is a major field of research nowadays, mainly due to the increased consumer awareness in relation to the origin, safety and quality of the foodstuffs. The industrial field of poultry production as well as the poultry processing industry have recently faced an increased law stringency in relation to the detailed analysis, authentication and labelling of these products. Towards this fact, hundreds of studies have been or are currently carried out with the aim of differentiating between different poultries, or for detecting adulteration of poultries by different animal species or chemical adulterants (antibiotics, bactericides, etc.). Considering the huge range of poultry production methods and the quantities produced and processed around the world, it is obvious that the development of cost-effective, rapid and easily applicable techniques for assessing all the factors that may affect poultry authenticity is a huge task. Extensive research is also required for the improvement of many of the existent techniques and their gradual systematic adoption by the industry as well as national and international regulations.

References

Abeni, F. and G. Bergoglio. 2001. Characterization of different strains of broiler chicken by carcass measurements, chemical and physical parameters and NIRS on breast muscle. Meat Science. 57: 133–137.

Aerts, R.M., I.M. Egberink, C.A. Kan, H.J. Keukens and W.M. Beek. 1991. Liquid chromatographic multicomponent method for determination of residues of ipronidazole, ronidazole, and dimetridazole and some relevant metabolites in eggs, plasma, and feces and its use in depletion studies in laying hens. Association of Official Analytical Chemists Journal. 74: 46–55.

Alonso, M.Z., N.L. Padola, A.E. Parma and P.M.A. Lucchesi. 2011. Enteropathogenic *Escherichia coli* contamination at different stages of the chicken slaughtering process. Poultry Science. 90(11): 2638–2641.

Anderson, A., E. Goodall, G.W. Bliss and R.N. Woodhouse. 1981. Analysis of the anti-coccidial drug, halofuginone, in chicken tissue and chicken feed using high-performance liquid chromatography. Journal of Chromatography. 212: 347–355.

Andrée, S., K. Altmann, R. Binke and F. Schawägele. 2004. Animal species identification and quantification in meat and meat products by means of traditional and real-time PCR. Fleischwirtschaft. 85 (1): 96–99.

Awad, A.M., E.-H.H.S. Abd, R.A.A. Abou and H.H. Ibrahim. 2010. Detection of reticuloendotheliosis virus as a contaminant of fowl pox vaccines. Poultry Science. 89: 2389–2395.

Baéza, E. 2004. Measuring quality parameters. pp. 304–331. *In*: Poultry meat processing and quality. Woodhead Publishing in Food Science and Technology, USA.

Balamatsia, C.C., A. Patsias, M.G. Kontominas and I.N. Savvaidis.2007. Possible role of volatile amines as quality-indicating metabolites in modified atmosphere-packaged chicken fillets: Correlation with microbiological and sensory attributes. Food Chemistry. 104: 1622–1628.

Beier, R.C., T.J. Dutko, S.A. Buckley, M.T. Muldoon, C.K. Holtzapple and L.H. Stanker. 1998. Detection of Halofuginone Residues in Chicken Liver Tissue by HPLC and a Monoclonal-Based Immunoassay. Journal of Agricultural and Food Chemistry. 46: 1049–1054.

Barbut, S. 1996. Estimates and detection of the PSE problem in young turkey breast meat. Canadian Journal of Animal Science. 76: 455–457.

Berger, R.G., R.P. Mageau, B. Schwab and R.W. Johnston. 1988. Detection of poultry and pork in cooked and canned meat foods by enzyme-linked immunosorbent assays. Journal of the Association of Official Analytical Chemists. 71: 406–409.

Berzaghi, P., A.D. Zotte, L.M. Jansson and I. Andrighetto. 2005. Near-Infrared Reflectance Spectroscopy as a Method to Predict Chemical Composition of Breast Meat and Discriminate Between Different n-3 Feeding Sources. Poultry Science Association, Inc. 128–136.

Bonnefoi, M., G. Benard and C. Labie. 1986. Gel Electrophoresis: A Qualitative Method for Detection of Duck and Goose Liver in Canned Foie Gras, Journal of Food Science. 51: 1362–1363.

Bowker, B.C. and H. Zhuang. 2013. Relationship between muscle exudate protein composition and broiler breast meat quality. Poultry Science. 92: 1385–1392.

Brown, M.L., D.M. Holbrook, E.F. Hoerning, M.G. Legendre and A.J. St. Angelo. 1986. Volatile Indicators of Deterioration in Liquid Egg Products. Poultry Science. 65(10): 1925–1933.

Er, B., F.K. Onurdag, B. Demirhan, S.Ö. Ozgacar, A.B. Oktem and U. Abbasoglu. 2013. Screening of quinolone antibiotic residues in chicken meat and beef sold in the markets of Ankara, Turkey. Poultry Science. 92(8): 2212–2215.

Byrne, D.V., W.L.P. Bredie, D.S. Mottram and M. Martens. 2002. Sensory and chemical investigations on the effect of oven cooking on warmed-over flavour development in chicken meat. Meat Science. 61: 127–139.

Calvo, J.H., P. Zaragoza and R. Osta. 2001. Random Amplified Polymorphic DNA Fingerprints for Identification of Species in Poultry Pâté. Poultry Science. 80: 552–524.

Cannavan, A. and D.G. Kennedy. 1997. Determination of dimetridazole in poultry tissues and eggs using liquid chromatography-thermospray mass spectrometry. Analyst. 122: 963–966.

Chao, K., C.C. Yang, Y.R. Chen, M.S. Kim and D.E. Chan. 2007. Hyperspectral-Multispectral Line-Scan Imaging System for Automated Poultry Carcass Inspection Applications for Food Safety. Poultry Science. 86(11): 2450–2460.

Chartrin, P., K. Méteau, H. Juin, M.D. Bernadet, G. Guy, C. Larzul, H. Rémignon, J. Mourot, M.J. Duclos and E. Baéza. 2006. Effects of Intramuscular Fat Levels on Sensory Characteristics of Duck Breast Meat. Poultry Science. 85: 914–922.

Chen, H. and B.P. Marks. 1997. Evaluating Previous Thermal Treatment of Chicken Patties by Visible/ Near-Infrared Spectroscopy. Journal of Food Science. 62: 753–756.

Chen, Y.R. and D.R. Massie. 1993. Visible/Near-Infrared Reflectance and Interectance Spectroscopy for Detection of Abnormal Poultry Carcasses. Transactions of the ASAE. 36: 863–869.

Chen, G. and M.J. Schneider. 2003. A Rapid Spectrofluorometric Screening Method for Enrofloxacin in Chicken Muscle. Journal of Agricultural and Food Chemistry. 51: 3249–3253.

Chiaochan, C., U. Koesukwiwat, S. Yudthavorasit and N. Leepipatpiboon. 2010. Efficient hydrophilic interaction liquid chromatography–tandem mass spectrometry for the multiclass analysis of veterinary drugs in chicken muscle. Analytica Chimica Acta. 682: 117–129.

Chisholm, J., A. Sánchez, J. Brown and H. Hird. 2008. The Development of Species-specific Real-time PCR Assays for the Detection of Pheasant and Quail in Food. Food Analytical Methods. 1: 190–194.

Chon, J.W., J.Y. Hyeon, J.H. Park, K.Y. Song and K.H. Seo. 2012. Comparison of 2 types of broths and 3 selective agars for the detection of Campylobacter species in whole-chicken carcass-rinse samples. Poultry Science. 91(9): 2382–2385.

Codony, R., R. Compañó, M. Granados, J.A. Garcia-Regueiro and M.D. Prat. 2002. Residue analysis of macrolides in poultry muscle by liquid chromatography-electrospray mass spectrometry. Journal of Chromatography A. 959: 131–141.

Colombo, F., R. Viacava and M. Giaretti. 2000. Differentiation of the species ostrich (Struthio camelus) and emu (Dromaius novaehollandiae) by polymerase chain reaction using an ostrich-specific primer pair. Meat Science. 56: 15–17.

Colombo, F., E. Marchisio, A. Pizzini and C. Cantoni. 2002. Identification of the goose species (Anser anser) in Italian "Mortara" salami by DNA sequencing and a Polymerase Chain Reaction with an original primer pair. Meat Science. 61: 291–294.

Cronly, M., P. Behan, B. Foley, E. Malone, S. Earley, M. Gallagher, P. Shearan and L. Regan. 2010. Development and validation of a rapid multi-class method for the confirmation of fourteen prohibited medicinal additives in pig and poultry compound feed by liquid chromatography–tandem mass spectrometry. Journal of Pharmaceutical and Biomedical Analysis. 53: 929–938.

Crooks, S.R., T.L. Fodey, G.R. Gilmore and C.T. Elliott. 1998. Rapid screening for monensin residues in poultry plasma by a dry reagent dissociation enhanced lanthanide fluoroimmunoassay. Analyst. 123: 2493–2496.

Cruz, V.C., P.C. Araújo, J.R. Sartori, A.C. Pezzato, J.C. Denadai, G.V. Polycarpo, L.H. Zanetti and C. Ducatti. 2012. Poultry offal meal in chicken: Traceability using the technique of carbon ($^{13}C/^{12}C$)- and nitrogen ($^{15}N/^{14}N$)-stable isotopes. Poultry Science. 91: 478–486.

De Ruyck, H., H. De Ridder, R. Van Renterghem and F. Van Wambeke. 1999. Validation of HPLC method of analysis of tetracycline residues in eggs and broiler meat and its application to a feeding trial. Food Additives and Contaminants. 16: 47–56.

De Marchi, M., M. Penasa, M. Battagin, E. Zanetti, C. Pulici and M. Cassandro. 2011. Feasibility of the direct application of near-infrared reflectance spectroscopy on intact chicken breasts to predict meat color and physical traits. Poultry Science. 90(7): 1594–1599.

Dey, B.P., Y.R. Chen, C. Hsieh and D.E. Chan. 2003. Detection of septicemia in chicken livers by spectroscopy. Poultry Science. 82(2): 199–206.

De Wasch, K., L. Okerman, S. Croubels, H. De Brabander, J. Van Hoof and P. De Backer. 1998. Detection of residues of tetracycline antibiotics in pork and chicken meat: correlation between results of screening and confirmatory tests. Analyst. 123: 2737–2741.

Dickens, J.A., N.A. Cox and J.S. Bailey. 1986. Evaluation of a Mechanical Shaker for Microbiological Rinse Sampling of Turkey Carcasses. Poultry Science. 65(6): 1100–1102.

Ding, H., R.-J. Xu and D.K.O. Chan. 1999. Identification of broiler chicken meat using a visible/near-infrared spectroscopic technique. Journal of the Science of Food and Agriculture. 79: 1282–1388.

Dooley, J.J., K.E. Paine, S.D. Garrett and H.M. Brown. 2004. Detection of meat species using TaqMan real-time PCR assays. Meat Science. 68: 431–438.

Donoghue, D.J., H. Hairston, S.A. Gaines, M.J. Bartholomew and A.M. Sonoghue. 1996. Modeling Residue Uptake by Eggs. 1. Similar Drug Residue Patterns in Developing Yolks Following Injection with Ampicillin or Oxytetracycline. Poultry Science. 75(3): 321–328.

Draisci, R., L. Lucentini, P. Boria and C. Lucarelli. 1995. Micro high-performance liquid chromatography for the determination of nicarbazin in chicken tissues, eggs, poultry feed and litter. Journal of Chromatography A. 697: 407–414.

Er, B., F.K. Onurdag, B. Demirhan, S.Ö. Ozgacar, A.B. Oktem and U. Abbasoglu. 2013. Screening of quinolone antibiotic residues in chicken meat and beef sold in the markets of Ankara, Turkey. Poultry Science. 92: 2212–2215.

Erel, Y., N. Yazici, S. Özvatan, D. Ercin and N. Cetinkaya. 2009. Detection of irradiated quail meat by using DNA comet assay and evaluation of comets by image analysis. Radiation Physics and Chemistry. 78: 776–781.

Ferguson, J., A. Baxter, P. Young, G. Kennedy, C. Elliott, S. Weigel, R. Gatermann, H. Ashwin, S. Stead and M. Sharman. 2005. Detection of chloramphenicol and chloramphenicol glucuronide residues in poultry muscle, honey, prawn and milk using a surface plasmon resonance biosensor and Qflex® kit chloramphenicol. Analytica Chimica Acta. 529: 109–113.

Franke, B.M., R. Hadorn, J.O. Bosset, G. Gremaud and M. Kreuzer. 2008. Is authentication of the geographic origin of poultry meat and dried beef improved by combining multiple trace element and oxygen isotope analysis? Meat Science. 80: 944–947.

Fumiére, O., M. Dubois, D. Grégoire, A. Théwis and G. Berben, . 2003. Identification on Commercialized Products of AFLP Markers Able To Discriminate Slow- from Fast-Growing Chicken Strains. Journal of Agricultural and Food Chemistry. 51: 1115–1119.

Gast, R.K., P.S. Holt, M.S. Nasir, M.E. Jolley and H.D. Stone. 2003. Detection of Salmonella enteritidis in incubated pools of egg contents by fluorescence polarization and lateral flow immunodiffusion. Poultry Science. 82(4): 687–690.

Girish, P.S., A.S.R. Anjaneyulu, K.N. Viswas, F.H. Santhosh, K.N. Bhilegaonkar, R.K. Agarwal, N. Kondaiah and K. Nagappa. 2007. Polymerase Chain Reaction–Restriction Fragment Length Polymorphism of Mitochondrial 12S rRNA Gene: A Simple Method for Identification of Poultry Meat Species. Veterinary Research Communications. 31: 447–455.

Haasnoot, W., H. Gerçek, G. Cazemier and M.W.F. Nielen. 2007. Biosensor immunoassay for flumequine in broiler serum and muscle. Analytica Chimica Acta. 586: 312–318.

Hagren, V., L. Connolly, C.T. Elliott, T. Lövgren and M. Tuomola. 2005. Rapid screening method for halofuginone residues in poultry eggs and liver using time-resolved fluorometry combined with the all-in-one dry chemistry assay concept. Analytica Chimica Acta. 529: 21–25.

Haider, N., I. Nabulsi and B. Al-Safadi. 2012. Identification of meat species by PCR-RFLP of the mitochondrial COI gene. Meat Science. 90: 490–493.

Hashim, I.B., A.V.A. Resurreccion and K.H. McWatters. 1995. Descriptive Sensory Analysis of Irradiated Frozen or Refrigerated Chicken. Journal of Food Science. 60: 664–666.

Haunshi, S., R. Basumatary, P.S. Girish, S. Doley, R.K. Bardoloi and A. Kumar. 2009. Identification of chicken, duck, pigeon and pig meat by species specific markers of mitochondrial origin. Meat Science. 83: 454–459.

Hemmen, F., A. Paraf and S. Smith-Gill. 1993. Lysozymes in Eggs and Plasma from Chicken, Duck and Goose: Choice and Use of mAbs to Detect Adulterants in «Foie Gras». Journal of Food Science. 58: 1291–1293.

Herman, L. 2004. Species Identification of Poultry Egg Products. Poultry Science. 83: 2083–2085.

Hird, H., R. Goodier and M. Hill. 2003. Rapid detection of chicken and turkey in heated meat products using the polymerase chain reaction followed by amplicon visualisation with vistra green. Meat Science. 65: 1117–1123.

Hird, H., J. Chisholm and J. Brown. 2005. The detection of commercial duck species in food using a single probe multiple species-specific primer real-time PCR assay. European Food Research and Technology. 221: 559–563.

Holt, P.S., H.D. Stone, R.K. Gast and C.R. Greene. 2000. Application of the agar gel precipitin test to detect antibodies to Salmonella enterica serovar enteritidis in serum and egg yolks from infected hens. Poultry Science. 79(9): 1246–1250.

Horvatovich. P., M. Miesch, C. Hasselmann, H. Delincée and E. Marchioni. 2005. Determination of Monounsaturated Alkyl Side Chain 2-Alkylcyclobutanones in Irradiated Foods. Journal of Agricultural and Food Chemistry. 53: 5836–5841.

Huang, L., Y. Wang, Y. Tao, D. Chen and Z. Yuan. 2008. Development of high performance liquid chromatographic methods for the determination of cyadox and its metabolites in plasma and tissues of chicken. Journal of Chromatography B. 874: 7–14.

Huet, A.-C., L. Mortier, E. Daeseleire, T. Fodey, C. Elliott and P. Delahaut. 2005. Development of an ELISA screening test for nitroimidazoles in egg and chicken muscle. Analytica Chimica Acta. 534: 157–162.

Imai, C. and J. Saito. 1983. Detection of Spoiled Eggs Using a New-Type Black Light Egg Inspection Unit. Poultry Science. 64(10): 1891–1899.

Jonker, K.M., J.J.H.C. Tilburg, G.H. Hägele and E. de Boer. 2008. Species identification in meat products using real-time PCR. Food Additives and Contaminants. 25: 527–533.

Kelly, S. 2007. The development of methods to determine the geographical origin of poultry. Institute of Food Research Enterprises Ltd. Norwich. pp. 1–89.

Kesmen, Z., Y. Celebi, A. Güllüce and H. Yetim. 2013. Detection of seagull meat in meat mixtures using real-time PCR analysis. Food Control. 34: 47–49.

Kim, H. and L.A. Shelef. 1986. Characterization and Identification of Raw Beef, Pork, Chicken and Turkey Meats by Electrophoretic Patterns of their Sarcoplasmic Proteins. Journal of Food Science. 51: 731–735.

Kiess, A.S., H.M. Parker and C.D. McDaniel. 2010. Evaluation of different selective media and culturing techniques for the quantification of Campylobacter ssp. from broiler litter. Poultry Science. 89(8): 1755–1762.

Lawrance, P., M. Woolfe and C. Tsampazi. 2010. The Effect of Superchilling and Rapid Freezing on the HADH Assay for Chicken and Turkey. Journal of the Association of Public Analysts. 38: 13–23.

Liu, Y., X. Fan, Y.-R. Chen and D.W. Thayer. 2003. Changes in structure and color characteristics of irradiated chicken breasts as a function of dosage and storage time. Meat Science. 63: 301–307.

Liu, Y., F.E. Barton, B.G. Lyon, W.R. Winham and C.E. Lyon. 2004. Two-Dimensional Correlation Analysis of Visible/Near-Infrared Spectral Intensity Variations of Chicken Breasts with Various Chilled and Frozen Storages. Journal of Agricultural and Food Chemistry. 52: 505–510.

Lockley, A.K. and R.D. Bardsley. 2002. Intron variability in an actin gene can be used to discriminate between chicken and turkey DNA. Meat Science. 61: 163–168.

Marchi, M.De., M. Penasa, M. Battagin, E. Zanetti, C. Pulici and M. Cassandro. 2011. Feasibility of the direct application of near-infrared reflectance spectroscopy on intact chicken breasts to predict meat color and physical traits. Poultry Science. 90: 1594–1599.

Martín, I., T. García, V. Fajardo, I. López-Calleja, M. Rojas, M.A. Pavón, P.E. Hernández, I. González and R. Martín. 2007. Technical note: Detection of chicken, turkey, duck, and goose tissues in feedstuffs using species-specific polymerase chain reaction. Journal of Animal Science. 85: 452–458.

Montowska, M. and E. Pospiech. 2013. Species-specific expression of various proteins in meat tissue: Proteomic analysis of raw and cooked meat and meat products made from beef, pork and selected poultry species. Food Chemistry. 136: 1461–1469.

Montowska, M. and E. Pospiech. 2011. Differences in two-dimensional gel electrophoresis patterns of skeletalmuscle myosin light chain isoforms between *Bostaurus, Sus scrofa* and selected poultry species. Journal of the Science of Food and Agriculture. 91: 2449–2456.

Nakamura, A., K. Kino, M. Minezawa, K. Noda and H. Takahashi. 2006. A Method for Discriminating a Japanese Chicken, the Nagoya Breed, Using Microsatellite Markers. Poultry Science. 85: 2124–2129.

Natonek-Wiśniewska, M., P. Krzyścin and A. Piestrzyńska-Kajtoch. 2013. The species identification of bovine, porcine, ovine and chicken components in animal meals, feeds and their ingredients, based on COX I analysis and ribosomal DNA sequences. Food Control. 34: 69–78.

Ng, J., J. Satkoski, A. Premasuthan and S. Kanthaswamy. 2012. A nuclear DNA-based species determination and DNA quantification assay for common poultry species. Journal of Food Science and Technology. DOI: 10.1007/s13197-012-0893-7.

Park, B., Y.R. Chen and R.W. Huffman. 1996. Integration of Visible/NIR Spectroscopy and Multispectral Imaging for Poultry Carcass Inspection. Journal of Food Engineering. 30: 197–207.

Pegels, N., I. González, I. López-Galleja, S. Fernández, T. García and R. Martín. 2012. Evaluation of a TaqMan real-time PCR assay for detection of chicken, turkey, duck, and goose material in highly processed industrial feed samples. Poultry Science. 91: 1709–1719.

Peippo, P., T. Lövgren and M. Tuomola. 2005. Rapid screening of narasin residues in poultry plasma by time-resolved fluoroimmunoassay. Analytica Chimica Acta. 529: 27–31.

Rhodes, C.N., J.H. Lofthouse, S. Hird, P. Rose, P. Reece, J. Christy, R. Macarthur and P.A. Brereton. 2010. The use of stable carbon isotopes to authenticate claims that poultry have been corn-fed. Food Chemistry. 118: 927–932.

Rikimaru, K. and H. Takahashi. 2007. A Method for Discriminating a Japanese Brand of Chicken, the Hinai-jidori, Using Microsatellite Markers. Poultry Science. 86: 1881–1886.

Riovanto, R., M. De Marchi, M. Cassandro and M. Penasa. 2012. Use of near infrared transmittance spectroscopy to predict fatty acid composition of chicken meat. Food Chemistry. 134: 2459–2464.

Rodríguez, M.A., T. García, I. González, L. Asensio, P.E. Hernández and R. Martin. 2003. Qualitative PCR for the detection of chicken and pork adulteration in goose and mule duck foie gras. Journal of the Science of Food and Agriculture. 83: 1176–1181.

Rojas, M., I. González, V. Fajardo, I. Martín, P.E. Hernández, T. García and R. Martín. 2009. Authentication of meats from quail (*Coturnix coturnix*), pheasant (*Phasianus colchicus*), partridge (*Alectoris* spp.), and guinea fowl (*Numida meleagris*) using polymerase chain reaction targeting specific sequences from the mitochondrial 12S rRNA gene. Food Control. 20: 896–902.

Rojas, M., I. González, M.A. Pavón, N. Pegels, P.E. Hernández, T. García and R. Martín. 2010. Polymerase chain reaction assay for verifying the labeling of meat and commercial meat products from game birds targeting specific sequences from the mitochondrial D-loop region. Poultry Science. 89: 1021–1032.

Rojas, M., I. González, M.Á. Pavón, N. Pegels, P.E. Hernández, T. García and R. Martín. 2011. Application of a real-time PCR assay for the detection of ostrich (*Struthio camelus*) mislabelling in meat products from the retail market. Food Control. 22: 523–531.

Rojas, M., I. González, T. García, P.E. Hernández and R. Martín 2012. Authentication of meat and commercial meat products from common pigeon (*Columba livia*), woodpigeon (*Columba palumbus*) and stock pigeon (*Columba oenas*) using a TaqMan real-time PCR assay. Food Control. 23: 369–376.

Saunal, H., F. Hemmen, A. Paraf and M.H.V. Van Regenmortel. 1995. Cross-reactivity and Heat Lability of Antigenic Determinants of Duck and Goose Lysozymes. Journal of Food Science. 60: 1019–1021.

Soares S., J.S. Amaral, I. Mafra and M.B. Oliveira. 2010. Quantitative detection of poultry meat adulteration with pork by a duplex PCR assay. Meat Science. 85: 531–536.

Soares, S., J.S. Amaral, M.B. Oliveira and I. Mafra. 2013. A SYBR Green real-time PCR assay to detect and quantify pork meat in processed poultry meat products. Meat Science. 94: 115–120.

Stamoulis, P., C. Stamatis, T. Sarafidou and Z. Mamuris. 2010. Development and application of molecular markers for poultry meat identification in food chain. Food Control. 21: 1061–1065.

Stedman, T.M. Jr., N.H. Booth, P.B. Bush, R.K. Page and D.D. Goetsch. 1980. Toxicity and bioaccumulation of pentachlorophenol in broiler chickens. Poultry Science. 59(5): 1018–1026.

Tewfik, I. 2008. Inter-laboratory Trial to Validate the Direct Solvent Extraction Method for the Identification of 2-dodecylcyclobutanone in Irradiated Chicken and Whole Liquid Egg. Food Science and Technology International. 14(3): 277–283.

Xu, C., X. Chu, C. Peng, L. Liu, L. Wang and Z.Y. Jin. 2006. Comparison of enzyme-linked immunosorbent assay with liquid chromatography–tandem mass spectrometry for the determination of diethylstilbesterol residues in chicken and liver tissues. Biomedical Chromatography. 20: 1056–1064.

Zhang, S., Z. Zhang, W. Shi, S.A. Eremin and J. Shen. 2006. Development of a Chemiluminescent ELISA for Determining Chloramphenicol in Chicken Muscle. Journal of Agricultural and Food Chemistry. 54: 5718–5722.

5

Fish and Seafood Authenticity

*Ioannis S. Arvanitoyannis** and *Konstantinos V. Kotsanopoulos*

5.1 Introduction

Authentication of fish and seafood is of high importance nowadays, mainly as a result of the increase in international trade of these products. Authentication can be either performed through generation of species-specific protein profiles or DNA profiles. The development and application of PCR-based techniques has been extensively examined due to their reliability, reproducibility and low cost. These methods enable the development of worldwide databanks. Techniques like FINS (forensically informative nucleotide sequencing), RFLP (restriction fragment length polymorphism), SSCP (single-stranded conformational polymorphism), RAPD and LP-RAPD (long-primer random amplified polymorphic DNA) and AFLP (amplified fragment length polymorphism) can be used to establish authentication methods (Bossier, 1999). Fish species cannot be easily identified when their common identifying characteristics are removed on processing and when only a portion of flesh can be examined. The examination of raw or cooked (under normal conditions) species can be easily performed by electrophoresis of the muscle proteins. Canned fish treated with heat sterilisation, cannot be examined using this procedure since the proteins are getting highly denatured. Degradation of DNA is also equally common. After amplification, the analysis of the samples can be performed using various techniques, some of which are suitable for food control laboratories (Mackie et al., 1999).

Protein and DNA analyses are probably the most suitable methods for species identification. Nevertheless, the development of new techniques for full authentication of freshness, production method, geographic origin, processing parameters, etc. should be accomplished. Natural isotope distribution, trace element and magnetic resonance (MR) analyses could potentially be used for this purpose. With advances in the fields of genomics, proteomics and metabolomics, and increased international concern about

School of Agricultural Sciences, Department of Agriculture, Ichthyology and Aquatic Environment, University of Thessaly, Fytoko St., 38446 Nea Ionia Magnesias, Volos, Hellas, Greece.
* Corresponding author

food quality and safety issues, new kits and equipment for ensuring the identity and safety of foods are continuously developed. Thousands of analyses could therefore be automatically performed in a much shorter time than required at present (Martinez et al., 2005). Furthermore, the substitution of fresh fish with frozen-thawed fish is a commonly met fraud. In this case, not only the qualitative characteristics of thawed meat are degraded during freezing, but also safety concerns are generated, since thawed meat is more susceptible to microbial growth. Although numerous techniques have been suggested for authenticating fresh fish, the classification efficiency is highly affected by the fish species examined (Ottavian et al., 2013). Table 5.1 presents some characteristic examples of techniques used for authenticating different species of fish and seafood.

This chapter is intended to provide an overview of the applications of the most common as well as of some innovative techniques used to ensure fish and seafood authenticity, including the authenticity of quality.

5.1.1 Application of PCR and Other DNA-based Methods Towards the Authentication of Fish and Seafood

Traceability of canned abalone is required for the protection of the qualitative characteristics and the safety of the product, as well as to uncover black market or fraudulent practices, such as canning other gastropods sold as abalone, 'abalone style', or showing an abalone shell picture on the label. The development of a forensic method to genetically differentiate three species of abalone from other gastropods was performed by analyzing 18S rDNA using PCR-SSCP. The development of a lysin gene marker was also performed towards the identification of species in Mexico. Both markers were effective for analysis of raw, frozen, and canned products. In a forensic analysis, two out of seven canned brands that claimed to contain abalone were found to contain other gastropods. None of three other brands showing the legend 'Abalone' or an abalone shell picture were found to contain abalone. The Chilean 'loco' *Concholepas concholepas* is most commonly used as an abalone substitute. The lysine gene effectively separated canned abalone based on species. Both methods effectively identified abalone and can be applied for certifying the authenticity of the Mexican commercial product or for identifying commercial fraud (Aranceta-Garza et al., 2011).

Moreover, the development of a conventional and a realtime multiplex PCR was performed for detecting fraudulent substitutions of Bianchetto (juvenile form of *Sardina philcardus*) and Rossetto (*Aphia minuta*) with Icefish (*Neosalanx* spp.), species that can be morphologically similar. *Engraulis encrasicolus* samples were also analyzed since this species is the major by-catch species. The development of a common reverse primer and forward species-specific primer was performed on the mitochondrial cytochrome b gene for amplifying sequences of various lengths using conventional PCR. The differentiation of each species was performed by providing specific peaks after melting temperature analysis in real-time PCR. The validation of the two PCR methods was carried out on fresh and processed products after the preparation of four dishes: two marinades (from raw or lightly boiled fish), a pasta sauce, and batter-fried fish cakes. All samples were correctly identified, but a portion of DNA was degraded after processing (Armani et al., 2012).

Table **5.1.** Examples of techniques used for the authentication of different species in several types of fresh and processed fish and seafood products.

Type of fish/seafood product	Treatment/Package	Type of fish to be detected	QC method	Efficiency	References
Sea cucumbers	Raw	*Aposticnopus japonicus, Cucumaria frondosa, Thelenota ananas,* and *Parastichopus californicus*	Multiplex-PCR	100% correct classification	Zuo et al., 2012
Surimi products	Admixed fresh surimi/ chewed	*Nibea albiflora*	Species-specific PCR	Detection limit < 0.5%	Zhao et al., 2013
Fresh Atlantic salmon products	Raw	*Oncorhynchus mykiss*	DGGE and AFLP-derived SCAR	AFLP-derived SCAR: at least 1% rainbow trout DNA in Atlantic salmon DNA is required; DGGE: at least 20% rainbow trout DNA in Atlantic salmon DNA is required.	Zhang et al., 2007
Salmon products	Frozen	Differentiation between Atlantic and Pacific salmon species	PCR-RFLP	Very effective, fragments < 100 bp cannot be identified individually.	Wolf et al., 2000
Commercial sea cucumber products	Frozen/dried	Identification of *Actinopyga lecanora, A. echinites, Bohadschia argus, Holothuria leucospilota, H. scabra, H. fuscogilva* and *H. fuscopunctata*	FINS methodology	Very specific, all species were differentiated	Wen et al., 2011
Various fish species	Raw, salted, dried, flavoured, peppered, peppered in oil, hot smoked, cold smoked, pickled and in tomato sauce	*Scomber scombrus*	Real Time-PCR	Efficiency 92.41%, 100% specificity, no cross reactivity	Velasco et al., 2013
Various fish species	Cooked or frozen only or frozen in 99% ethanol or 50% isopropanol	Grouper	RT-NASBA	Very fast, low cross reactivity	Ulrich et al., 2013

Table 5.1. contd....

Table 5.1. contd.

Type of fish/seafood product	Treatment/Package	Type of fish to be detected	QC method	Efficiency	References
Canned sardine, sild, herring, mackerel	Canning	*Sardina pilchardus*	Non-competitive indirect ELISA	Relatively low cross reactivity	Taylor and Jones, 1992
Fresh fish/processed seafood	Fresh, frozen or processed	*Genypterus*	PCR followed by FINS	100% specificity	Santaclara et al., 2014
Decapoda	Frozen	Species identification	Native isoelectric focusing	Differentiation of all species	Ortea et al., 2010
Fushi	Dried	Identification of different fish species	Immunostaining using anti-myosin light chain antiserum	Quite effective but unclear protein staining patterns	Ochiai and Watabe, 2003
Commercial smoked and gravad fish products	Smoked and gravad	Various species	SDS-PAGE	Not effective for all species	Mackie et al., 2000
Canned abalone	Canning	Other gastropods	PCR-SSCP	100% correct classification	Araneta-Garza et al., 2011
Sardina pilchardus, *Aphia minuta*	Fresh, marinated and cooked	Icefish	RM-PCR	100% correct identification	Armani et al., 2012
Commercial fish fillets of *Epinephelus marginatus*	Raw fillet	*Lates niloticus*, *Polyprion americanus*	ELISA	Very reliable and specific	Asensio et al., 2009
Epinephelus guaza	Raw fillet	Other fish (authentication)	Indirect ELISA	Accurate, very specific	Asensio et al., 2003
Perch products	Raw fillet	Differentiation of European perch, Nile perch, European pikeperch, Sunshine bass	IEF, 2-DE	Very reliable, 100% correct identification	Berrini et al., 2006
Fresh and canned sardine and anchovy-type products	Fresh/Canned	*Sardina pilchardus*, *Sardinella aurita* and *Engraulis encrasicolus* (differentiation)	PCR-RFLP	Very specific and reliable	Besbes et al., 2012
Raw and canned tuna	Raw, canned	*Thunnus alalunga*, *Thunnus albacares*, *Thunnus obesus*, *Thunnus thynnus*, *Katsuwonus pelamis* (differentiation)	PCR	Simultaneous and unambiguously differentiate between the examined species	Bottero et al., 2007

Sea cucumbers are traditional Asia type of seafood and are widely consumed. The different species of cucumbers cannot be easily identified by solely relying on their morphological characteristics and therefore a multiplex-PCR was used for the identification of four sea cucumber species (*Apostichopus japonicus*, *Cucumaria frondosa*, *Thelenotaananas* and *Parastichopus californicus*). After the construction of the 692 bp fragments of mitochondrial cytochrome oxidase I (COI) gene, four sets of species-specific primers were generated. After amplification, the fragments had the following lengths: 212 bp for *A. japonicus*, 301 bp for *C. frondosa*, 261 bp for *T. ananas* and 358 bp for *P. californicus* and could be easily differentiated using DNA electrophoresis. The four sets of species-specific primer were mixed and applied for the simultaneous detection of sea cucumber species. As a result, multiplex-PCR is a quite easy, useful, and highly specific method for authenticating sea cucumber species. The method can also be applied for traceability purposes (Zuo et al., 2012). Furthermore, since sea cucumbers are usually provided to consumers as dried or frozen goods and both treatments can significantly modify the flavour, the species identification cannot be performed based on morphological characteristics. Two molecular methods for the identification of six species of the Stichopodidae family in commercial products were evaluated in the study of Wen et al. (2010). The first one is PCR–RFLP using PCR amplification of a common 570 bp fragment of the 16S rRNA gene and subsequent co-digestion with two restriction enzymes (*XmnI* and *MwoI*), while the second one is FINS of the 16S r-RNA gene, which uses phylogenetic analysis of 16S r-RNA gene sequences. The PCR–RFLP methodology can be applied for the authentication of species tested in the present study and can be used for the detection of fraudulent labelling of frozen *A. japonicus* foodstuffs, while the FINS technique can be used to confirm that the species substituted for *A. japonicus* belongs to the genera Isostichopus. As a result, the methods could be successfully used for the authentication of the species of commercial sea cucumber foodstuffs. Also, according to Wen et al. (2011), the most common processing technique applied to sea cucumbers includes gutting, boiling and roasting and subsequent preservation through drying, smoking or freezing, leading to extreme alterations of their morphological characteristics. A FINS technology was therefore used to validate labelling of sea cucumbers belonging to family Holothuriidae in commercial seafood. It was proved that 63.6% of the samples were incorrectly labelled (63.6%). Furthermore, it was shown that the same methodology can be used for the genetic identification of more than 40 species from all over the world.

Food control policies controlling seafood label authenticity are now of very high importance due to increased incidence of species substitution or mislabelling. Proper species-level identification in the field of processed scallop foodstuffs is very difficult due to the absence of distinct morphological characteristics. The identification of four commercially popular scallop species (*Argopecten purpuratus*, *A. irradians*, *Mizuhopectenyessoensis*, *Pecten albicans*) was carried out using a species-specific multiplex PCR reaction. Novel reverse species-specific primers combined with one universal forward primer generated for the amplification of a partial region of the mitochondrial 16S rRNA gene were applied for the analysis of fresh and processed scallop samples. It was proved that all PCR reactions were highly specific, thus permitting the identification and authentication of unambiguous species (Marín et al., 2013). The wedge clam *Donax trunculus* is a commercially important bivalve species in Portugal. However, it presents similar morphological characteristics with three other species (*D. semistriatus, D. vittatus* and *D. variegatus*), which are provided at a lower market price. This fact may potentially lead to fraudulent substitution of the product, targeting at higher profits. Therefore, the development of reliable analytical techniques that can be applied to certify the authenticity of these species is required. In the study of Pereira et al. (2012), two DNA extraction techniques were developed together with a simple PCR method to accurately identify the *D. trunculus* based on the amplification of the nuclear marker 5S rDNA. The results of the PCR amplification indicated that the technique can be effectively used for differentiating the *D. trunculus* and *D. variegatus* from the remaining Donax species, since the resulted fragments of *D. trunculus* ranged from 275 to 300 bp while in the case of *D. variegatus* the corresponding fragments were about ~450 bp, clearly characterised by lower molecular weight in comparison to the DNA fragments of the other two species (~500 bp).

In another study, an effort was made towards the authentication of fish fillets of grouper (*Epinephelus marginatus*) in an attempt to prevent the substitution of the product with Nile perch (*Latesniloticus*) and wreck fish (*Polyprion americanus*). APCR technique relied on the use of specific primers of these species was used for this purpose. The oligonucleotides designed from the 12S ribosomal RNA gene, led to the generation of PCR fragments of 100, 138 and 169 bp length for grouper, wreck fish and Nile perch, respectively. The specificity degree of the primers was evaluated against more than 50 different fish species. In addition, the analysis of 70 commercial fish fillets samples was performed following the same methodology, and it was proved that 58 out of them were not correctly labelled. It was therefore proved that this method might be a reliable tool for detecting grouper adulteration. Furthermore, it could be a useful technique for strengthening traceability systems in the seafood industry (Asensio et al., 2009). In the study of Catanese et al. (2010), a novel multiplex-PCR assay was evaluated for its potential to authenticate mackerels of the genus Scomber in processed food. Two *Scomber japonicus*- (104 bp) and *S. australasicus*-specific (143 bp) amplicons were employed corresponding to the mitochondrial control region. A *S. colias*-specific product (159 bp) corresponding to the 5S ribosomal DNA, the *S. scombrus*-specific fragment (123 bp) from the mitochondrial NADH dehydrogenase subunit 5, and a positive amplification control corresponding to the small 12S rRNA subunit (188 bp), were also used. Both fresh and 40 commercial samples were examined, including 28 different canned products and 12 unprocessed fresh fillets. All samples were correctly labelled. It was proved that the method could be a useful molecular tool that could be directly applied for authenticating Scomber mackerels. Also, in another study, a RT-PCR method using a TaqMan technology was applied for identifying *S. scombrus*. A system of specific primers and a Minor Groove Binding (MGB) TaqMan probe relied on sequences of the mitochondrial cytochrome b region was generated. The technique was successfully applied in 81 specimens of *S. scombrus* and related species and validation was carried out in 26 different commercial samples. An average Threshold Cycle (Ct) value of 15.3 was found for *S. scombrus* DNA. As regards the other species tested, no fluorescence signal was detected and the Ct was significantly higher ($P < 0.001$). The efficiency of the assay was 92.41% and 100% specific, while no cross reactivity with any other species was detected. It was therefore proved that this method is speedy and efficient for effectively identifying *S. scombrus* and could be used for preventing fraud or mislabelling in mackerel products (Velasco et al., 2013).

Tuna are high cost fishes commonly used in processed foodstuffs. To establish an effective fishery management and protect consumers' rights, a molecular method was developed for identifying the species of the tuna products. The method was based on the development of a real-time PCR (RT-PCR) technology that can rapidly identify four tuna species. Four species specific TaqMan probes were generated to authenticate bigeye tuna (*Thunnus obesus*), Pacific bluefin tuna (*T. orientalis*), southern bluefin tuna (*T. maccoyii*), and yellowfin tuna (*T. albacares*). A SYBR green system was also generated to increase the efficiency of *T. obesus* authentication. Both systems can effectively and efficiently differentiate target species from others and can be used to identify tuna species in commercial products (Chuang et al., 2012). The TaqMan real-time PCR is characterised by the highest potential for automation and is therefore the most suitable method for screening, effectively detecting fraudulent or unintentional mislabelling of species. AnRT-PCR system was developed for detecting and identifying octopus (*Octopus vulgaris*) and main substitute species (*Eledone cirrhosa* and *Dosidicus gigas*). Specifically, sequences of the cytochrome oxidase subunit III (COIII) gene were used. From the alignment, specific primers/probe set were generated and a second specific primers/probe set as well as amplifications of the partial 16S RNA gene fragment were performed. PCR products were sequenced in both directions. DNA automatic sequencing in an ABI Prism was subsequently carried out. Thermal cycle sequencing reaction and the subsequent sequencing product cleanup by ethanol precipitation were performed. Raw data were subjected to analysis and the nucleotide sequences obtained were submitted to the GeneBank database of the NCBI. From the sequences of the 16S RNA gene, two internal primers and probe were generated inside this region. The method is very simple, rapid, and specific, while it can be used for all kinds of products: fresh frozen and processed, including those intensively processed and transformed. Its validation was performed to check how the degree of food processing affects the method and the detection of each species. Furthermore, it was used for the evaluation of the labelling of 34 commercial samples. It was proved that this method can be used to check that the labelling regulation for seafood are followed as well as for traceability verification in commercial trade and fisheries control (Espiñeira and Vieites, 2012).

The development of a real-time PCR assay based on LNA TaqMan probe technology was performed to detect and identify Atlantic salmon (*Salmo salar*). The method was proved to be simple and rapid, while it can potentially be automated. As previously discussed, the TaqMan real-time PCR is currently the most effective method of screening, thus allowing the detection of fraudulent or unintentional mislabelling of these products. The validation of the methodology, using a specific primers-probe set, was performed by application to 20 commercial samples labelled as salmon or *S. salar* aiming at determining whether the species contained in the product correspond to what was stated on the label. It was shown that the technique could be used to verify the abidance by the law as well as for traceability purposes (Herrero et al., 2011). The European sole (*Solea solea*) is a species with high-quality characteristics and thus commercially valuable. In the study of Herrero et al. (2013), a highly specific and fast real-time PCR was used to authenticate *S. solea* (distinguishing it from other related species) and prevent intentional and unintentional substitution with other species. The technique is relied on the use of a species-specific set of primers and MGB Taqman probe which is employed for the amplification of a 116-bp fragment of the internal transcribed spacer 1 (ITS 1) ribosomal DNA region. This assay is both highly specific and sensitive due to the use of RT-PCR, but it is also rapid, thus effectively detecting *S. solea* in a short period of time. The validation of the method was successfully carried out for application to all types of commercial products that contain *S. solea*. It was also used for the analysis of 40 commercial samples in an effort to determine whether correct labelling had been employed in the market. It was proved that the assay could be successfully applied to monitor and verify food-labelling regulations. In the study of Santaclara et al. (2014), a method of authentication of Genypterus and their substitute species was reported. This method included the use of PCR and phylogenetic analysis (FINS). All species examined were identified using the mitochondrial cytochrome oxidase subunit I (CoxI) as molecular marker. The species that belong to Genypterus genera could be easily substituted by other species of lower value. The morphological characteristics of these animals are not readily distinguishable, especially after processing. Thus numerous methodological strategies were implemented and all effectively authenticated the examined species in any kind of products (from fresh, frozen or ready-to-eat fish meals). As a result, the method can be used to routinely prevent mislabelling of Genypterus species, as well as for examining their traceability potential.

When PCR is applied without species-specific primers, an additional discriminatory technique, such as RFLP, is essential. RFLP analysis of PCR products has been widely applied for species discrimination and a single primer pair can lead to the production of a fragment that can be used for identifying multiple species with judicious choice of restriction enzymes (Gil, 2007). In the study of Ardura et al. (2010), the development of three genetic methods for identifying species of Amazonian marketed fish was reported. The methods were relied on: polymerase chain reaction-restriction fragment length polymorphism (PCR-RFLP), FINS technique, and PCR-fragment size visualization of 5S rDNA in agarose gels. The above-mentioned techniques were used to identify 20 species belonging to nine taxonomic families and could be implemented to authenticate fish products and labelling of Amazonian fisheries. Samples of commercial products collected from restaurants were successfully identified, even when salted or frozen, thus confirming the effectiveness of these methods for routinely controlling Amazonian fish food. After assessing the time required for the completion of the methods, the equipment required and the cost, it was shown that they could be effectively applied in industrial quality control in developing countries. Wolf et al. (2000) evaluated a method that could be used for identifying fish species. The method is relied on the amplification of a specific part of the mitochondrial genome (tRNAGlu/cytochrome b) using PCR. The differentiation between several fish species was performed by cutting the obtained 464 bp long PCR-products using different restriction endonucleases (RE) thus leading to species-specific RFLP. The technique can be successfully applied for the identification of, e.g., Atlantic and several Pacific salmon species, while it is also characterised by simplicity, allowing detection of falsely declared fish or fish products made up of a single species. Four different restriction enzymes (e.g., Nla III, Dde I, Hae III and Mbo II) were only required to identify all 23 investigated species. One particular species can be identified within a mixture of several fish, preferably by using species-specific PCR systems.

In the study of Zhao et al. (2013), the development of a species-specific PCR assay authenticating and enhancing the traceability potential of Yellow Drum (*Nibea albiflora*) from surimi products was performed using a primer pair design based on mt-DNA 12S rDNA sequences for amplification of 193 bp DNA fragments. The PCR assay was examined against DNA of eleven species (Yellow Drum, Large Yellow Croaker, Hairtail, Silver carp, Grass Carp, Bighead carp, Eel, chicken, pork, corn starch and wheat starch), commonly used in surimi products in China, to evaluate its specificity. It was proved that positive amplification was only possible for *N. albiflora*. The detection limits of the PCR assay were lower than 0.5% in admixed fresh surimi products containing Yellow Drum and Large Yellow Croaker. Even in chewed and vomitted samples, authentication of *N. albiflora* could be performed and traced effectively. It was therefore concluded that the technique is characterised by high specificity and sensitivity and can rapidly authenticate and trace *N. albiflora* in extensively processed surimi products. Proper identification of fish species (including origin determination) can be performed only by using recognized material for verifying the conformity of the product with the labelling requirements. In this study, the discrimination between Tunisian small pelagic fish: *S. pilchardus*, *Sardinellaaurita* and *E. encrasicolus* was performed through DNA extraction from fresh fish, 12 canned sardine and two anchovy-type products and subsequent amplification of a 252 bp fragment of the cytochrome b gene. The new sequences, which were deposited in the NCBI GenBank, were tested against a nucleotide sequence database (GenBank) and the phylogenetic analysis was carried out using the MEGA software. Multiple alignments of three analyzed reference samples, belonging to Clupeomorpha species, were performed versus the canned samples. Low levels of intraspecific variability was detected for canned anchovy and sardine (< 0.05), whereas mean interspecific variability was 0.23. The construction of a phylogenetic tree was performed, and the calculated bootstrap values (BP, 71–100%) indicated the correct assignment of unknown canned samples to reference species. Post-amplification digestion with *HaeIII* and *ALuI* restriction enzymes, led to the generation of specific profiles that allowed the differentiation between the sardine species. This PCR-RFLP very effectively identified fish species in canned products (Besbes et al., 2012).

The distribution of Cyprinidae fish is very wide in Taiwan and therefore an effective species identification technique based on PCR-RFLP was developed by Chen et al. (2012). After completion of the mitochondrial cytochrome b genes (cytb, 1141 bp) of 8 Cyprinidae, a rapid PCR-RFLP method was implemented for the amplification of a 426 bp fragment of cytb using primer set Lnew1/Hnew1, while the digestion of the fragment was performed with NlaIII and BstXI. It was proved that the use of this PCR-RFLP technique was effective for identifying the 8 Cyprinidae. The method was also used for the investigation of 16 commercial samples of processed Cyprinidae products from traditional markets. It was shown that nine of the samples were *Carassius auratus auratus*, *Cyprinus carpio carpio*, or *C. idella*. The other six samples were not of the Cyprinidae family according to the results extracted by using PCR-RFLP. These samples were subjected to further analysis using FINS and BLAST methods. It was found that these six processed products belonged to *Oreochromis* spp., a monophyletic fish group. The last commercial product could not be amplified to obtain a 426 bp fragment, and, as a result, no identification was performed. It was therefore shown that PCR-RFLP and FINS can be applied for the identification of processed products fraudulently labelled as Cyprinidae.

According to Pascoal et al. (2012), genomic and proteomic techniques for species identification of meat and seafood products are extensively used nowadays and, therefore, a genomic approach was selected to differentiate *Pandalus borealis* (the Northern shrimp), which belongs to the superfamily Pandaloidea, from 30 crustaceans, consisting of 19 commercially relevant prawns/shrimps species, that belong to the superfamily Penaeoidea (which includes the families Penaeidae and Solenoceridae), and 11 other crustacean species, including prawns, shrimps, lobsters, and crabs. Specifically, a PCR–RFLP method was generated by amplifying the 16S rRNA/ tRNAVal/12S rRNA mitochondrial regions using the primers 16S-CruF and 16S-CruR. 966-bp PCR products were generated and cleaved using the restriction enzymes AluI, TaqI, and HinfI, providing species-specific restriction patterns. Furthermore, a proteomic technique using matrix-assisted laser desorption/ionization time-of-flight (MALDI–TOF) and electrospray ionization–ion trap (ESI–IT) mass spectrometry, was applied for the identification and characterisation of new *P. borealis*-specific peptides that could potentially be used as markers of this species in protein-based detection methods. This is probably the first study that describes a molecular method that can be effectively used for the identification of a vast variety of prawn and shrimp species, including *P. borealis*, for either whole individuals or processed products. Nevertheless, the methods should be validated by using them for the examination of more samples from different populations and geographic origins.

In another study, five different tuna species were differentiated from commercial canned tuna using Nested Primer PCR–RFLP. The identification of canned tuna at species level cannot be easily performed using PCR methods due to the use of additives, as well as due to the fact that severe degradation of the DNA commonly occurs. The application of Nested Primer PCR, significantly enhances the specificity and sensitivity of the reactions, thus allowing the generation of an amplicon of 276 bp (TUN276) from commercial canned tuna, in spite of the presence of additives. As a consequence, this is a very effective method for authenticating tuna in brine, oil, or for the authentication of pickled, sauced and spiced tuna products. The 276 bp amplicon obtained in this study, is probably the longest fragment obtained so far from canned tuna (Pardo and Pérez-Villareal, 2004). The fish roe is a product that originates from Messolongi, Greece with designation of origin. This processed seafood product is produced from the whole ovaries of the fish *M. cephalus*. A PCR–RFLP based method was used to authenticate the fish roe of Messolongi. The PCR was applied for amplifying a mitochondrial 16S rRNA gene segment. The PCR products were digested using the restriction enzymes BstNI, TaqI and HinfI and subsequent agarose gel electrophoresis was applied, yielding species specific restriction patterns and clearly discriminating the product from the fish roe produced from other Mugilidae species coexisting in the same area (Klossa-Kilia et al., 2002). Moreover, Infante et al. (2004) described a novel procedure that could be used for authenticating frigate tunas (*Auxis thazard* and *A. rochei*) found in commercial canned products. Simultaneous amplification of three mitochondrial regions was carried out using multiplex-Polymerase Chain Reaction, one corresponding to the small r-RNA 12S subunit as a positive amplification control and two species-specific fragments corresponding to cytochrome b for *A. rochei* and ATPase 6 for *A. thazard*, respectively. By testing two different detection systems, it was shown that the fluorescence-based technique was characterised by the highest sensitivity. It was therefore suggested that this rapid, low-cost method could be used to reliably authenticate canned products at a molecular level.

Another multiplex-PCR assay for authenticating the Atlantic mackerel *S. scombrus* in commercial canned products was described by Infante et al. (2006). The method employes a *S. scombrus*-specific fragment (123 bp) which corresponds to the mitochondrial NADH dehydrogenase subunit 5, and a positive amplification control which corresponds to the small rRNA 12S subunit (188 bp). The assay was carried out on six different canned products labelled as *S. scombrus,* thus proving that the technique can be applied as a molecular tool to directly authenticate *S. scombrus* canned products. The identification of mislabelling or fraudulent substitution of toxic puffer fish in thermally treated fish products was performed using PCR which includes restriction sites and sequence analysis. A 376-bp fragment of the cytochrome b gene was generated after PCR amplification. The preparation of the samples was performed through autoclaving the fishes at 121°C for 10–90 min at 10 min intervals. No DNA fragments could be found after 90 min of autoclaving at 121°C. The PCR products were digested using BsaJ I, Aci I, Hinf I, Taq I and Sap I endonucleases, yielding species-specific profiles that could be used to identify puffer fish species derived from 60 commercial market samples. It was proved that the restriction fragment length polymorphism technique could be applied for the identification of 17 puffer fish species obtained from commercial products even after extreme heat treatment (Hsieh et al., 2010). A molecular method was applied to discriminate between eight species widely used in the production of codfish. Since only whole fish can be visually identified, a PCR method was applied to obtain a short fragment of the cytochrome b (cytb) gene that was subsequently subjected to analysis using RFLP, SSCP and DGGE (denaturing gradient gel electrophoresis). While RFLP and SSCP led to discrimination of only some of the species examined, DGGE led to the production of patterns that were used to identify the species considered. The molecular methods for identifying the species in this work were both rapid and reliable (Comi et al., 2005).

Bottero et al. (2007) described a novel methodology that employs analysis of mtDNA-cytb diagnostic sites for differentiating between four closely related species of Thunnus (*T. alalunga, T. albacares, T. obesus* and *T. thynnus*) and one species of Euthynnus (*Katsuwonus pelamis*) genus in raw and canned tuna. The primers applied in the preliminary PCR led to successful amplification of a 132 bp region from the cytb gene of all the species examined. The sites of diagnosis have been interrogated simultaneously using a multiplex primer-extension assay (PER) and the method was validated using fragment sequencing. The effectiveness of the multiplex PER assay on the analysis of commercial canned tuna products was also shown. This technique can be used for detecting frauds as well as for seafood traceability purposes.

Another study based on the use of cytochrome b was carried out by Espiñeira et al. (2009) for authenticating scombroid products. FINS technique (Polymerase Chain Reaction (PCR) followed by phylogenetic analysis) was employed and the method allowed the identification of commercially significant scombroid species using the mitochondrial cytochrome b as a molecular marker. Due to the various commercial values of the different species of this family, high-cost species could potentially be substituted by lower-cost ones. Numerous methods have been used to authenticate scombroid species in several types of foods, from fresh fish to canned products. This analytical tool was validated and was then used for the analysis of 20 commercial samples labelled as tuna, bonito, mackerel and frigate tuna, indicating that three of them (15%) were not correctly labelled. As a result, this technique can be routinely applied to prevent or detect mislabelling of products containing scombroid species as well as for the assessment of traceability systems.

Rajidae family is among the most diverse families within Batoidea super-order including more than 220 described species. These species are mainly marketed as fresh and frozen wings. Thus, they are not identifiable based on their morphology. As a result, a method was suggested to genetically identify different skate species. The method is based on sequencing of a fragment of 555 bp from amplified DNA, COI gene, using PCR and subsequent phylogenetic analysis (FINS). Using this technique, more than 40 skate species were determined in skate products. The applied method is very effective since, at present, there is no work about the genetic identification that includes so many skate species. As a result, this molecular technique could be used to simplify the controls of the labelling of skate commercial products, enhance the traceability of raw materials, and the control of imported skates (Lago et al., 2012).

Espiñeira et al. (2010) reported the development of a molecular method for authenticating cephalopods products by allowing the genetic identification of about 30 species belonging to the families Octopodidae, Sepiidae and Sepiolidae. The technique is performed by phylogenetically analyzing DNA sequences. The molecular marker examined was the cytochrome b gene (cytb), which was subjected to amplification using PCR and subsequent sequencing. The validation of the method was followed by application to 20 commercial samples. Six incorrectly labelled products (30%) were detected, indicating that the method can be used to enhance the correct labelling, traceability, and import control of foodstuffs that contain the examined taxonomic groups.

The establishment of specific PCR-RFLP profiles of 10 different Cyprinidae fish species was attempted by employing capillary electrophoresis (CE). The primer set CypbL/CypbH was applied for the amplification of a partial mitochondrial cytochrome b gene (331 bp), subsequent co-digestion of the sample was carried out using endonuclease (NlaIII and MseI). The determination of the PCR-RFLP profiles of the 10 Cyprinidae fish species was performed using CE and successful differentiation of the different fish species was achieved. The technique was also used for the detection of the species used in 24 commercial Cyprinidae-related products. Four samples were identified as *C. carpio carpio*, eight samples as *C. auratusauratus*, and 12 samples as *Ctenopharyngodon idella*. It was proved that application of PCR-RFLP in a CE system can be used for the identification of fish species and therefore it could potentially be established as an analytical tool for large and routine food authentication checks (Chen et al., 2013).

The development of PCR-based methods for the differentiation of four species of commercially important eels (*Anguilla anguilla*, *A. rostrata*, *A. japonica* and *A. australis*) was reported. A 464 base pair (bp) section of the cytochrome b gene was subjected to amplification. Sequencing of the amplicon was performed and restriction enzymes were chosen for RFLP analysis. A fragment (123 bp) of the 464 bp section was subjected to amplification using another set of primers and used for SSCP analysis. The effectiveness of RFLP and SSCP on differentiating hot-smoked eel was verified by a collaborative study. When mixtures were analyzed, however, it was shown that one of the other components may not be detected if *A. anguilla* is present (Rehbein et al., 2002).

In another study, forty-one cod steaks, fifteen escolar steaks, and fifteen salted escolar roe products obtained from the Taiwan market were examined for determining biogenic amine, histamine-forming bacteria, and for identifying the fish species. The levels of pH value, salt content, Aw, TVBN and APC in all samples ranged from 5.3 to 7.0, 0.7 to 5.6%, 0.80 to 0.99, 0.8 to 59.9 mg/100 g and 2.5 to 7.3 log cfu/g, respectively. No samples were found to contain coliform bacteria or *Escherichia coli*. The average level of histamine in all samples examined was lower than 5 mg/100 g. Nine histamine-forming bacteria found in cod, escolar, and salted escolar roe products led to the production of 2.0–62.3 ppm of histamine in trypticase soy broth (TSB) supplemented with 1.0% L-histidine (TSBH). Application of DNA direct sequence and PCR-RFLP proved that 31.6% (13/41) of products were mislabelled. Among them, seven samples (17%) and six samples (14.6%) were *Ruvettus pretiosus* (oilfish) and *Reinhardtius hippoglossoides* (Greenland halibut), respectively. Furthermore, the majority of escolar steaks and salted escolar roe foodstuffs were identified as *Lepidocybium flavobrunneum* (escolar), while other samples were identified as *R. pretiosus* (Hwang et al., 2012).

According to Bossier et al. (2004), Artemia cysts are currently commercially exploited and their quality is highly dependent on their intrinsic nutritional quality, harvesting and processing conditions, diapause characteristics and size. In an effort to mediate authentication of commercial Artemia cyst samples, a database with eight RFLP patterns of a 1500 bp mitochondrial rDNA fragment was generated. The database includes 53 samples from almost all geographical areas that Artemia can be found. PCR was used in a Hybaid PCR Express (Labsystemsk, Belgium) and the primer combinations 12SA/16Sbr, 12S-R/16S-F and 12S-SP/16S-SP were applied for the amplification of the fragment of DNA of interest-a 1500 bp fragment of mitochondrial rDNA. On the basis of a band-sharing index the above-mentioned samples could be categorised into five groups. Within each of these groups, high diversity between samples can still be found, demonstrating the genetic diversity within each species. The examined method is useful for the assignment of the samples to these clusters, making the authentication of species easier.

The development of DNA-based methods for identifying fish species is of high importance for fisheries research and control, as well as for detecting unintentional or intentional adulteration. The generation of a comprehensive reference database of DNA sequences from the mitochondrial 16S and 12S ribosomal RNA (rRNA) genes was carried out for 53 commercial fish species in South Africa. The effectiveness of these genetic markers on identifying fish, at species level, was assessed. Extraction of DNA from the species of interest was followed by amplification using universal primers, which targeted both rRNA gene regions. A ca. 570 bp fragment of the 16S rRNA gene was subjected to amplification using PCR from all DNA extracts and the universal primers 16SarL (5'-CGC CTG TTT ATC AAA AAC AT-3') and 16SbrH (5'-CCG GTC TGA ACT CAG ATC ACG T-3'). Submission of sequences from the 16S and 12S rRNA genes to GenBank was carried out for 34% and 53% of the fish species, respectively. Cumulative analysis of the 16S rRNA gene sequences revealed mean con-specific, con-generic and con-familial Kimura two parameter (K2P) distances of 0.03%, 0.70% and 5.10% and the corresponding values at the 12S level were 0.03%, 1.00% and 5.57%. In general, K2P neighbour joining trees based on both sequence datasets led to clustering of species according to their taxonomic classifications. The nucleotide variation in both the 16S and 12S sequences was effective for the identification of most of the examined fish specimens to, at least, the level of genus, but was not as effective for differentiating certain congeneric fish species. It was suggested that the analysis of one or more faster-evolving DNA regions could be used for confirming the identities of closely related fish species in South Africa (Cawthorn et al., 2012). The giant grouper (*E. lanceolatus*) is a fish species of high commercial value in Taiwan. A RAPD, inter-simple sequence repeat (ISSR) and the CoxI of mitochondrial DNA were applied for the analysis of the genetic variation in *E. lanceolatus*. Specific primers originating from these methods were generated. Wild and cultivated giant grouper were obtained from the Penghu, Pingtung, and Kaohsiung regions of Taiwan and identified to species levels.

Extraction of DNA was followed by analysis using 95 primers of RAPD and 59 primers of ISSR in an effort to examine the genetic variation among samples. It was shown that RAPD analysis using 21 primers yielded 279 bands and 86 polymorphic bands (31% polymorphism). Both RAPD115 and RAPD73 primers effectively discriminated wild and cultivated giant groupers. Seventeen of the 59 ISSR primers (29.3% of total primers) led to the production of 166 bands and 58 polymorphic bands (34.9%). In CoxI analysis, the sequences could lead to differentiation of wild and cultivated *E. lanceolatus*. In an attempt to develop sequence characterised amplified region (SCAR) markers, a 747 bp amplicon was formed by the ISSR method and subjected to sequencing. Primer pairs, which targeted the SCAR marker sequence, were generated, enabling species-specific detection and preventing any samples of other species of Serranidae fish from being cross-amplified. Specific primers (SCAR1/SCAR2) effectively amplified products of cultured giant grouper but not of the wild population. The specific DNA fragment generated from SCARF/SCAR2 primers was subjected to amplification in all giant grouper samples but was not amplified in other Serranidae fish. The development of specific ISSR-SCAR markers was performed from distinguishing wild and cultivated populations of giant grouper as well as for distinguishing them from other Serranidae and giant grouper fish. The development of DNA molecular marker techniques and CoxI sequences for generating information for species identification, trace genetic variation between different individuals in aquaculture, and authenticate fish and fishery products, was finally reported (Chiu et al., 2012).

In another study, a method based on SSCP was applied for the identification of potential species-specific markers for Pangasiidae species. Seven primers were generated using AFLP. It was proved that only one of them had conserved variation within species and could be distinguished among species. Use of *PL8* differentiated five out of nine Pangasiidae species, among which species of high value such as

Pangasianodon gigas, P. hypopthalmus, Pangasius bocourti and *P. larnaudii.* Confirmation of the effectiveness of this marker on identifying species was obtained by examining samples of larvae, although no unambiguous differentiation between hybrids of *P. hypophthalmus* and *P. gigas,* or *P. bocourti* from their parents, could be achieved. *PL8* could be successfully used for identifying species of Pangasiid catfishes, as well as for authenticating species in food products (Sriphairoj et al., 2010). According to Zhang et al. (2007): "Commercial frauds in fresh Atlantic salmon (*S. salar*) products due to adulteration and substitution with rainbow trout (*Oncorhynchus mykiss*) can be performed using two molecular tools, AFLP and DGGE. The development of a species-specific SCAR marker for rainbow trout (*O. mykiss*) was performed based on AFLP analysis. Furthermore, DGGE was applied for the discrimination of Atlantic salmon (*S. salar*) from rainbow trout, by employing a fragment of the mitochondrial cytochrome b gene. Analysis of experimental mixtures showed that AFLP-derived SCAR is more sensitive than DGGE."

The 5S ribosomal RNA can be easily used to rapidly and efficiently identify fish species mainly due to its structural characteristics (a conserved region followed by a species-specific non-coding region, the 'non-transcribed spacer'). This species-specificity based in length and sequence was evaluated in an effort to differentiate between fish species, which can be adulterated in the fish markets. The targeted portions of the 5S rRNAs of different fish species were sequenced and aligned, and the primer pairs required for PCR amplification on the DNA traits and a primer pair on conserved regions, were generated. It was proved that the method was feasible, simple and reliable for detecting mislabelling or fraudulent substitution of fish species (Tognoli et al., 2011). Grouper is the raw material for many commercial products of high importance in the state of Florida in US and is commonly used as a high-end restaurant dish across the country. Unfortunately, a high number of mislabelling incidents connected with the substitution of the product with cheaper fish species are continuously reported. The U.S. Food and Drug Administration has already named 56 species of fish that can be labelled as "grouper" for commercial purposes. Due to the fact that this group of fish consists of species from ten different genera, the species cannot be easily identified after removal of characteristics such as skin, head and tail. Thus, regulatory agencies are moving towards the use of genetic identification techniques, characterised by higher species-level resolution than phenotypic methods. The common methods used to genetically modify these products are highly technical and expensive lab-based equipment is demanded to carry out the analysis. As a result, a generic grouper assay was developed that can be used for detecting most of the grouper species listed on the 2011 FDA Seafood List, as well as the species found in Florida waters. This assay relied on the use of real-time nucleic acid sequence-based amplification (RT-NASBA) that targets mitochondrial 16S rRNA to accurately detect grouper. This assay can be carried out in less than 90 min, while very low levels of cross-reactivity from non-target species is observed (Ulrich et al., 2013).

Authentication of commercial marine foodstuffs is of high importance nowadays. Crustaceans are benthic marine organisms extensively consumed in Chile. For the commercial point of view, Brachyuran are the most important crustaceans and in Chile are offered as various processed foodstuffs. In general, Chilean brachyuran crab meat products are not labelled as per their taxonomic origin which in addition to the absence of recognisable characteristics, due to treatment, makes the identification of species very difficult. Therefore, a study was published at which determination of brachyuran in the Chilean market was performed using DNA Barcoding and phylogenetic analysis (Haye et al., 2012). DNA Barcoding, based on the sequencing of a standardised region of the COI gene, is used more and more frequently to accurately identify animal species (Cawthorn et al., 2012). To authenticate commercial crab meat and DNA Barcoding a partial sequence of the COI gene of seven commercialized brachyuran species in Chile was employed. The authentication was carried out by obtaining products of seven marketed types from the local market in Coquimbo, Chile. The majority of packages contained more than one species of crab. The species found in order of frequency are *Metacarcinus edwardsii*, *Romaleon polyodon*, *Cancer porteri*, *Cancer plebejus*, and *Homalaspis plana*. Mislabelling was detected in the case of one of the commercial formats, which was labelled as meat from Cancer species. The methodology applied in the study relied on standard DNA Barcoding and phylogenetic analyses, and can be effective for the normative control of crab meat processed foodstuffs (Haye et al., 2012).

5.1.2 Application of Near-Infrared Spectroscopic Methods towards the Authentication of Fish and Seafood

Several studies have examined the use of Near-Infrared Spectroscopy (NIR) for food authentication purposes. An evaluation of the effectiveness on the determination of moisture, protein and fat content of the famous Greek dish taramosalata was carried out by Adamopoulos and Goula (2004). Additionally, the relative performance of the calibration model in the context of varying calibration data size and number of used wavelengths was also examined. Numerous calibration data sizes (n . 40; 50; . . . ; 80) and number of wavelengths (1, 2,. . ., 6) were taken into consideration. The analysis of calibration samples was performed using traditional chemical methods and scanning was carried out by employing an Instalab 600-Dickey-John NIR apparatus. Calibrations were performed using multilinear regression between chemical and spectral data from each calibration data set. The optimal prediction errors [root mean square error of prediction (RMSEP)], found by models using six wavelengths and a calibration data size of 80, were 0.115%, 0.023% and 0.088% for moisture, protein and fat, respectively. The prediction errors decreased by increasing the calibration size and number of wavelengths but were stable when a calibration data size of 60 and 3 wavelengths were applied. Due to the high cost of the calibration set and the fact that calibrations using many wavelengths tend to over fit the data, accurate calibrations using three wavelengths and a calibration data size of 60 seem to be the best option. The RMSEP values acquired using the above-mentioned models were 0.271% for moisture, 0.115% for protein and 0.222% for fat. NIR measurements as conducted using the Dickey John Analyzer were used to rapidly and accurately analyze taramosalata and could potentially replace the conventional expensive and time-consuming wet chemistry methods.

In another study, short-wavelength near-infrared (SW-NIR) spectroscopy (700–1100 nm) was employed for the prediction of the crude lipid content of intact, whole rainbow trout (weight range 66.5–883 g). Non-invasive measurements were taken in a diffuse reflectance mode by conveying the light to the intact, whole fish over an optical fibre bundle. Collection of the backscattered light was achieved using a second fibre bundle, concentrically positioned against the first, and focused onto the entrance slit of the monochromator. Two multivariate calibration models, a multiple linear regression (MLR) model and a partial least-squares (PLS) model, were employed for correlating chemical with spectral data. Optimum correlation between the laboratory obtained measurements and predicted values from spectral measurements of crude lipid content was achieved at a point midway between the dorsal fin and the adipose fin above the lateral line. The PLS cross-validation model was based on the use of three latent variables, yielding a standard error of prediction of cross-validation [SEPCV = 2.27% (w/w), R = 0.811. According to the authors, "comparable results were achieved with a MLR cross-validation model using three wavelengths (934, 957, and 845 nm) [SEPCV = 2.48% (w/w), R = 0.761]". It was proved that both a full spectrum and a discrete wavelength SW-NIR methodology can be effective for predicting the crude lipid level of fish muscle in whole trout without scale damaging the animal and therefore without affecting its qualitative characteristics (Lee et al., 1992). NIR was also used to non-destructively obtain measurements from 100 live, anaesthetized farmed Atlantic salmon. Two NIR instruments were used: a grating monochromator instrument equipped with a fibre optic interactance probe, and a diode array instrument measuring diffuse reflectance in a non-contact mode. The determination of the crude fat level was performed by employing PLS regression. In the case of the fibre optic instrument (wavelength range from 800 to 1098 nm), correlation coefficient of 0.90 and an RMSEP equal to 14 g kg^{-1} fat were found. The diode array instrument with wavelength ranging from 900 to 1700 nm was equally accurate. The measurement times were determined at 21 and 3 s respectively. It was shown that both instruments could be applied for determining the crude fat content of live Atlantic salmon (Solberg et al., 2003).

Reflectance spectroscopy in the NIR region (900–1800 nm) was employed for the prediction of the proximate composition of fresh and frozen rainbow trout muscle. The RMSEP for moisture was 1.1% (over the range 71.5–76.7% moisture), for crude lipid, 3.1% [range 8.8–17.0% dry weight basis (dwb)], and for protein nitrogen (Kjeldahl, % N X 6.25), 5.4% [over the range 69.3–87.4% protein (dwb)]. PLS and MLR were employed for the development of models capable of relating absorbance to moisture, lipid and protein contents. This NIR reflectance spectroscopic technique could be used for the approximate estimation of the moisture and protein levels of fish muscle as well as for facilitating quality control. No significant sample preparation is required and no homogenization drying or extraction is necessary for this type of analysis (Rasco et al., 1991).

The lower price of frozen fish compared to fresh product is frequently the reason for adulteration incidents. As a result, a NIR spectroscopic method was used for differentiating between frozen-thawed and fresh products. Fresh and frozen-thawed samples of horse mackerel (n = 162) were assessed. Dry extract spectroscopy by infrared reflection (DESIR) was used to analyse the meat juices and subsequent discrimination was carried out using principal component analysis (PCA) and MLR. In DESIR spectra, the overall absorbance level decreased in frozen-thawed samples, as an indication of the distinct chemical composition of juice, amount of dry matter, particle size, and their scattering properties. The spectral changes between fresh and frozen-thawed samples were easily discriminated in the 1920- to 2350-nm region. The spectra were characterised by numerous peaks, which represent proteins (1510, 1700, 1738, 2056, 2176, 2298 and 2346 nm). It was clearly shown that the separation of fresh and frozen-thawed fish was performed with 100% accuracy using the DESIR technique (Uddin Okazaki, 2004). Near infrared diffuse reflectance spectra were also obtained to analyse the skin of 115 thawed whole cod (*Gadus morhua*) of different quality grades. In fish samples obtained from the same fish, four frozen fish quality parameters (water holding capacity, concentration of total volatile nitrogen bases, dimethylamine, and formaldehyde) were examined. By using PCA, it was proved that the above-mentioned qualitative attributes were highly connected with each other. Comparison of PLS regression models relied on full spectra and on principal variables, respectively, was performed to evaluate the prediction ability towards the water holding capacity. PLS models based on only seven principal variables of the 459 measured variables generated models characterised by the same prediction abilities as the full spectrum model.

The high water content of the samples is the most important factor that limits the potential use of NIR analysis for determining chemical quality attributes. Nevertheless, NIR-reflectance analyses can be used to effectively determine the water holding capacity as well as other properties (Bechmann and Jørgensen, 1998). Furthermore, NIR spectra were examined for 105 samples of cod mince obtained from thawed cod fillets of different quality grades packed under modified atmosphere packaging (MAP) and chilled temperature. Traditional chemical, physical, microbiological and sensory techniques used to assess fresh fish products were also implemented to evaluate the same cod fillets. The purpose of the study was the evaluation of the potential of NIR spectroscopy to estimate (i) frozen storage temperature, (ii) frozen storage period and (iii) chill storage period of thawed-chilled MAP Barents Sea cod fillets. Moreover, the ability of taking measurements of specific qualitative characteristics such as dip loss, water holding capacity and content of dimethylamine by NIR, was assessed. The presentation of the extracted data was performed using multivariate modelling methods such as PLS regression and discriminant PLS regression. Systematic differences in the NIR measurements in the minced cod fillets were mainly attributed to the chill storage duration (days at 2°C). PLS regression models using wavelengths selected by a new Jack-knife method gave a correlation coefficient of 0.90 between measured and predicted duration of chill storage period (days at 2°C). The root-mean-square error of cross-validation (RMSECV) was 3.4 d at 2°C. NIR analyses resulted in useful data for evaluating the freshness of thawed-chilled MAP cod fillets, enhancing the reliability of the results extracted by traditional quality methods. Nevertheless, the examination of the effect of, e.g., sample preparation, season, fishing ground and cod size as well as the use of more sophisticated pre-treatments of NIR spectra is essential, prior to integration of the NIR method as a method for reliably evaluating thawed-chilled MAP cod fillets (Bøknæs et al., 2002). According to Nielsen et al. (2002): "The parameter of freshness, expressed as storage time, in cod (*G. morhua*) and salmon (*S. salar*) was determined using visible/near infrared (VIS/NIR) spectroscopy. Modelling of the correlation between spectral data and storage time was carried out using multivariate statistics. In the case of cod, the best-fit model was provided using the visible wavelength range (correlation of prediction of 0.97 with an error value of

1.04 d), while as regards salmon, the best-fit model was given using data from the NIR range (correlation of prediction of 0.98 with an error value of 1.20 d). Therefore, VIS/ NIR spectroscopy proved to be an effective tool toward evaluating fish freshness."

NIR spectroscopy in the spectral range of 1000–2500 nm, was used to analyse brine from barrel salted herring, in an effort to assess the use of NIR as a rapid technique for determining protein levels. A PCA performed at the NIR spectra separated two groups, samples stored for up to 100 days and samples stored for more than 100 days. A partial least-squares regression model between pre-determined regions of the NIR spectra and the protein content resulted in a correlation coefficient of 0.93 and a prediction error (RMSECV) of 0.25 g/100 g. It was therefore shown that NIR spectroscopy can be used to rapidly and non-invasively assess the protein content in brine from barrel salted herring, which can be used to indicate the ripening quality of barrel salted herring (Svensson et al., 2004). Huang et al. (2002) examined the determination of salt and moisture levels in cold-smoked salmon using SW-NIR reflectance spectroscopy (600 to 1100 nm). PLS regression models gave the best results between three linear regression methods examined. Back-propagation neural networks (BPNN) was found to be slightly better in modelling salt and moisture concentrations (Salt: $R^2 = 0.824$, RMS = 0.55; Moisture: $R^2 = 0.946$, RMS = 2.44) than PLS (Salt: R2 = 0.775, RMS = 0.63; Moisture: $R^2 = 0.936$, RMS = 2.65). By selecting samples from different axial locations of the fish, no effects on the prediction error for salt or WPS was observed, but it did affected the prediction error for moisture. In another study, the determination of moisture (49.70 to 74.20% w/w) and salt (0.13 to 12.30% w/w) levels in cured Atlantic salmon (*S. salar*) or teijin was performed using SW-NIR reflectance spectroscopy (600 to 1100 nm) using PLS regression and artificial neural networks (ANN) calibration methods. ANN and PLS gave similar results (Salt: ANN RMS = 1.43% w/w, PLS RMS = 1.37% w/w; Water, ANN RMS = 2.08% w/w, PLS RMS = 2.04% w/w). Selection of samples from the dorsal or ventral portion of the fish had no effect on the prediction error of the salt or moisture models (Huang et al., 2003). In the study of Isaksson et al. (1995), NIR diffuse spectroscopy was employed for determining the fat, moisture and protein levels of whole and ground farmed Atlantic salmon fillets. A remote fibre-optic probe was applied for NIR measurements on 50 whole salmon fillets. The results extracted ranged from 91 to 205 g kg^{-1} for fat, 599 to 709 g kg^{-1} for moisture and 186 to 209 g kg^{-1} for protein. Principal component regression resulted in the following prediction errors for ground salmon fillets, expressed as root mean square error of cross validation: 6.6 g kg^{-1} fat, 3.8 g kg^{-1} moisture and 2.0 g kg^{-1} protein. On the other hand, the prediction errors for non-destructively measuring whole salmon fillets were 104 g kg^{-1} fat, 8.5 g kg^{-1} moisture and 3.7 g kg^{-1} protein. Regression models of 760–1100 nm range were characterised by lower prediction errors than models using the 1100–2500 nm or 760–2500 nm ranges. It was shown that fibre-optic probe NIR instruments could be used for the accurate determination of fat and moisture in whole salmon fillets.

NIR spectroscopy analysis was employed for predicting the chemical composition and identifying the rearing system of 236 European sea bass obtained from four Italian fish farms (extensive ponds, semi-intensive ponds, intensive tanks and intensive sea-cages). The samples were prepared in three different ways: intact fillet portions, whole fresh minced fillet and freeze-dried minced fillet. It was proved that NIR spectroscopy was very reliable in predicting the chemical composition of sea bass fillets but not as effective for predicting crude protein. NIR spectroscopy predicted the chemical composition of fresh minced fillets more accurately than the composition of intact fillets (Xiccato et al., 2004).

Lin et al. (2003) reported the development of a rapid method for determination of the iodine value (IV) and saponification value (SV) of fish oils by employing NIR spectroscopy. The PLS calibration model was relied on a spectral range of 7560 to 9100 cm^{-1} due to CH bond. The validation of the methodology was performed through comparison of the IV and SV of a series of fish oils, predicted by the PLS model, to the results extracted by the titration methods of the Japan Oil Chemists' Society. The NIR-predicted IV and SV were characterised by complete consistency with the chemically found IV and SV. The NIR method was more accurate and reproducible than the titration method. It could be used to successfully determine IV and SV of fish oils, as well as for the accurate and rapid determination of vegetable oils (2 min/sample) (Endo et al., 2005).

Lin et al. (2003) reported the development of PLS based SW-NIR prediction models for salt levels in commercial hot smoked fillets of king Chinook (*O. tshawytscha*) (N = 140; 212–468 g) and chum salmon (*O. keta*) (N = 120; 137–356 g). Spectra were obtained from the diffuse reflectance mode (600–1100 nm). The salt level ranged from 1.66 to 5.95% w/w and 2.15 to 5.69% w/w for king and chum salmon, respectively. The moisture ranged from 50.7 to 71.6% w/w for king and 55.5 to 69.7% w/w for chum salmon. The optimum PLS model for salt employed eight latent variables for king salmon (R^2 = 0.83, SEP = 0.32% w/w) and eight latent variables for chum salmon (R^2 = 0.82, SEP = 0.25% w/w).

Near infrared transmittance (NIT) spectroscopy (850–1048 nm) was used to measure fat, protein and dry matter in wet homogenized Atlantic halibut (*Hippoglossus hippoglossus*) fillet. A total of 155 fillet samples were obtained at the start of the experiment (survey I) and 98 fillet samples were obtained from eight fish groups six months later (survey II). Separate testing of the multivariate calibration models was performed for each survey and after mixing the survey samples. Testing of the models was also performed after taking into account the samples' temperature, and their weights, or after reducing to spectra of eight wavelengths. The samples of survey II resulted in calibration models with optimum prediction abilities for fat and dry matter, while by combining samples from both surveys, optimum prediction potential for protein was obtained. The constituent ranges (w/w) were as follows: 10–123 g/kg fat, 165–274 g/kg protein and 234–335 g/kg dry matter. The partial least squares regression gave prediction errors (expressed as root-mean-square error of cross validation) of 2.7 g/kg fat, 5.2 g/kg protein and 4.2 g/kg dry matter. The temperature adjustment positively affected the test set validation of protein and dry matter, while the fish-weight adjustment positively affected the test set validation of fat. Calibration models using eight wavelengths presented deviations, but were promising in terms of applicability in simpler instruments for future online quality monitoring of the product (Nortvedt et al., 1998).

5.1.3 Application of Fluorescence Spectroscopy for Authenticating Fish and Seafood

Fluorescence spectroscopy is generally characterised by 1–3 orders of magnitude higher sensitivity in comparison to absorption spectroscopy techniques. In fluorescence spectroscopy, the analyzed signal is electromagnetic radiation emitted from the analyte as it relaxes from an excited electronic energy status to its ground state. The higher energy level of the analyte is achieved through absorbing radiation in the UV or Vis range. The activation and deactivation procedures are performed at the same time during a fluorescence measurement. Each molecular system will have an optimum radiation wavelength for sample excitation and another, longer wavelength, for monitoring fluorescence emission. The wavelengths for excitation and emission are dependent on the chemistry of the system of interest (Penner, 2010). It is a rapid, sensitive, and non-destructive analytical technique, which can provide spectral signatures in just a few seconds. These spectra can be used as fingerprints of the foods examined. The use of fluorescence in food analysis has been significantly enhanced during the last years, mainly due to the increased use of chemometrics (Sádecká and Tóthová, 2007).

Evaluation of the intrinsic fluorescence of fish muscle was performed to rapidly and non-destructively monitor fish freshness. The fluorescence emission spectra of aromatic amino acids and nucleic-acids (excitation: 250 nm, emission: 280–480 nm), tryptophan residues (excitation: 290 nm, emission: 305–400 nm) of proteins and NADH (excitation: 336 nm, emission: 360–600 nm) were analyzed for cod, mackerel, salmon and whiting fillets after a period of 1, 5, 8 and 13 days of storage. PCA and Mahalanobis distance methods were employed. As regards the mackerel, the similarity map obtained by the principal components 1 and 2 indicated that the aromatic amino acids and nucleic acids spectra obtained after one day of storage could be discriminated from the spectra obtained after 5 and 8 days of storage in accordance with the principal component 1. Similar results were found for all the fish species and the fluorophores examined. According to this study, "the intrinsic fluorescence spectra could be potentially used to discriminate between fresh and aged fish fillets (Dufour et al., 2003)."

Inductively coupled plasma double focusing sector field mass spectrometry (ICP-SFMS) and multi-collector inductively coupled plasma mass spectrometry (MC-ICP–MS) were employed to differentiate the origin of vendace whitefish caviars and brackish freshwater caviars. Differences in elemental concentrations or sample-specific isotopic composition (Sr and Os) were identified. The determination of the concentrations of 72 elements was performed using ICP-SFMS following microwave-based digestion. Vendace and whitefish caviar products originated from Sweden (from both brackish and freshwater), Finland and USA, as well as untreated vendace roe and salt used in caviar production were examined. This data set can be used to identify elements whose contents in caviar can be altered due to salt addition and contamination issues during production and packaging. The long-term reproducibility of all analytes was evaluated based on replicate caviar preparations/analyses, and the variations in element levels in caviar derived from different harvests were assessed. The highest degree of differentiation was shown for elements with varying levels between brackish and freshwaters (e.g., As, Br, Sr). Elemental ratios (Sr/Ca, Sr/Mg and Sr/Ba) can be successfully employed to authenticate vendace caviar obtained from brackish water roe, as a result of the differences between caviar from different sources, low between-harvest variations and existence in relatively high levels in samples, allowing them to be accurately determined using modern analytical techniques. Differences in the $^{87}Sr/^{86}Sr$ ratio for vendace caviar from different harvests (on the order of 0.05–0.1%) is more than 10 times lower in comparison to differences between caviar processed from brackish and freshwater roe. As a result, Sr isotope ratio determinations using either ICP-SFMS or MC-ICP–MS, can be effectively used to differentiate the origin of the different samples. On the other hand, the differentiation between Swedish caviar obtained from brackish water roe and Finnish freshwater caviar was not possible by solely relying on $^{187}Os/^{188}Os$ ratios (Rodushkin et al., 2007).

5.1.4 Use of Electrophoretic Techniques for the Authentication of Fish and Seafood

The determination of the origin of foods can be carried out using electrophoretic procedures. These techniques can be very useful when the electropherograms of the protein extracts reveal protein zones or bands specific for the protein source. The analyses are carried out through extraction of the sarcoplasm proteins using water and conduction of the electrophoretic separation mainly on polyacrylamide gels (but also starch and agarose gels may be used). The application of a pH gradient [Isoelectric focusing (IEF)] can lead to great protein patterns (Belitz et al., 2009). IEF and two-dimensional electrophoresis (2-DE) were employed for distinguishing four freshwater fish marketed as "perch": *Perca fluviatilis* (European perch), *L. niloticus* (Nile perch), *Stizostedion lucioperca* (European pikeperch) and *Morone chrysops* × *saxatilis* (sunshine bass). The above-mentioned species cannot be easily identified in their commercial form since they are usually sold in the form of fillets. IEF of the water-soluble proteins extracted from the samples was resolved in species-specific patterns. It was shown that the intra-species polymorphism was quite low, and did not significantly affect the bands characterising each species. Furthermore, 2-DE maps presented various species-specific protein spots. It is important to mention that while no IEF band was common to all four species, numerous major 2-DE spots presented high similarity. As a result, IEF of water-soluble sarcoplasmic proteins can be effectively used for the accurate discrimination between the four species examined. Analysis using 2-DE, which is characterised by higher resolution, but also higher cost and higher time requirements, may be used to obtain further knowledge on the proteome of poorly characterised species (Berrini et al., 2006). In the study of Mackie et al. (2000), sodium dodecylsulphate polyacrylamide gel electrophoresis (SDS-PAGE), urea-isoelectric focusing (urea-IEF) and native isoelectric focusing were employed in an effort to identify species of smoked salmonids, gravad salmonids and smoked eels. The use of SDS-PAGE led to minor changes in the profiles of the samples rendering the identification of closely related species extremely difficult. The use of urea-IEF led to fewer changes in the profiles thus enhancing the species-discrimination power of the samples. The profiles of the eel species, as obtained on SDS-PAGE or urea-IEF, were not affected by smoking. Urea-IEF was characterised by higher discrimination power than SDS-PAGE for the eel species. Native isoelectric focusing could be successfully used to provide supplementary identification on species, whose identification is difficult using SDS-PAGE or urea-IEF.

High resolution 2-DE has also been employed in an effort to differentiate between wild and farmed cod (*G. morhua*) as well as for the assessment of the protein composition of klipfish. Tris and CHAPS-urea extracts obtained from wild and farmed cod muscle and rehydrated cod klipfish fillets were subjected to analysis using 1DE and 2DE. It was proved that 2DE maps of tris extracts from farmed cod differed from the wild in a series of spots of Mw 35 and 45 kDa. The CHAPS-urea extracts obtained from farmed cod presented numerous spots of Mw between 100 and 45 kDa, which could not be easily detected in wild cod, while were very prominent in klipfish. Klipfish differed significantly from the other samples. Specifically, the myosin heavy chain could not be easily detected and the tris extracts and the CHAPS-urea were characterised by fewer and more spots, respectively, in comparison to the corresponding extracts obtained from raw samples. Further identification of these potentially diagnostic spots will facilitate the differentiation of farmed and wild cod and the assessment of klipfish processing on the protein level of the samples (Martinez et al., 2007).

In another study, the effectiveness and reliability of urea IEF and SDS-PAGE for identifying cooked fish flesh were evaluated in a collaborative study among nine laboratories. Urea IEF was carried out using CleanGels and ImmobilineGels, while ExcelGels were applied in the case of SDS-PAGE, allowing the three types of gels to be run in the same flatbed electrophoresis chamber. The use of urea IEF led to 34 out of 35 correct identifications, while the use of SDS-PAGE was similarly successful. It was proved that the above-mentioned methods could be effective for validating the accuracy of the labelling of fishery products (Rehbein et al., 1999).

5.1.5 Implementation of Multivariate Analysis towards Authenticity of Fish and Seafood

Nowadays authenticity of foods and fish in particular is becoming extremely important due to the high number of adulteration incidents. However, at the same time, authenticity control is becoming easier due to the development of different rapid physicochemical and microbiological techniques, which can be used to distinguish one species from another using scientifically approved methodology. It is approved, however, that despite the fact that analytical and protein and DNA-based techniques may be very precise and accurate, detection of authenticity could not be claimed without resorting to multivariate analysis (Arvanitoyannis et al., 2005).

In the study of Balladin et al. (1998), analysis of seventy-eight samples of *Scomberomorus brasiliensis* (carite) muscle tissue was performed for determining the fat, hypoxanthine (determined by a flow injection method), total volatile acids and bases contents, the total bacterial count, and several textural characteristics such as firmness, hardness, fracturability, cohesiveness, chewiness, and elasticity. Discriminant and factor analyses were carried out and the apparent error rate obtained using discriminant analysis with all the parameters in the analysis was about 5.6%. However, classification did not improve (apparent error rate 6.9%) when cohesiveness and chewiness were not included in the analysis. The stepwise methods of selection (Wilks and Minresid) H × FI ranked second, and fourth when the Mahal and Maxminf methods were applied. In the factor analysis, the use of PCA led to the extraction of five factors. Factor loadings relate factor 1 (tastiness) with hypoxanthine concentration, total volatile acids and bases: factors 2, 3, 5 (texture) with the textural attributes; and factor 4 (rancidity) with the % fat content. Rotation of the reference axes indicated that factors 1, 2 and 5 did not change, while factors 3 and 4 were interchanged.

According to Bechmann et al. (1998), several physicochemical and organoleptic characteristics were determined for 115 cod (*G. morhua*) samples stored under different frozen storage conditions. Five different process parameters (period of frozen storage, frozen storage temperature, place of catch, season for catching and state of rigour) varied systematically at two levels. The evaluation of the extracted data was performed using the multivariate methods, PCA and PLS regression. The PCA models were employed for the identification of the most significant factors in terms of quality of the frozen cod. Generation of PLS models used for predicting the physicochemical and organoleptic quality parameters from the process parameters of the frozen raw material were carried out. The prediction abilities of the PLS were sufficient for providing reasonable results even when the process parameters were characterised by ones and zeroes only. In another study, the freshness factor of sea bass stored at 4°C (and 1°C with ice covering) was assessed using sensory analysis, descriptors and instrumental parameters [rigour index (RI), dielectric properties, K_I, compactness (IC) and resilience (IEL), pH, free water (FW) and cooking losses (CL)]. The sensory scores were highly repeatable (≥ 0.70). The PCA of sensorial parameters indicated that Factor 1, accounting for a 95% of the variation and linked to storage duration, was relatively homogenous. The existence of an antithesis between descriptors of external characteristics of fish and descriptors of internal parts was found, especially in the intermediate phase of the shelf life—which was highly affected by temperature. The PCA of the organoleptic and physicochemical characteristics together indicated that the characteristics most highly correlated with the first factor (83% of overall variance) were sensorial [(loading scores > 0.96), K_I freshness index, rigour index and dielectric properties (loading scores = −0.96, 0.73 and 0.71, respectively)]. It was therefore proved that the external characteristics of the fish alone could be used as indicators of freshness. The K_I, rigour index and dielectrics properties can be employed to objectively measure freshness. Sensory ratings are less capable of analysing freshness at the later stages of shelf life (Parisi et al., 2002).

5.1.6 Use of Nuclear Magnetic Resonance for Ensuring the Authenticity of Fish and Seafood

It is likely that DNA-based techniques will be the most frequently applied ones for identifying species, mainly due to the fact that they are easy to use. Even after severe degradation of the DNA, both identification and quantification could be performed using quantitative PCR. However, DNA-based methods are not effective in determining the geographical origin of samples. For this purpose, spectroscopic techniques will most likely be used. For the differentiation between wild and cultivated fish, however, either trace element analyses or nuclear magnetic resonance (NMR) techniques are the most effective (Martinez et al., 2005). The use of MR for authenticating fish products is mentioned in many studies. Some of the most important efforts to use this technique are described below.

In the study by Gribbestad et al. (2005), the spectra from high-resolution ^{1}H MR spectroscopy of extracts, muscles, and whole Atlantic salmon, were interpreted. For the analysis of whole fish, an MR image localisation technique was employed. The individual components were identified through comparing the published values of chemical shifts, using the knowledge of the biochemical composition of salmon skeletal muscle and spiking samples, with authentic compounds. Single chemical substances, such as hypoxanthine, amino acids, anserine, lactate and some fatty acids were identified in extracts, whole muscle and whole fish. These methods could be applied for selecting live specimens for breeding and for classifying both live specimens and fillets in accordance with the quality and quantity of lipids and small molecules, connected with the nutritional value of fish, as well as for authentication purposes (Gribbestad et al., 2005). Bonny et al. (2001) employed an MR imaging technique relied on susceptibility-induced contrast for the visualisation of the spatial distribution of connective tissue in meat. MRI of bovine meat samples was performed using a high-field 4.7T imager. Comparison of the MR images obtained with spin-echo and gradient-echo sequences was performed aiming at elucidating how the connective tissue affects the additional signal losses found in the gradient-echo images. The reconstruction of the T_2^* maps was carried out from the multiple gradient-echo images, which give quantitative information. By comparing them with histological pictures, it was shown that these T_2^* maps exhibit the overall organisation of the primary perimysium at the scale of the whole muscle. The distinct perimysial organization found between the Gluteo biceps and *Pectoralis profundis* muscles indicates that MRI could be used for the characterisation of the muscle connective tissue structure.

In another study, gas chromatography (GC), isotope ratio mass spectrometry, and high-resolution ^{2}H site-specific natural isotope fractionation/nuclear MR spectroscopy were applied to examine several types of fish oils and lipids obtained from muscle samples of wild and farmed salmon (Norway, Scotland). The fatty acid contents, overall ^{2}H and ^{13}C isotope ratios, and molar fractions of the isotopomeric deuterium clusters were statistically analyzed in an effort to select the most efficient variables to distinguish the different groups of salmons and fishes examined. A classification analysis using four fatty acid compositions, three deuterium molar fractions, and the overall $(D/H)_{tot}$ isotope ratio of fish oils accurately classified the oils to the right group (Aursand et al., 2000).

Phosphorus-Nuclear MR (^{31}P-NMR) was applied for the evaluation of the freshness of loach muscle depending on metabolic changes of high energy phosphate compounds. The phosphocreatine ([PCr])/inorganic phosphate ([Pi]) ratio was proved to be an accurate index of early metabolic hypofunction. Ratios of [PCr]/P-phosphate of ATP ([β-ATPI), and [Pi]/[β-ATP] were successfully used for estimating such metabolic changes in fresh fish. The intramuscular levels of creatine phosphate, ATP and pH remained higher in blood-drained loach in comparison to the untreated fish. The freshness of samples bled and washed under lower-temperature conditions, was better preserved (Chiba, 1991).

The investigation of changes in the muscle samples of cod (*Gadus rnorhua*) and haddock (*Melanogrammus aeglejnus*) using high-resolution NMR and magnetic resonance imaging (MRI) was performed by Howell et al. (1996). Water- and salt-soluble extracts obtained from samples preserved at $-20°C$ and $-30°C$ were subjected to analysis. Specifically, high-resolution proton NMR was used enabling accurate identification of metabolites such as trimethylamine oxide, trimethylamine (TMA) and dimethylamine. The detection of formaldehyde could not be carried out using NMR neither in the stored fish samples nor in spiked water or salt extracts even when high quantities of formaldehyde were added, probably due to polymerisation. The use of systematic and controlled storage trials revealed that dimethylamine was present at around nine months for samples stored at $-20°C$, but no such changes were found in the case of the control storage temperature of $-30°C$. Cod and haddock fillets preserved for a period of one year at -20 and $-30°C$ were compared, confirming that dimethylamine was produced only in cod stored at $-20°C$. It is important to mention that 'fresh' cod and haddock obtained from a supermarket presented high levels of TMA, which indicates that trimethylamine oxide had been broken down to TMA by bacteria. No detection of TMA was carried out in the fish fillets obtained for the storage trials. MRI of fresh cod and fish stored at -8 and $-30°C$ clearly showed that the fish half stored at $-8°C$ was characterized by the formation of dense lines or arches which indicate the existence of gaps in the tissue due to possible breakdown of the connective tissue. The images of fish stored at $-30°C$ did not present any differences upon comparison with fresh samples. MRI also indicated the existence of frozen and unfrozen areas in the fish in a non-destructive way.

In the study of Igarashi et al. (2000), a 500 MHz proton NMR (^1H NMR) spectrometer was used for developing a quantitative method for the determination of the levels of docosahexaenoic acid (DHA) in fish oils (mg/g), the molar proportions (mol%) of DHA to all other fatty acids which form parts of the fish oils, and the molar proportions of total n-3 fatty acids to all other non-n-3 fatty acids in the fish oils. The examination of ethylene glycol dimethyl ether (EGDM), methanol, and 1,4-dioxane as internal standards allowed the optimization of the experimental conditions. EGDM was basically employed as internal standard. The pulse repetition time was set at 30 s, which was five times longer than the longest T1 of the ^1H NMR signals of fish oils, allowing the generation of reproducible data and analytical times shorter than 10 min. By using the internal standard, quantification of DHA was also carried out on a weight basis (mg/g). Good agreement was found between the ^1H NMR data and data obtained using GC. The sample preparation prior to using ^1H NMR measurements is carried out through weighing the sample and preparing an internal standard solution. Due to the fact that this method was highly reproducible and simple, it was suggested as a promising alternative to the GC method for quantifying DHA and n-3 fatty acids in fish oils (Igarashi et al., 2000). In another study, the potential of low field NMR (LF NMR) for rapidly estimating the quality of hake (*Merluccius merluccius*) stored at $-10°C$ for upto six months was assessed. LF NMR identified three types of water: water strongly bound to macromolecules (T_{2b}), trapped water (T_{21}) and free water (T_{22}). By increasing the storage time, and concomitant with an increase in the T_{22} and a decrease in the T_{21}, the water holding capacity and apparent viscosity values decreased while the shear strength increased, demonstrating the loss of juiciness and the development of tougher texture. The construction of two mathematical models was also reported: a simple regression using the bi-exponential analysis of the relaxation times (T_{21}, T_{22}) and amplitudes (A_{21}, A_{22}) and a PLS regression of CONTIN analysis. Both models were judged to be effective for estimating the quality of the samples (Sánchez-Alonso et al., 2012).

According to Standal et al. (2010), evaluation of the potential use of phospholipid profiles obtained by ^{13}C NMR spectroscopy for separating species of lean gadoid fish was carried out. ^{13}C NMR data were extracted by analyzing muscle lipids of five types of lean gadoid fish [north-east arctic cod and Norwegian coastal cod (*G. morhua*), haddock (*Melanogrammus aeglifinus*), saithe (*Pollachiusvirens*), and pollack (*P. pollachius*)]. Analysis of 27 fish caught at the same location of the Norwegian coast was performed. Investigation of the sn-2 position specificity of 22:6n-3 (docosahexaenoic acid, DHA) in phosphatidyl choline (PC) and phosphatidyl ethanolamine (PE) for the various species/stocks was also performed, and the full ^{13}C NMR spectra were used for the multivariate analysis. Stereospecific distribution calculations indicated that the different species significantly differed in the distribution of 22:6n-3 in PC and PE, and the pollack group demonstrated the lowest values for 22:6n-3 in sn-2 position, both in PC and PE. This first screening proved that by applying ^{13}C NMR fingerprint of muscle lipids, linear discriminant analysis correctly classified 78% of samples in accordance with the five categories of lean gadoid fish, while Bayesian belief networks (BBN) successfully classified all samples.

In the study of Nott et al. (1999), MRI was employed for visualising the main organs and muscular-skeletal framework of fresh rainbow trout (*S. gairdneri*) in two dimensions, and for identifying the spatial distribution of lipid and collagen-rich tissues. Quantitative MRI can be used for providing the MR parameters [T_1, T_2, M_0, T_1^{sat}, M_{sat}/M_0, and the Magnetisation Transfer (MT) rate] and the tissue water variations in these can be used for differentiating among freshly killed and frozen–thawed trout. This study aimed at demonstrating the type of anatomical resolution that can be achieved between individual organs of intact trout by using MRI and also at exploring the use of MRI as a technique for authenticating fresh trout and differentiating it from frozen-thawed products. It was found that MRI could be successfully used for the visualisation of the anatomical characteristics of rainbow trout by highlighting different soft tissues through combining protocols that provided different contrasts. This is highly important for monitoring fish quality since the fat distribution and the connective tissue levels highly affect quality. Application of quantitative MRI resulted in parameters sensitive to the effects of freeze–thawing as well as to the method of freezing and duration of frozen-storage. No significant changes in the MR factors were observed when liquid nitrogen was used for freezing the samples, while the MT rate significantly changed when slow freezing was used. For the trout steaks T_1, T_1^{sat} and MT rate were the most prominent effects. In the case of previously frozen samples, the change in the MR parameters after repeat freeze–thawing was lower in comparison to the changes found during the first two day freeze–thaw of fresh fish, thus proving that the method could be suitable for examining fresh trout. It was also suggested that the proposed methodology can be used for seawater fish, crustaceans and mollusks, whose qualitative characteristics degrade through mistreatment but which are not as commonly available in the 'fresh' state.

Support vector machines (SVMs) were employed in the study of Masoum et al. (2007) in an effort to authenticate the origin of salmon. SVMs offer the advantage of using an extensively studied theory and have already successfully tested in various practical applications. An innovative method was suggested for discriminating between wild and farm salmon, eliminating any possibility of fraud through misrepresentation of the country of origin of salmon. The samples are prepared following a simple technique that includes extraction of the fish oils from the white muscle. ^{1}H NMR spectroscopic analysis can be applied for the effective analysis of the fatty acid components of the fish oils. The SVM correctly distinguished between wild and farmed salmon, but it was finally proved that 5% of the country of origin was not classified correctly.

5.1.7 Application of Immunological and Protein-based Methods for the Authentication of Fish and Seafood

The immunological detection of fish species used in boiled and dried fish products (dried fish sticks) can be performed through examination of the availability of antiserum against myosin light chains. Specifically, the samples are dissolved in the presence of 8 M urea and 1% sodium dodecyl sulphate (SDS), and solubilized matters are used for the SDS–polyacrylamide gel electrophoresis and electroblotted onto PVDF membranes. Subsequently, the protein bands of interest are stained using anti-myosin light chain (alkali light chain 1) rabbit antiserum and horseradish peroxidase-conjugated anti-rabbit immunoglobulin. Using fingerprinting of immunostained patterns, the majority of the fish species can be identified, even after drastic processing, but the protein staining patterns are not clear enough to be used for species identification (Ochiai and Watabe, 2003).

The development of an immunoassay was performed with the aim of facilitating the discrimination between canned whole sardine (*S. pilchardus* Walbaum) and other fish species (Pacific pilchard, mackerel, herring, sild and anchovy) that may be used as its substitutes in canned products. The non-competitive indirect ELISA employs an antiserum, which is raised against a crude water-soluble extract of canned sardine. Sufficient antiserum specificity was obtained by using extracts of those fish, which should not cross-react with, to prevent any post-production undesirable cross-reactions Taylor and Jones (1992). According to Asensio et al. (2003), a monoclonal antibody (MAb) generated against soluble muscle proteins from grouper (*E. guaza*) was applied in two indirect ELISA formats (microtiter plates and immunostick tubes) to rapidly authenticate grouper fillets. The 3D12 MAb was generated using the hybridoma technique and testing was performed using ELISA widely consumed fish species. The 3D12 MAb effectively identified grouper samples and therefore could be effective for discriminating grouper among other fish species of lower cost.

In another study, native isoelectric focusing of water-soluble sarcoplasmic proteins was used to identify 14 commercially important shrimp species, which belong to the order Decapoda. These species were characterised by different commercial values as well as many phenotypic similarities. Due to the fact that since the carapace is also removed during their industrial processing, incorrect food labelling and deliberate or unintentional adulteration can occur. Each of the 14 tested species presented species-specific protein band profiles and low intra-specific polymorphism, thus allowing for the accurate differentiation between them. Therefore, IEF of water-soluble sarcoplasmic proteins could be used to identify the 14 species examined (*Penaeusmonodon,*

P. semisulcatus, Farfantepenaeus notialis, F. aztecus, F. brevirostris, F. merguiensis, F. indicus, Litopenaeus vannamei, Parapenaeus longirostri, Marsupenaeus japonicus, Melicertus latisulcatus, Pleoticus muelleri, Solenocera agassizii and *P. borealis*). Furthermore, sarcoplasmic calcium-binding proteins (SCPs) were identified by tandem mass spectrometry (MS/MS) as the most important species-specific proteins of the examined species, indicating the need for further studies in order to examine their potential use as specific biomarkers (Ortea et al., 2010).

5.1.8 Further Technologies with Potential Uses in Fish Authentication

Other techniques that have been used for authenticating fish and seafood include HPLC techniques, Fourier transform infrared (FTIR) spectroscopy and the use of portable electronic nose. Description of a method for separating sarcoplasmic fish proteins using Reversed Phase-High Performance Liquid Chromatography (RP-HPLC) was made in the study of Knuutinen and Harjula (1998). It was proved that there were significant differences that could be used to reliably identify fish species. The differentiation of sixteen of the most common Finnish freshwater fish species was performed by employing species-specific High Performance Liquid Chromatography (HPLC) chromatograms obtained with the use of photodiode array detection (PAD) at 200–350 nm. The analytical column was a Hi-Pore RP-304 reversed-phase column. Separation was carried out using a linear gradient of acetonitrile and water with low levels of trifluoracetic acid (TFA). The construction of star-symbol plots was performed using chromatograms with the aim of visualising the data. Clearly different HPLC protein profiles were produced for the majority of the fish species. The chromatograms of salmonoids presented some similarities, whereas the protein profiles of cyprinids were highly different. Minor differences between the same species were found for three types of powan (*Coregonus lavaretus*).

Due to the fact that some vegetable oils such as canola (CaO), corn (CO), soybean (SO), and walnut (WO) oils are characterised by colour similar to that of cod liver oil (CLO), the potential presence of these oils cannot be easily detected using naked eye. Thus, Fourier transform infrared (FTIR) spectroscopy using horizontal attenuated total reflectance (HATR) as sampling accessory and combined with chemometrics was used to detect and quantify these vegetable oils as adulterants in CLO. The vegetable oils were quantified using multivariate calibrations of PLS and principal component regression, while pure CLO and CLOs adulterated with CaO, CO, SO, and WO were classified by employing discriminant analysis (DA). PLS with FTIR normal spectra was more effective in comparison to principal component regression for quantification purposes with coefficient of determination (R2) higher than 0.99 and root mean square error of calibration (RMSEC) ranging from 0.04 to 0.82% (v/v). The PLS model was also applied for the prediction of the quantities of these vegetable oils in independent samples in an effort to validate the methodology. The RMSEP values obtained were: 1.75% (v/v) (CaO), 1.39% (v/v) (CO), 1.35% (v/v) (SO), and 1.37% (v/v) (WO), respectively. It was proved that DA can be used for the classification of CLO as well as CLO mixed with these vegetable oils using nine principal components (Rohman and Che Man, 2011).

The development of a portable electronic nose was reported in the study of O'Connell et al. (2001) who used it to determine the fish freshness determinations of Argentinean hake. The methodology used is different from the methods, which are relied on the headspace method or glass syringe manipulation. Specifically, in this work, the introduction of a weighed piece of fish was performed into a sensor's chamber and the signals of the sensors due to the fish emissions are recorded as function of time. Commercial gas sensors, based on tin dioxide, were employed. The individual sensor signals and the array's pattern were analyzed. An increase of the signals was reported by increasing storage days and the mass of the hake pieces (up to 50 g). The obtained response patterns did not depend on the weight and changed storage, indicating the rottenness of the sample. Two different patterns were reported, which were related to rotten and non-rotten samples, respectively. These patterns did not depend on storage conditions. For example, the pattern of a rotten sample would be the same regardless whether the sample was stored for some days in a refrigerator or whether it was stored for one day at ambient temperature. Therefore, it was proved that this is a rapid and easy method for differentiating between fresh and old samples. The results were confirmed using PCA.

5.2 Conclusions

Substitution of fish species is becoming more and more important within the food industry and the need for rapid, reliable, and reproducible tests for fish authentication is now dramatically increasing. Complex international trade routes and seafood consumption, along with fluctuations in the supply chain and different consumers' requirements may lead to intentional product mislabelling. The effects of this phenomenon may include economic fraud, health hazards, and illegal trade of protected species (Rasmussen and Morrissey, 2008). Authentication of fish and seafood, including authentication of the quality of the products, can be performed using a vast variety of methods including different PCR methods, NIR, fluorescence spectroscopy, electrophoretic techniques, magnetic resonance, multivariate analyses, etc.

The illegal mislabelling of fish and seafood can detrimentally affect the industry and the consumers. These effects could be prevented through developing species authentication techniques that can rapidly and reliably authenticate these foodstuffs (Rasmussen and Morrissey, 2008). Although the techniques currently used are usually very effective and reproducible, the development of even more reliable and easily applicable techniques is required. Further studies should examine the applications of techniques that could be practically used in commercial procedures and could systematically be applied by authorities and the industry to verify the authenticity of fish and seafood.

References

Adamopoulos, K.G. and A.M. Goula. 2004. Application of near-infrared reflectance spectroscopy in the determination of major components in taramosalata. Journal of Food Engineering. 63: 199–207.
Aranceta-Garza, F., R. Perez-Enrique and P. Cruz. 2011. PCR-SSCP method for genetic differentiation of canned abalone and commercial gastropods in the Mexican retail market. Food Control. 22: 1015–1020.

Ardura, A., I.G. Pola, A.R. Linde and E. Garcia-Vazquez. 2010. DNA-based methods for species authentication of Amazonian commercial fish. Food Research International. 43: 2295–2302.

Armani, A., L. Castigliego, L. Tinacci, D. Gianfaldoni and A. Guidi. 2012. Multiplex conventional and real-time PCR for fish species identification of Bianchetto (juvenile form of *Sardina pilchardus*), Rossetto (*Aphia minuta*), and Icefish in fresh, marinated and cooked products. Food Chemistry. 133: 184–192.

Arvanitoyannis, I.S., E.V. Tsitsika and P. Panagiotaki. 2005. Implementation of quality control methods (physico-chemical, microbiological and sensory) in conjunction with multivariate analysis towards fish authenticity. International Journal of Food Science and Technology. 40: 237–263.

Asensio, L., I. González, M.A. Rodríguez, B. Mayoral, I. López-Calleja, P.E. Hernández, T. García and R. Martín. 2003. Development of a Specific Monoclonal Antibody for Grouper (*Epinephelus guaza*) Identification by an Indirect Enzyme-Linked Immunosorbent Assay. Journal of Food Protection. 66: 886–889.

Asensio, L., I. González, M. Rojas, T. García and R. Martín. 2009. PCR-based methodology for the authentication of grouper (*Epinephelus marginatus*) in commercial fish fillets. Food Control. 20: 618–622.

Aursand, M., F. Mabon and G.J. Martin. 2000. Characterization of Farmed and Wild Salmon (*Salmo salar*) by a Combined Use of Compositional and Isotopic Analyses. Journal of the American Oil Chemists' Society. 77: 659–666.

Balladin, D.A., D. Narinesingh, V.A. Stoute and T.T. Ngo. 1998. Multivariate statistical (using factor and discriminant) analyses of some selected chemical and physical freshness indicators of fish [*Scomberomorus brasiliensis* (carite)] muscle tissue samples. Chemometrics and Intelligent Laboratory Systems. 40: 175–192.

Bechmann, E.I. and B.M. Jørgensen. 1998. Rapid Assessment of Quality Parameters for Frozen Cod Using Near Infrared Spectroscopy. Lebensmittel—Wissenschaft & Technologie. 31: 648–652.

Bechmann, E.I., H.S. Jensen, N. Bøknæs, K. Warm and J. Nielsen. 1998. Prediction of Chemical, Physical and Sensory Data from Process Parameters for Frozen Cod using Multivariate Analysis. Journal of the Science of Food and Agriculture. 78: 329–336.

Belitz, H.D., W. Grosch and P. Schieberle. 2009. Meat. pp. 563–616. *In*: Food Chemistry. Springer, Germany.

Berrini, A., V. Tepedino, V. Borromeo and C. Secchi. 2006. Identification of freshwater fish commercially labelled "perch" by isoelectric focusing and two dimensional electrophoresis. Food Chemistry. 96: 163–168.

Besbes, N., S. Fatouuch and S. Sadok. 2012. Differential detection of small pelagic fish in Tunisian canned products by PCR-RFLP: An efficient tool to control the label information. Food Control. 25: 260–264.

Bøknæs, N., K.N. Jensen, C.M. Andersen and H. Martens. 2002. Freshness Assessment of Thawed and Chilled Cod Fillets Packed in Modified Atmosphere Using Near-infrared Spectroscopy. Lebensmittel—Wissenschaft & Technologie. 35: 628–634.

Bonny, J.M., W. Laurent, R. Labas, R. Taylor, P. Berge and J.P. Renou. 2001. Magnetic resonance imaging of connective tissue: a non-destructive method for characterising muscle structure. Journal of the Science of Food and Agriculture. 81: 337–341.

Bossier, P., W. Xiaomei, F. Catania, S. Dooms, G.V. Stappen, E. Naessens and P. Sorgeloos. 2004. An RFLP database for authentication of commercial cyst samples of the brine shrimp *Artemia* spp. (International Study on *Artemia* LXX). Aquaculture. 231: 93–112.

Bossier, P. 1999. Authentication of Seafood Products by DNA Patterns. Journal of Food Science. 64: 189–193.

Bottero, M.T., A. Dalmasso, M. Cappelletti, C. Secchi and T. Civera. 2007. Differentiation of five tuna species by a multiplex primer-extension assay. Journal of Biotechnology. 129: 575–580.

Catanese, G., M. Manchado, A. Fernández-Trujillo and C. Infante. 2010. A multiplex-PCR assay for the authentication of mackerels of the genus Scomber in processed fish products. Food Chemistry. 122: 319–326.

Cawthorn, D.-M., H.A. Steinman and R.C. Witthuhn. 2012. DNA barcoding reveals a high incidence of fish species misrepresentation and substitution on the South African market. Food Research International. 46: 30–40.

Cawthorn, D.M., H.A. Steinman and R.C. Witthuhn. 2012. Evaluation of the 16S and 12S rRNA genes as universal markers for the identification of commercial fish species in South Africa. Gene. 491: 40–48.

Chen, C.H., C.H. Hsieh and D.F. Hwang. 2012. Species identification of Cyprinidae fish in Taiwan by FINS and PCR-RFLP analysis. Food Control. 28: 240–245.

Chen, C.H., C.H. Hsieh and D.F. Hwang. 2013. PCR-RFLP analysis using capillary electrophoresis for species identification of Cyprinidae-related products. Food Control. 33: 477–483.

Chiba, A., M. Hamaguchi, M. Kosaka, T. Tokuno, T. Asai and S. Chichibu. 1991. Quality evaluation of fish meat by phosphorus-nuclear magnetic resonance. Journal of Food Science. 56(3): 660–664.

Chiu, T.S., Y.C. Su, J.Y. Pai and H.C. Chang. 2012. Molecular markers for detection and diagnosis of the giant grouper (*Epinephelus lanceolatus*). Food Control. 24: 29–37.

Chuang, P.S., M.I. Chen and J.C. Shiao. 2012. Identification of tuna species by a real-time polymerase chain reaction technique. Food Chemistry. 133: 1055–1061.

Comi, G., L. Iacumi, K. Rantsiou, C. Cantoni and L. Cocolin. 2005. Molecular methods for the differentiation of species used in production of cod-fish can detect commercial frauds. Food Control. 16: 37–42.

Dufour, E., J.P. Frencia and E. Kane. 2003. Development of a rapid method based on front-face fluorescence spectroscopy for the monitoring of fish freshness. Food Research International. 36: 415–423.

Endo, Y., M. Tagiri-Endo and K. Kimura. 2005. Rapid Determination of Iodine Value and Saponification Value of Fish Oils by Near-Infrared Spectroscopy. Journal of Food Science. 70(2): 127–131.

Espiñeira, M. and J.M. Vieites. 2012. Rapid method for controlling the correct labeling of products containing common octopus (*Octopus vulgaris*) and main substitute species (*Eledone cirrhosa* and *Dosidicus gigas*) by fast real-time PCR. Food Chemistry. 135: 2439–2444.

Espiñeira, M., N. Gonzalez-Lavín, J.M. Vieites and F.J. Santaclara. 2009. Development of a method for the identification of scombroid and common substitute species in seafood products by FINS. Food Chemistry. 117: 698–704.

Espiñeira, M., J.M. Vieites and F.J. Santaclara. 2010. Species authentication of octopus, cuttlefish, bobtail and bottle squids (families Octopodidae, Sepiidae and Sepiolidae) by FINS methodology in seafoods. Food Chemistry. 121: 527–532.

Gil, L.A. 2007. PCR-based methods for fish and fishery products authentication. Trends in Food Science & Technology. 18: 558–566.

Gribbestad, I.S., M. Aursand and I. Martinez. 2005. High-resolution ^{1}H magnetic resonance spectroscopy of whole fish, fillets and extracts of farmed Atlantic salmon (*Salmo salar*) for quality assessment and compositional analyses. Aquaculture. 250: 445–457.

Haye, P.A., N.I. Segovia, R. Vera, M. Ángeles Gallardo and C. Gallardo-Escárate. 2012. Authentication of commercialized crab-meat in Chile using DNA Barcoding. Food Control. 25: 239–244.

Herrero, B., J.M. Vieites and M. Espiñeira. 2011. Authentication of Atlantic salmon (*Salmo salar*) using real-time PCR. Food Chemistry. 127: 1268–1272.

Herrero, B., F.C. Lago, J.M. Vieites and M. Espiñeira. 2012. Real-time PCR method applied to seafood products for authentication of European sole (*Solea solea*) and differentiation of common substitute species. Food Additives & Contaminants: Part A. 29(1): 12–18.

Howell, N., Y. Shavila, M. Grootveld and S. Williams. 1996. High-Resolution NMR and Magnetic Resonance Imaging (MRI) Studies on Fresh and Frozen Cod (*Gadusmorhua*) and Haddock (*Melanogrammus aeglefinus*). Journal of the Science of Food and Agriculture. 72: 49–56.

Hsieh, C.H., W.T. Chang, H.C. Chang, H.S. Hsieh, Y.L. Chung and D.F. Hwang. 2010. Puffer fish-based commercial fraud identification in a segment of cytochrome *b* region by PCR–RFLP analysis. Food Chemistry. 121: 1305–1311.

Huang, Y., A.G. Cavinato, D.M. Mayes, G.E. Bledsoe and B.A. Rasco. 2002. Nondestructive Prediction of Moisture and Sodium Chloride in Cold Smoked Atlantic Salmon (*Salmo salar*). Journal of Food Science. 67(7): 2543–2547.

Huang, Y., A.G. Cavinato, D.M. Mayes, L.J. Kangas, G.E. Bledsoe and B.A. Rasco. 2003. Nondestructive Determination of Moisture and Sodium Chloride in Cured Atlantic Salmon (*Salmo salar*) (Teijin) Using Short-wavelength Near-infrared Spectroscopy (SW-NIR). Journal of Food Science. 68(2): 482–486.

Hwang, C.C., C.M. Lin, C.Y. Huang, Y.L. Huang, F.C. Kang, D.F. Hwang and Y.H. Tsai. 2012. Chemical characterisation, biogenic amines contents, and identification of fish species in cod and escolar steaks, and salted escolar roe products. Food Control. 25: 415–420.

Igarashi, T., M. Aursand, Y. Hirata, I.S. Gribbestad, S. Wada and M. Nonaka. 2000. Nondestructive Quantitative Determination of Docosahexaenoic Acid and n-3 Fatty Acids in Fish Oils by High-Resolution ^{1}H Nuclear Magnetic Resonance Spectroscopy. Journal of the American Oil Chemists' Society. 77(7): 737–748.

Infante, C., G. Catanese, M. Ponce and M. Manchado. 2004. Novel Method for the Authentication of Frigate Tunas (Auxis thazard and Auxis rochei) in Commercial Canned Products. Journal of Agricultural and Food Chemistry. 54: 7435–7443.

Infante, C., A. Crespo, E. Zuasti, M. Ponce, L. Pérez, V. Funes, G. Catanese and M. Manchado. 2006. PCR-based methodology for the authentication of the Atlantic mackerel *Scomber scombrus* in commercial canned products. Food Research International. 39: 1023–1028.

Isaksson, T., G. Tøgersen, A. Iversen and K.I. Hildrum. 1995. Non-Destructive Determination of Fat, Moisture and Protein in Salmon Fillets by Use of Near-Infrared Diffuse Spectroscopy. Journal of the Science of Food and Agriculture. 69: 95–100.

Klossa-Kilia, E., V. Papasotiropoulos, G. Kilias and S. Alahiotis. 2002. Authentication of Messolongi (Greece) fish roe using PCR–RFLP analysis of 16s rRNA mtDNA segment. Food Control. 13: 169–172.

Knuutinen, J. and P. Harjula. 1998. Identification of fish species by reversed-phase high-performance liquid chromatography with photodiode-array detection. Journal of Chromatography B. 705: 11–21.

Lago, F.C., J.M. Vieites and M. Espiñeira. 2012. Development of a FINS- based method for the identification of skates species of commercial interest. Food Control. 24: 38–43.

Lee, M.H., A.G. Cavinato, D.M. Mayes and B.A. Rasco. 1992. Noninvasive Short-Wavelength Near-Infrared Spectroscopic Method To Estimate the Crude Lipid Content in the Muscle of Intact Rainbow Trout. Journal of Agricultural and Food Chemistry. 40: 2176–2181.

Lin, M., A.G. Cavinato, Y. Huang and B.A. Rasco. 2003. Predicting sodium chloride content in commercial king (*Oncorhynchus tshawytscha*) and chum (*O. keta*) hot smoked salmon fillet portions by short-wavelength near-infrared (SW-NIR) spectroscopy. Food Research International. 36: 761–766.

Mackie, I.M., S.E. Pryde, C. Gonzales-Sotelo, I. Medina, R. Peréz-Martín, J. Quinteiro, M. Rey-Mendez and H. Rehbein. 1999. Challenges in the identification of species of canned fish. Trends in Food Science & Technology. 10: 9–14.

Mackie, I., A. Craig, M. Etienne, M. Jérôm, J. Fleurence, F. Jenssen, A. Smelt, A. Kruijt, I.M. Yman, M. Ferm, I. Martinez, R. Peréz-Martín, C. Piñeiro, H. Rehbein and R. Kündiger. 2000. Species identification of smoked and gravad fish products by sodium dodecylsulphate polyacrylamide gel electrophoresis, urea isoelectric focusing and native isoelectric focusing: a collaborative study. Food Chemisty. 71: 1–7.

Marín, A., T. Fujimoto and K. Arai. 2013. Rapid species identification of fresh and processed scallops by multiplex PCR. Food Control. 32: 472–476.

Martinez, I., D. James and H. Loréal. 2005. Application of modern analytical techniques to ensure seafood safety and authenticity. FAO Fisheries Technical Paper, 455, Rome.

Martinez, I., R. Šližytė and E. Daukšas. 2007. High resolution two-dimensional electrophoresis as a tool to differentiate wild from farmed cod (*Gadus morhua*) and to assess the protein composition of klipfish. Food Chemistry. 102: 504–510.

Masoum, S., C. Malabat, M. Jalalo-Heravi, C. Guillou, S. Rezzi and D.N. Rutledge. 2007. Application of support vector machines to ¹H NMR data of fish oils: methodology for the confirmation of wild and farmed salmon and their origins. Analytical and Bioanalytical Chemistry. 387: 1499–1510.

Nilsen, H., M. Esaiassen, K. Heia and F. Sigerncs. 2002. Visible/Near-Infrared Spectroscopy: A New Tool for the Evaluation of Fish Freshness. Journal of Food Science. 67(5): 1821–1826.

Nortevdt, R., O.J. Torrissen and S. Tuene. 1998. Application of near-infrared transmittance spectroscopy in the determination of fat, protein and dry matter in Atlantic halibut fillet. Chemometrics and Interlligent Laboratory Systems. 42: 199–207.

Nott, K.P., S.D. Evans and L.D. Hall. 1999. Quantitative magnetic resonance imaging of fresh and frozen-thawed trout. Magnetic Resonance Imaging. 17(3): 445–455.

O'Conell, M., G. Valdora, G. Peltzer and R.M. Negri. 2001. A practical approach for fish freshness determinations using a portable electronic nose. Sensors and Actuators B. 80: 149–154.

Ochiai, Y. and S. Watabe. 2003. Identification of fish species in dried fish products by immunostaining using anti-myosin light chain antiserum. Food Research International. 36: 1029–1035.

Ortea, I., B. Cañas, P. Calo-Mata, J. Barros-Velázquez and J.M. Gallardo. 2010. Identification of commercial prawn and shrimp species of food interest by native isoelectric focusing. Food Chemistry. 121: 569–574.

Ottavian, M., L. Fasolato, P. Facco and M. Barolo. 2013. Foodstuff authentication from spectral data: Toward a species-independent discrimination between fresh and frozen–thawed fish samples. Journal of Food Engineering. 119: 765–775.

Pardo, M.A. and B. Pérez-Villareal. 2004. Identification of commercial canned tuna species by restriction site analysis of mitochondrial DNA products obtained by nested primer PCR. Food Chemistry. 86: 143–150.

Parisi, G., O. Franci and B.M. Poli. 2002. Application of multivariate analysis to sensorial and instrumental parameters of freshness in refrigerated sea bass (*Dicentrarchus labrax*) during shelf life. Aquaculture. 214: 153–167.

Pascoal, A., I. Ortea, J.M. Gallardo, B. Cañas, J. Barros-Velázquez and P. Calo-Mata. 2012. Species identification of the Northern shrimp (Pandalus borealis) by polymerase chain reaction–restriction fragment length polymorphism and proteomic analysis. Analytical Biochemistry. 421: 56–67.

Penner, M.H. 2010. Ultraviolet, Visible, and Fluorescence Spectroscopy. pp. 387–405. *In*: S.S. Nielsen (ed.). Food Analysis. Springer, USA.

Pereira, A.M., J. Fernández-Tajes, M.B. Gaspar and J. Méndez. 2012. Identification of the wedge clam Donax trunculus by a simple PCR technique. Food Control. 23: 268–270.

Rasco, B.A., C.E. Miller and T.L. King. 1991. Utilization of NIR Spectroscopy To Estimate the Proximate Composition of Trout Muscle with Minimal Sample Pretreatment. Journa of Agricultural and Food Chemistry. 39: 67–72.

Rasmussen, R.S. and M.T. Morrissey. 2008. DNA-Based Methods for the Identification of Commercial Fish and Seafood Species. Comprehensive Reviews in Food Science and Food Safety. 7: 280–295.

Rehbein, H., R. Kündiger, I.M. Yman, M. Ferm, M. Etienne, M. Jerome, A. Craig, I. Mackie, F. Jessen, I. Martinez, R. Mendes, A. Smelt, J. Luten, C. Pineiro and R. Perez-Martin. 1999. Species identification of cooked fish by urea isoelectric focusing and sodium dodecylsulfate polyacrylamide gel electrophoresis: a collaborative study. Food Chemistry. 67: 333–339.

Rehbein, H., C.G. Sotelo, R.I. Perez-Martin, M.J. Chapela-Garrido, G.L. Hold, V.J. Russell, S.E. Pryde, A.T. Santos, C. Rosa, J. Quinteiro and M. Rey-Mendez. 2002. Differentiation of raw or processed eel by PCR-based techniques: restriction fragment length polymorphism analysis (RFLP) and single strand conformation polymorphism analysis (SSCP). European Food Research and Technology. 214: 171–177.

Rodushkin, I., T. Bergman, G. Douglas, E. Engström, D. Sörlin and D.C. Baxter. 2007. Authentication of Kalix (N.E. Sweden) vendace caviar using inductively coupled plasma-based analytical techniques: Evaluation of different approaches. Analytica Chimica Acta. 583: 310–318.

Rohman, A. and Y.B. Che Man. 2011. Application of Fourier transform infrared (FT-IR) spectroscopy combined with chemometrics for authentication of cod-liver oil. Vibrational Spectroscopy. 55: 141–145.

Sádecká, J. and J. Tóthová. 2007. Fluorescence spectroscopy and chemometrics in the food classification – a review. Czech Journal of Food Sciences. 25(4): 159–173.

Sánchez-Alonso, I., I. Martinez, J. Sánchez-Valencia and M. Careche. 2012. Estimation of freezing storage time and quality changes in hake (Merluccius merluccius, L.) by low field NMR. Food Chemistry. 135: 1626–1634.

Santaclara, F.J., R. Pérez-Martín and C.G. Sotelo. 2014. Developed of a method for the genetic identification of ling species (*Genypterus* spp.) in seafood products by FINS methodology. Food Chemistry. 143: 22–26.

Solberg, C., E. Saugen, L.P. Swenson, L. Bruun and T. Isaksson. 2003. Determination of fat in live farmed Atlantic salmon using non-invasive NIR techniques. Journal of the Science of Food and Agriculture. 83: 692–696.

Sriphairoj, K., S. Klinbu-nga, W. Kamonrat and U. Na-Nakorm. 2010. Species identification of four economically important *Pangasiid* catfishes and closely related species using SSCP markers. Aquaculture. 308: S47–S50.

Standal, I.B., D.E. Axelson and M. Aursand. 2010. ¹³C NMR as a tool for authentication of different gadoid fish species with emphasis on phospholipid profiles. Food Chemistry. 121: 608–615.

Svensson, V.T., H.H. Nielsen and R. Bro. 2004. Determination of the protein content in brine from salted herring using near-infrared spectroscopy. Lebensmittel-Wissenschaft & Technologie. 37: 803–809.

Taylor, W.J. and J.L. Jones. 1992. An immunoassay for verifying the identity of canned sardines. Food and Agricultural Immunology. 4(3): 169–175.

Tognoli, C., M. Soraglia, G. Terova, R. Gornatti and G. Bernardini. 2011. Identification of fish species by 5S rRNA gene amplification. Food Chemistry. 129: 1860–1864.

Uddin, M. and E. Okazaki. 2004. Classification of Fresh and Frozen-thawed Fish by Near-infrared Spectroscopy. Journal of Food Science. 69(8): 665–668.

Ulrich, R.M., D.E. John, G.W. Barton, G.S. Hendrick, D.P. Fries and J.H. Paul. 2013. Ensuring seafood identity: Grouper identification by real-time nucleic acid sequence-based amplification (RT-NASBA). Food Control. 31: 337–344.

Velasco, A., A. Sánchez, I. Martínez, F.J. Santaclara, R.I. Pérez-Martín and C.G. Sotelo. 2013. Development of a Real-Time PCR method for the identification of Atlantic mackerel (*Scomber scombrus*). Food Chemistry. 141: 2006–2010.

Wen, J., C. Hu, L. Zhang, P. Luo, Z. Zhao, S. Fan and T. Su. 2010. The application of PCR–RFLP and FINS for species identification used in sea cucumbers (*Aspidochirotida: Stichopodidae*) products from the market. Food Control. 21: 403–407.

Wen, J., C. Hu, L. Zhang and S. Fan. 2011. Genetic identification of global commercial sea cucumber species on the basis of mitochondrial DNA sequences. Food Control. 22: 72–77.

Wolf, C., M. Burgener, P. Hübner and J. Lüthy. 2000. PCR-RFLP Analysis of Mitochondrial DNA: Differentiation of Fish Species. Lebensmittel—Wissenschaft & Technologie. 33: 144–150.

Xiccato, G., A. Trocino, F. Tulli and E. Tibaldi. 2004. Prediction of chemical composition and origin identification of European sea bass (*Dicentrarchus labrax* L.) by near infrared reflectance spectroscopy (NIRS). Food Chemistry. 86: 275–281.

Zhang, J., H. Wang and Z. Cai. 2007. The application of DGGE and AFLP-derived SCAR for discrimination between Atlantic salmon (*Salmo salar*) and rainbow trout (*Oncorhynchus mykiss*). Food Control. 18: 672–676.

Zhao, W., Y. Zhao, Y. Pan, X. Wang, Z. Wang and J. Xie. 2013. Authentication and traceability of *Nibea albiflora* from surimi products by species-specific polymerase chain reaction. Food Control. 31: 97–101.

Zuo, T., Z. Li, Y. Lv, G. Duan, C. Wang, Q. Tang and C. Xue. 2012. Rapid identification of sea cucumber species with multiplex-PCR. Food Control. 26: 58–62.

6

Milk and Dairy Products Authenticity

*Ioannis S. Arvanitoyannis** and *Konstantinos V. Kotsanopoulos*

6.1 Introduction

Both consumers and the food industry recognise the need for the development of measurement tools that would allow the effective characterisation of raw materials or final food. Dairy products such as milk, ice cream, yogurt, butter, cheese, etc., are in considerable demand, have premium prices and thus can be subject to economic adulteration. Authenticity of these products is of high significance for food processors, retailers, regulatory authorities and consumers. It is also very important for ensuring fair competition as well as for the protection of the consumers against fraud due to mislabelling (Karoui and De Baerdemacker, 2007). The authentication of dairy products has become major issue, attracting the attention of scientists, producers, consumers, and policymakers. Among various others, some of the practices considered to lead to adulteration of milk and dairies include substituting a part of the fat or proteins, mixing milk of different species, adding low-cost dairy products (such as whey derivatives), or mislabelling of foodstuffs protected by denomination of origin. Various analytical techniques have been used for detecting frauds and have been repeatedly modified, and reassessed to be one step ahead of individuals/companies who pursue the above illegal activities. Some traditional techniques that are used for the assessment of the authenticity of dairy products include chromatography, electrophoretic, and immunoenzymatic techniques (Fuente and Juárez, 2005). Techniques have also been developed for the detection of contaminants in dairy products including methods for detecting artificial colours, hypochlorite and antibiotics (Kirk and Sawyer, 1991). New techniques such as capillary electrophoresis (CE), polymerase chain reaction (PCR), and isotope ratio mass spectrometry have also been recently employed (Fuente and

School of Agricultural Sciences, Department of Agriculture, Ichthyology and Aquatic Environment, University of Thessaly, Fytoko St., 38446 Nea Ionia Magnesias, Volos, Hellas, Greece.
* Corresponding author

Juárez, 2005). The establishment of chromatographic and electrophoretic methods has been performed to effectively determine cheese ripening and for detecting milk adulteration (Veloso et al., 2004). According to Moatsou and Anifantakis (2003) a high number of different cheeses are made from ovine milk, or from its mixtures with caprine milk, and are highly acceptable by consumers worldwide. The organoleptic and physicochemical characteristics of the cheeses are affected by the composition of the milk used as raw material. The seasonal production and the higher prices of caprine milk and ovine milk compared to bovine milk are the basic motive for the admixture of cheese milk with bovine milk. Moreover, the higher price of ovine milk and the existence of mixed flocks of goats and ewes can lead to the accidental or fraudulent substitution of ovine milk by caprine. However, the need for genuine products and accurate labels requires the establishment of certain protective measures against adulteration of milk species in dairy products. As a result, numerous analytical techniques, such as chromatographic, electrophoretic, immunological and, more recently, DNA-based techniques, have been implemented for detecting ovine and caprine milk adulteration and the majority of them rely on the analysis of milk protein fractions.

6.2 Use of Electrophoresis for Ensuring the Authenticity of Dairy Products

In the study of Amigo et al. (1991), valuable information was extracted when High Performance Liquid Chromatography (HPLC) and urea–polyacrylamide gel electrophoresis (PAGE) were used for examining the proteolysis of casein during a 30-day ripening period of ovine milk cheeses, ovine milk cheeses with 10% and 20% bovine milk and bovine milk cheeses, which were manufactured in accordance with the traditional Terrincho technology. In the case of ovine cheeses, a-casein demonstrated the highest degree of degradation during cheese ripening. High degradation was also observed in the case of ovine milk cheese with 10% bovine milk. The profile of ovine milk cheese with 20% bovine milk was more similar to that obtained for bovine cheese. As regards bovine milk cheeses, electrophoresis appeared to have the highest sensitivity when used for evaluating the proteolysis of casein. The detection of 10 and 20% of bovine milk in ovine milk cheeses could be performed using urea–PAGE and HPLC, respectively, even after 30 days of ripening (Veloso et al., 2004). However, when PAGE, isoelectric focusing (IEF), and radial immune-diffusion (RIO) were used for determining cow's and goat's milks in Serra da Estrela cheeses, it was proved that the qualitative results were similar for all three methods, apart from two samples in which the immunological method failed to detect cow's milk and three samples in which it failed to detect goat's milk. Adulteration was mainly in the form of adding goat's milk in the product. The quantitative results, obtained by electrophoresis and IEF, were not significantly different ($P < 0.05$) (Amigo et al., 1991).

A comparison of cellulose acetate electrophoresis and PAGE techniques was performed when they were used for fractionating and quantifying milk proteins. Staining of protein bands was carried out using Ponceau-S and aniline blue black in the case of cellulose acetate electrophoresis and PAGE, respectively. α_{s1}-casein and

β-casein absorbed almost equal quantities of Ponceau-S per unit weight, whereas β-casein absorbed higher quantities of aniline blue black per unit in comparison to α_{s1}-casein. β-lactoglobulin, α-lactalbumin, and bovine serum albumin absorbed equal amounts of Ponceau-S per unit, but different amounts of aniline blue black. It was proved that cellulose acetate electrophoresis was the best technique for rapidly fractionating and quantifying milk proteins (Deshmukh and Donker, 1989). Similarly, in the study of Lin et al. (2010), native-PAGE was employed to simultaneously, qualitatively and quantitatively, analyze whey proteins of raw, commercial and laboratory heat-treated bovine milks. The separation of four whey protein bands, including β-lactoglobulin variants (β-LG A and B), was effectively performed in the gel. It was shown that the levels of the major whey proteins were reduced by approximately 23% in the pasteurized milks and by more than 85% in the UHT milks, in comparison to raw milk. The α-lactalbumin proved to be the most heat-tolerant with about 32% of its remaining in its native state after heating the milk at 100°C for 10 min, while about 42% of β-LG A and 53% of β-LG B were lost by heating the milk at 75°C for 30 min. Blood serum albumin (BSA) was lost almost completely by heating the milk (pH 5.0) at a temperature of 75°C or higher. The β-LGA and β-LGB were characterised by higher stability at low pH than in neutral conditions. Also, in the study of Kaminarides and Koukiassa (2002), the potential for detecting and determining bovine milk in adulterated yoghurt using cationic PAGE of yoghurt caseins, treated with rennet, was assessed. Yoghurts derived from bovine and ovine milk, as well as from their mixtures, were used. The bovine milk in yoghurt was evaluated by using the optical density of the bovine para-κ-casein band, which was clearly separated and linearly related to the amount of bovine milk in the mixtures. By using PAGE of bovine para-κ-casein, the detection of amounts of bovine milk as low as 1% was easily performed in ovine yoghurt. Furthermore, Pesic et al. (2011) employed a native-PAGE to simultaneously analyze, both qualitatively and quantitatively, bovine milk adulteration in caprine and ovine milk using whole milk samples as well as their whey protein fraction. The quantification was relied on the measurement of the band intensity of bovine β-lactoglobulins in all milk mixtures and bovine α-lactalbumin in caprine/bovine milk blends. Establishment of linear relationships was carried out between the band intensity of bovine β-lactoglobulins and α-lactalbumin versus volume percentage of added bovine milk in all samples examined, with the correlation coefficient being from 0.9950 to 0.9998. These correlations were used to quantify the bovine milk level within a wide range from 3% and 5% to 90% in caprine/bovine and ovine/bovine milk blends, respectively. The differences between the actual amounts of bovine milk added in the samples and those found using the regression lines were lower or equal to 5% for all samples. This method can be used to rapidly determine and unequivocally identify the bovine whey proteins in almost every caprine/bovine or ovine/bovine milk mixtures.

Izco et al. (1999) found that CE could be used instead of other techniques such as PAGE or SDS-PAGE to quantitatively analyze and separate the different casein fractions in cow's and ewe's milk. However, there has been no clarification yet on whether that method can achieve good quantifications. In this experiment, commercial whole ovine casein product used as standard and a mixture of the standard and whole casein extracted from ewe's milk cheese were employed for testing how reliable this method is. It was demonstrated that CE could be successfully used for the quantification of the ewe's milk caseins. The areas under four of the most representative peaks of the electrophoretogram for two α- and two β-caseins (designated α-casein1$_{CE}$, α-casein$_{2CE}$, β-casein$_{1CE}$, and β-casein$_{2CE}$ in order of elution) were taken into account for validating the method. As regards linearity, coefficient of determination (r^2) values greater than 99% were obtained for the regressions of each of the caseins. Furthermore, each casein led to response factors with a relative standard deviation (RSD) lower or equal to 5. The coefficients found in the day-to-day reproducibility analysis were higher than those for the same-day repeatability, but all the values were found to be within acceptable limits. In terms of accuracy, the percentage recovery rates of the α-casein fractions were higher in comparison to those of β-casein fractions, and therefore quantification of the latter using this technique could be achieved with higher accuracy under the conditions applied.

In another study, the development of a microfluidic "lab-on-a-chip" technique for separating and quantifying milk proteins was reported and compared with traditional SDS-PAGE. Separation of all major milk proteins was achieved using this technique when standard protein solutions were used. In a milk system, separation of α-lactalbumin, β-lactoglobulin, α-casein, β-casein and κ-casein was easily achieved, while the resolution could be compared with that of SDS-PAGE. Nevertheless, the immunoglobulins, lactoferrin and bovine serum albumin could not be resolved from the background in the microfluidic chip technique, but were readily resolved using SDS-PAGE. Linear standard curves for the major whey proteins were found using both techniques and the amounts of the major proteins found in a milk sample were comparable using both microfluidic chip and SDS-PAGE. It was therefore suggested that the microfluidic chip technology could be used to rapidly separate and quantify proteins in milk products (Anema, 2009).

6.3 Use of Chromatographic Methods for Ensuring the Authenticity of Dairy Products

Several chromatographic techniques have been employed for detecting adulteration of dairy products (Chmilenko et al., 2011). For example, the development of a liquid chromatography–mass spectrometry (LC-MS) method that could be used for the detection of a fraudulent addition of cow's milk in water buffalo milk and mozzarella, was reported by Czerwenka et al. (2010). This method was based on the use of β-lactoglobulin as marker of adulteration. It can rapidly determine and unequivocally identify the marker protein in every run. After examining 18 commercial buffalo mozzarella samples, it was found that three products were adulterated with high levels of cow's milk. Similarly, in the study of Romero et al. (1996), HPLC was employed for the detection of added cows' milk in goat's and ewe's milk. Detection of the additions at levels lower than 1% was achieved through analyzing the whey proteins. The method does not distinguish between goat's and ewe's milk. Also, in the study of Abernethy and Higgs (2013), LC-MS was used for the rapid detection of economic adulterants such as small, nitrogen containing compounds (melamine, ammeline, ammelide, cyanuric acid, allantoin, thiourea, urea, biuret, triuret, semicarbazide, aminotriazine, 3- and

4-aminotriazole, cyanamide, dicyandiamide, guanidine, choline, hydroxyproline, nitrate, and a range of amino acids) in fresh milk. $^{15}N_2$-urea was used as an internal standard. The detection of the adulteration of milk with exogenous urea is not easy due to the variation in the naturally occurring levels of urea in milk. However, by examining the contaminants biuret and triuret, which comprise up to 1% of synthetic urea, the detection of the adulteration of milk with urea-based fertilizer can be achieved. It was estimated that an economically viable adulteration would mean that addition of 90–4000 ppm of the above adulterants would need to be carried out, and thus for the majority of the adulterants, an arbitrary detection threshold of 2 ppm would be sufficient. As regards biuret, a lower detection threshold, better than 0.5 ppm, is needed and the technique for biuret and triuret can become more sensitive using post-column addition of lithium, in order to create lithium adducts under electrospray ionization (EI). Sample handling involves a two-step solvent precipitation method that is deployed in a 96-well plate format, and the hydrophilic interaction liquid chromatography uses a rapid gradient (1.2 min). Three separate injections were performed for detecting positively and negatively charged compounds, as well as amino acids and finally the lithium adducts. This rapid and qualitative survey method could be used as a second tier screening method to reduce the number of samples indicated as irregular by an Fourier Transform Infrared spectroscopy (FTIR spectroscopy) based screening system, and for directing the examination to suitable quantification methods.

According to Tay et al. (2013), reports of infant milk formula adulteration by detergent powders in the form of both economic frauds and poisoning incidents are quite frequent since detergents are widely available and low-cost materials. LC-Qtrap and LC–hybrid quadrupole time-of-flight mass spectrometry (LC-QTOF–MS) combined with chemometrics were successfully used for detecting the presence of detergent powder adulterated in infant milk formula. Partial least square analysis (PLS) regression was also employed for the quantification of the level of detergent powder in adulterated infant milk formula without the use of any standards. The identification of dodecylbenzenesulfonate (C12-LAS) and its verification as the marker were carried out using LC-QTOF–MS. Quantification of the level of C12-LAS that was present in the admixture was effectively performed through standard addition method.

Caseinomacropeptide is a peptide derived from chymosin and released during cheese production. Since it remains in whey, it can be employed as a biomarker of fluid milk adulteration through whey addition. Analysis of CMP is usually carried out using reversed phase (RP-HPLC) or size-exclusion chromatography (SEC). However, some psychrotropic microorganisms (such as *Pseudomonas fluorescens*) if present in milk, can enzymatically produce a CMP-like peptide commonly known as pseudo-CMP. These two peptides are different from each other by only one amino acid. RP-HPLC and SEC methods are not effective in distinguishing these two peptides, thus the development of a confirmatory method characterised by high selectivity is required. Taking into account the different degrees of glycosylation and phosphorylation in CMP, in combination with possible genetic variation (CMP A and CMP B), the development of analytical methods for differentiating these peptides is very complex. As a result, a proteomic-like technique was proposed for separating and characterising these peptides, using LC-MS with electrospray ionization capable of differentiating and subsequently quantifying CMP and pseudo-CMP in milk samples, in an effort to detect adulteration or contaminants in the products. The method was quite precise with a detection limit of 1.0 μg mL^{-1} and a quantification limit of 5.0 μg mL^{-1} (Motta et al., 2014).

The development of improved techniques in the field of chemometrics in combination with the use of vibrational spectroscopy has been very important for identifying and quantifying food contaminants. The above-mentioned techniques are employed by regulatory agencies and can be easily used for monitoring food processing, quality control, and quality assurance processes, thus ensuring the authenticity of the foodstuff in relation to the variety, geographical origin, and presence or absence of contaminants (Domingo et al., 2014). For example, in the study of Santos et al. (2013), the application of attenuated total reflectance mid-infrared microspectroscopy (MIR-microspectroscopy) was assessed for its potential in rapidly detecting and quantifying milk adulteration. Milk samples were obtained from local grocery stores (Columbus, OH, USA) and spiked at different concentrations of whey, hydrogen peroxide, synthetic urine, urea and synthetic milk. The samples were placed on a 192-well micro-array slide, air-dried and the collection of the spectra was performed using MIR-microspectroscopy. Pattern recognition analysis by Soft Independent Modelling of Class Analogy (SIMCA) demonstrated tight and well-separated clusters thus permitting the easy discrimination of control samples from adulterated milk. PLS Regression gave standard error of prediction of ≈2.33, 0.06, 0.41, 0.30 and 0.014 g/L when used for estimating the adulteration with whey, synthetic milk, synthetic urine, urea and hydrogen peroxide, respectively. It was finally shown that MIR-microspectroscopy could be an alternative method for detecting economic adulteration of cow's milk.

6.4 Use of Polymerase Chain Reaction (PCR) for Authenticating Dairy Products

PCR allows the amplification of defined DNA-fragments in a very short time by a factor of up to some millions in three steps: ((1) denaturation for obtaining a single-stranded DNA, (2) annealing in which primers flank the molecular region of interest, and (3) extension where synthesis of the new strand by DNA polymerase occurs). Subsequent analysis of the DNA can be carried out using numerous molecular biological procedures, mainly by size fragment length polymorphism, and visualisation can then be performed with ethidium bromide followed by electrophoresis in agarose gel. The main advantage of PCR-analysis compared to chemical methods based on the use of proteins is that the whole DNA is always identically present in all organs of a species. Due to that, the determination of molecular genetic differences can be carried out very easily. Molecular biology techniques have been applied for the identification of the species of origin in foods, especially meat products. The use of these molecular techniques in dairy products has been historically used only for the detection of bacterial contaminants, and it has only very recently been used for controlling authenticity and differentiating between species related to the dairy industry (Fuente and Juárez, 2005).

It is commonly accepted that it is highly important to monitor adulterations of genuine cheeses in the dairy industry. As a result, the development of a PCR-based method was carried out for detecting bovine-specific mitochondrial DNA sequence in Italian water buffalo Mozzarella cheese. The isolation of DNA from cheese matrix was performed using organic extractions and kit purifications. A 134-bp fragment was amplified with a bovine-specific set of primers designed on the sequence alignment of bovine and buffalo mitochondrial cytochrome oxidase subunit I. The specificity degree of the primers was evaluated using DNA derived from the blood of water buffalo and bovine, which were present together in adulterated Italian Mozzarella cheese. The method was very reliable and effectively detected a level of 0.5% of bovine milk (Feligini et al., 2005).

According to Cheng et al. (2006), goats' milk adulteration with cows' milk is becoming a major issue. Historically, the urea-PAGE assay has been used for identifying cows' milk adulteration, with a detection sensitivity of 1.0%. In this study, a faster and more sensitive method was developed for detecting cows' milk, which may be present in adulterated goats' milk and goats' milk powder. The primer targeted highly conserved regions in bovine mitochondrial DNA (a 271 bp amplicon). This amplicon was cloned and sequencing was carried out to further confirm bovine specific sequence. The chelex-100 was employed for the separation of bovine somatic cells from goats' milk or goats' milk powder samples. Analysis of random samples for several brands of goats' milk powder and tablets from numerous regions of Taiwan indicated that the rate of adulteration was 20 out of 80 (25%) in goats' milk powders and 12 out of 24 (50%) in goats' milk tablets. Using this method, quantities of cows' milk or cows' milk powder as low as 0.1% could be detected in goat milk or goat milk powder. Therefore the method can detect adulterated goats' milk products rapidly and with high sensitivity and reproducibility. In another study, the development of a duplex-PCR method, with two pairs of primers specific to sequences of mitochondrial D-loop region was carried out for the identification of cows' milk in the milk of goats. The PCR was characterised by high specificity and sensitivity, thus allowing levels of cows' milk as low as 1% to be detected when added to the milk of goats. By simultaneously using a primer pair for goats' and cows' mitochondrial DNA fragment, false negative results could be prevented. The method was used for tracking the adulteration of goat milk with cow's milk in the Polish market. Examination of 54 milk samples from three Polish (34) and one foreign producer (20) was performed. Cow DNA was found in 33 samples, while 21 samples, including all 20 samples obtained from foreign producers, contained goat-derived milk only (Kotowicz et al., 2007).

Similarly, in the study of López-Calleja et al. (2005) the PCR was used to accurately detect goats' milk in sheep's milk by employing primers that target the mitochondrial 12S ribosomal RNA gene. By using goat-specific primers, a 122-bp fragment from goats' milk DNA was generated, while no amplification signal was detected from sheep's, cows', and water buffaloes' milk DNA. PCR analysis of raw and heat-treated milk binary mixtures of sheep/goat allowed the detection of goats' milk. The method was very sensitive (sensitivity threshold of 0.1%). It was therefore proved that the PCR assay was very useful for routinely authenticating milk products. Moreover, López-Calleja et al. (2007) reported the development of a method for quantifying goats' milk in sheep's milk mixtures. This method is based on the use of a RT-PCR technique, which relied on the amplification of a fragment of the mitochondrial 12S ribosomal RNA gene (rRNA). The technique takes advantage of the use of both goat-specific primers specific to the amplification of a 171 bp fragment from goat DNA, and mammalian-specific primers that amplify a 119 bp fragment of mammalian species DNA and are used as endogenous control. An internal fluorogenic probe (TaqMan) that leads to hybridization of "goat-specific" and "mammalian" DNA fragments was employed for monitoring the amplification of the target gene. By comparing the cycle number (C_t) at which mammalian and goat-specific PCR products are first detected with reference standards of pre-determined caprine content, the exact level of goats' milk in a milk mixture can be determined. The method was employed for the analysis of raw and heat-treated milk binary mixtures (goat/sheep), allowing the quantification of goats' milk when present at levels of 0.6–10%. The proposed PCR assay can rapidly and effectively authenticate milk and dairies and could be a valuable tool for ensuring the authentication of dairy products.

Another study was carried out by Darwish et al. (2009) for the assessment of a PCR-based method, which could be used for detecting cow's milk in water buffalo's milk. It employed primers specific to the mitochondrial 12S r-RNA gene. The detection limit of this method was 0.5% and its determination was allowed by using model samples derived from buffalo's milk, which contained known levels of cow's milk. An evaluation of the method was also performed through using it for the examination of 21 market milk samples labelled "buffalo milk". It was shown that ten out of the 21 examined milk samples were of pure buffalo's milk, while three samples were found to be of pure cow's milk and eight samples were mixtures of cow and buffalo milk. It was therefore suggested that this PCR method could be effective for the detection of cow's milk in water buffalo milk with a detection limit of 0.5%. Moreover, by analyzing market milk samples, it was shown that adulteration of buffalo milk by adding cow's milk or substituting it with cow's milk occurs very frequently in the dairy industry. A RT-PCR was also used in the study of Dąbrowska et al. (2010) to specifically detect and quantify goat's milk adulteration with cows' milk. A primer pair specific to conserved regions in goats and cow's mitochondrial genomes was employed for the normalisation of the total quantity of DNA. It was proved that the method was characterised by high specificity since the cow 300-bp amplicon was observed only when cow's milk and mixtures of cow and goat milk DNA were present. The effectiveness of the primers used to normalise the quantity of DNA was confirmed with a dilution series of goat's milk which contained cow's milk, indicating that the results were not affected by dilution of the DNA template. The standard curve had an R^2 value higher than 0.99 and therefore can be used to quantify milk addition in the range of 0.5–100%.

The usefulness of the PCR method was also highlighted by Mašková and Paulíčková (2006). Specifically, validation of a technique relied on the PCR principle was carried out for the detection of cow's milk in goat and sheep cheeses. Isolation of DNA from the cheeses was performed with the use of the isolation kit Invisorb Spin Food I by Invitek Co. The PCR method employed utilizes the sequence of the mitochondrial gene coding cytochrome β, specific to mammals. After electrophoresis, the characterisation of the cow's DNA was performed by using a fragment of 274 bp, while a DNA fragment of 157 bp, and a DNA fragment of 331 bp were used for goat's and sheep's DNA, respectively. The detection limit of the PCR method was 1% and its determination was performed using model samples made from pure goat cheese containing a defined level of cheese made from cow's milk. The technique was used for analyzing 17 goat cheeses and seven sheep cheeses obtained from retailers. Products of Czech, Slovak, French, Dutch, and Italian origin were studied. Detection of undeclared cow's milk was achieved in three types of goat cheese and in one type of sheep cheese. Furthermore, Rodrigues et al. (2012) examined the adulteration of goat milk obtained from small holders in semi-arid northeastern Brazil with bovine milk. This study was performed in an effort to evaluate the extent of adulteration phenomena and to move towards their inhibition, ensuring the quality and safety of goat milk. The development of a duplex PCR assay was performed. Further validation was carried out in 160 fresh bulk goat milk samples. The detection limit of the duplex PCR was 0.5% bovine milk in goat milk. It was found that 41.2% of the goat milk samples examined contained bovine milk. The effectiveness of this technique was also shown in the study of Díaz et al. (2007). Specifically, a PCR procedure was used for detecting caprine milk in ovine cheeses using primers which targeted the mitochondrial 12S ribosomal RNA gene. A primer pair, specific for goat, was employed in PCR analysis of cheese samples and a goat fragment of 122 bp was amplified. The specificity was about 1%, while no amplification of sheep's, cow's and water buffalo's milk DNA was observed. This study proved that the proposed PCR assay was very effective and could be used to qualitatively detect goats' milk in ewes' cheeses, allowing the easy and accurate authentication of cheese or other dairy products. Finally, Branciari et al. (2000) described a simple procedure that could be used for the detection of the species origin of milk used as raw material for producing cheese. DNA was isolated from Italian mozzarella or Greek feta using sequential organic extractions and resin purification and its analysis was carried out using PCR-restriction fragment length polymorphism. A clear differentiation was achieved between mozzarella derived from water buffalo milk and mozzarella derived from cheaper bovine milk as well as from feta cheeses made from bovine, ovine, and caprine milk.

6.5 Use of Immunological Methods for Ensuring the Authenticity of Dairies

Immunological procedures relied on antigen-antibody precipitation reactions can be used for the authentication of dairy products since they can effectively differentiate milk proteins from different species. These methods are highly specific due to the antigen-antibody reactions they employ. Additionally, they are characterised by low cost and high sensitivity, while relatively less time is required for the tests in comparison to conventional detection methods. Immunoassay kits can be easily used in the laboratory for the detection of the authenticity of dairy products and could be a valuable tool in the food industry. The most commonly used immunological technique for evaluating the authenticity of dairy products is enzyme immunoassay using Enzyme Linked Immunosorbent Assay (ELISA) (Fuente and Juárez, 2005).

The aim of the work of Song et al. (2011) was the development of an ELISA, which was able to detect adulteration of goat milk or goat milk products with bovine milk. Polyclonal antibodies were applied against bovine β-casein and were mixed with lyophilized Saanen goat β-casein (blocking) in an effort to enhance the specificity of

Table 6.1. Methods and techniques developed for detecting adulterants in dairy products.

Product(s)	Adulterant(s) to be detected	QC method	Effectiveness	References
Fresh milk	Nitrogen-containing adulterants	Liquid chromatography—tandem mass spectrometry	Rapid-qualitative	Abernethy and Higgs, 2013
Ewe's milk cheeses	Cow's milk	Electrophoretic analysis	$\geq$ 1% cow's milk can be detected	Amigo et al., 1991
Ewes'/goats' milk	Cows' milk	Indirect ELISA	$\geq$0–100 ml/1 adulterant can be detected	Anguita et al., 1995
Ewe's milk and cheese	Cow's milk	Immunostick ELISA—Monoclonal Antibody against Bovine β-Casein	1% cow's milk in ewe's milk/0.5% cow's cheese in ewe's cheese can be detected	Anguita et al., 1996
Ovine/Caprine milk	Bovine milk	Competitive ELISA—Monoclonal Antibody against Bovine β-Casein	0.5 to 25% of adulterant can be detected	Anguita et al., 1997
Liquid milk, infant formula, and milk powder	Melamine	MIR/NIR spectroscopy	0.76 ± 0.11 ppm of adulterant can be detected	Balabin and Smirnov, 2011
Ewes' and goats' milk cheese	Native/heat-denatured bovine β-lactoglobulin	Indirect competitive ELISA	100 ng/ml bovine native Ig or hd-β-Ig or 0.1–0.2% of cows' milk equivalent in cheese can be detected	Beer et al., 1996
Milk	Cheese whey	Proteomic-like sample preparation and liquid chromatography electrospray tandem mass spectrometry analysis	Satisfactory precision (<11%) with a detection limit of 1.0 μg ml^{-1} and quantification limit of 5.0 μg ml^{-1}	Motta et al., 2014
Raw milk	Determination of glycomacropeptide to detect liquid whey	Sensitive sandwich ELISA	Detection limit of 0.047% (v/v) and a quantification limit of 0.14% (v/v)	Chávez et al., 2012
Goats' milk	Cow's milk	PCR	$\geq$ 0.1% cows' milk or cows' milk powder in goat milk or goat milk powder can be detected	Cheng et al., 2006

Table 6.1. contd....

Table 6.1. contd.

Product(s)	Adulterant(s) to be detected	QC method	Effectiveness	References
Water buffalo milk and mozzarella	Cow's milk	LC-MS analysis of β-lactoglobulin	Unequivocal identification	Czerwenka et al., 2010
Water buffalo milk	Cow's milk	PCR	Detection limit of 0.5%	Darwish et al., 2009
Ovine cheeses	Caprine milk	PCR	Sensitivity threshold of approximately 1%	Diaz et al., 2007
Italian Mozzarella cheese	Bovine DNA	PCR-based	≥ 0.5% bovine milk can be detected	Feligini et al., 2005
Ovine milk	Caprine milk	Indirect ELISA	0.5–15% of goat's milk in ewe's milk (v/v) can be detected	Haza et al., 1996
Ovine milk	Caprine milk	Monoclonal antibody and ELISA	0.25–15% (v/v) of goat's milk in ewe's milk can be detected	Haza et al., 1997
Ewe's cheese	Goat's cheese	Monoclonal antibody and ELISA	0.5 to 25% (w/w) of goat's cheese in ewe's cheese can be detected	Haza et al., 1999
Milk	Melamine, urea, tetracycline and glucose	FTIR combined with 2D correlation spectroscopy	Detects the presence of adulterants	He et al., 2010
Goat's, sheep's, and buffalo's milk	Cows' milk	Indirect competitive ELISA	≥ 0.1% (v/v) adulteration with cows' milk can be detected	Hurley et al., 2004a
Ovine yoghurt	Bovine milk	PAGE	≥ 1% adulterant can be detected	Kaminarides and Koukiassa, 2002
Cow milk	Water/whey	NIR spectroscopy	Effective for detecting the adulterants and their contents	Kasemsumran et al., 2007
Goats' milk	Cows' milk	Duplex-PCR	The adulterant can be detected when 1% of cows' milk in goats' milk is present	Kotowicz et al., 2007
Goats'/ewes' milk	Cows' milk	Two-site ELISA using monoclonal antibodies	As little as 5 ng β-lg/ml or 1 part cows' milk per 100,000 parts goats' or ewes' milk can be detected	Levieux and Venien, 1994
Sheep's milk	Goats' milk	PCR	Sensitivity threshold of 0.1%	López-Calleja et al., 2005

Sample	Target	Method	Results	Reference
Sheep's milk	Goats' milk	Real time PCR	Quantification of goats' milk in the range 0.6–10%	López-Calleja et al., 2007
Goat and sheep cheeses	Cow's milk	PCR-based method	Detection limit of 1%	Mašková and Paulíčková, 2006
Caprine and ovine milks	Bovine milk	Native-PAGE	Percentages of bovine milk in caprine/bovine and ovine/bovine milk mixtures ≥ 3% or 5%, respectively can be detected	Pesic et al., 2011
Goats' and ewes' milk and cheese	Cows' milk and caseinate	Indirect competitive ELISA	Detection limit of 0.1%	Richter et al., 1997
Goat milk	Bovine milk	Duplex PCR	Detection limit of 0.5% bovine milk in goat milk	Rodrigues et al., 2012
Ewes' milk and cheese	Goat milk	Indirect ELISA	Adulterant amounts of 1% can be detected	Rodríguez et al., 1991
Ovine milk and cheese	Bovine milk	Polyclonal antibodies	Adulterant amounts of 0–125 to 64% (v/v) in sheep's milk and 0–5 to 25% (v/v) in cheese can be detected	Rolland et al., 1993
Ewe's/Goat's milk	Cow's milk	HPLC	Detection of 1% adulterant—no distinguishment between goat and ewe's milk	Romero et al., 1996
Milk	Tetracycline	FT-MIR and FT-NIR Spectroscopy	Fast, specific, simple, and easily automatable method	Sivakesava and Irudayaraj, 2002
Shaanxi goat's milk	Cow's milk	ELISA	Detection limit: 2%	Song et al., 2011

the antibodies. The blocked antibodies were specific to a number of specific epitopes of bovine β-casein and did not react with bovine α-casein, κ-casein and whey protein. The absorbance values of the indirect ELISA were linearly related with the concentration of the adulterant (bovine milk) when present in quantities of 2%–50%, and could be employed for the quantification of bovine milk. No significant difference was found in titration curves between Saanen and Guanzhong goat milk, and the detection limits were determined at 2% for goat caseins, raw milk and heat-treated milk. The low coefficients of variation (<10%) of the experiments indicated good repeatability. Therefore this ELISA assay could be potentially used to routinely detect cow's milk adulteration of Shaanxi goat's milk. In the study of Anguita et al. (1995), an indirect ELISA using a monoclonal antibody against β-casein was employed for detecting cow's milk in ewes' and goats' milks. The higher value of ewe and goat milk and the possibility of substituting it with cheaper cow's milk make this experiment very important. Screening of hybridoma cell supernatants using an indirect ELISA demonstrated that the cell line AH4 led to the production of a monoclonal antibody with high specificity against bovine caseins, while it did not cross react with ewes' and goats' caseins, bovine serum albumin, soyabean protein, gelatin or soluble muscle proteins from beef, pork, horse and chicken. By testing the monoclonal antibody class and subclass using a commercial kit (Sigma) in a sandwich ELISA system, it was proved that the hybridoma cell line AH4 produced IgG_1 monoclonal antibodies. According to the authors, the assay was characterised by high accuracy, sensitivity and specificity and is also quite simple. The use of a monoclonal antibody as the immunorecognition reagent can ensure the supply of homogeneous antibodies with consistent specificity and could therefore be used to commercially produce stable kits for species verification. Similarly, in the study of Haza et al. (1996), monoclonal antibodies were applied against purified caprine alpha S2-casein recovered from goat's milk casein by cation-exchange chromatography. The use of hybridoma technology led to identification and characterisation of eight monoclonal antibodies reactive against whole caprine caseins. No cross-reactivity was detected by testing the monoclonal antibodies against sheep and cow whole caseins and beef, horse, porcine, chicken, and soya proteins, or gelatin and bovine serum albumin.

One of the monoclonal antibodies generated by the hybridoma cell line B2B was found to be species monospecific, reacting only with the caprine alpha S2-casein fraction. Subsequently, monoclonal antibody B2B was employed in an indirect ELISA format in an effort to detect pre-determined quantities of goat's milk (0.5–15%) in ewe's milk (v/v) mixtures. Moreover, Hurley et al. (2006) reported the development of an assay for the detection of the adulteration of soft goat, sheep and buffalo milk cheese with bovine milk from cheaper sources. A sandwich ELISA was generated using a monoclonal antibody combined with a polyclonal goat anti-bovine IgG antibody. After optimization, the ELISA became highly specific. The detection limits were 0.001% for cows' milk in sheep or buffalo milk, and 0.01% for cows' milk in goat milk, while the detection limits in soft cheese were 0.001% in goat cheese and 0.01% in sheep or buffalo cheese. The assay had very good reproducibility with both intra- and inter-assay coefficient of variation <10%. It could therefore be successfully used to routinely examine milk and soft cheese. Also, in the study of Beer et al. (1996), an indirect competitive ELISA was generated for detecting native and heat-denatured bovine-β-lactoglobulin (hd-β-lg) in goats' and ewes' milk cheese. Purification of polyclonal antibodies raised in chicken against hd-β-lg was performed using immunadsorption chromatography on native bovine, ovine and caprine-β-lactoglobulins conjugated to CNBr-activated Sepharose. Despite the fact that lactoglobulins from ewe, goat and cow are characterised by quite similar amino acid sequences, no cross reactivity was observed. A goat-anti-chicken alkaline phosphatase conjugate was employed as a secondary antibody for the detection of the anti-hd-β-lg antibodies bound to immobilized hd-β-lg. The detection limit of the assay was 100 ng/ml bovine native lg or hd-β-lg or 0.1–0.2% of cows' milk equivalent in cheese. An indirect competitive ELISA method was also developed in the study of Richter et al. (1997) for detecting bovine milk and caseinate in goats' and ewes' milk and cheese. Polyclonal antibodies were raised in rabbits and chickens against bovine γ3-casein. Initially an affinity chromatography analysis was used. Absorption of antibodies that recognise caseins was performed on bovine casein-Sepharose. After the dialysis, antibodies presenting cross reactivity with ewes' and goats' milk protein were eliminated using immunoadsorption onto stationary phases that contained ovine casein and protein extracted from genuine ewes' and goats' milk cheese. The detection limit of the ELISA test was 0.1% and the method was used with success in a EU collaborative study aiming at evaluating methods for detecting cows' milk.

Similarly, the development of an indirect ELISA was reported by Rodríguez et al. (1991). The technique was used for detecting defined quantities of goats' milk (1–25%) in ewes' milk and cheese. The assay employed polyclonal antibodies raised in rabbits against goats' caseins (GC). The anti-GC antibodies were obtained from the crude antiserum using immunoadsorption and elution from a column containing immobilized goats' caseins. The anti-GC antibodies were biotinylated and their specificity to goats' milk was confirmed after mixing them with lyophilized bovine and ovine caseins. The development of the assay was carried out in a non-competitive ELISA format and included coating plates with extracts from samples. ExtrAvidin-peroxidase was employed for detecting the biotinylated anti GC antibodies connected with the goats' caseins. The enzymes converting the substrate led to the development of a distinct colour that allowed the differentiation of the optical densities when the assay was applied to mixtures of ewes' milk and cheese derived from different amounts of goats' milk. Also, Rodríguez et al. (1990) reported the development of an indirect ELISA for detecting cows' milk (1–50%) in sheeps' milk and cheese. The assay was based on the use of polyclonal antibodies raised in rabbits against bovine caseins (BC). The recovery of the anti-BC antibodies from the crude antiserum was performed using immunoadsorption and elution from a column containing BC. The biotinylation of the antibodies was followed by rendering them cows' milk specific, through mixing them with lyophilized ovine and caprine caseins. The detection of the extrAvidin-peroxidase was performed using biotinylated anti-BC antibodies bound to BC immobilized on 96-well plates. Subsequently, the substrate was enzymatically converted allowing the detection of clear absorbance differences when mixtures of sheeps' milk and cheese of different levels of cows' milk were assayed.

The use of a sandwich ELISA of the two-site type was suggested by Levieux and Venien (1994) as a method for detecting cows' milk in goats' or ewes' milk. The assay is based on the use of two monoclonal antibodies raised in mice against cows' β-lactoglobulin (β-lg). These monoclonal antibodies are capable of recognizing different epitopes of the β-lg, which are quite distinct thus allowing the simultaneous binding of the corresponding antibodies. According to the authors of this study, one of the monoclonal antibody recognizes a species-specific epitope of the bovine β-lg and was adsorbed to a plastic microtitration plate (capture antibody). Labelling of the other monoclonal antibody with peroxidase rendered it suitable for detecting the captured cows' β-lg. The parameters that could affect the effectiveness of the assay were examined. After optimization, the assay was characterised by high specificity, reproducibility (intra- and inter-assay CV were 8 and 13% respectively) and sensitivity. It was shown that levels as low as 5 ng β-lg/ml or one part cows' milk per 100,000 parts goats' or ewes' milk could be effectively detected. The technique is cheap, rapid and very reliable and could therefore be used in semi-automated and screening surveys.

In another study, a competitive ELISA was generated for detecting quantifying bovine milk in ovine and caprine milk and cheese through the use of a monoclonal antibody (AH4 monoclonal antibody) against bovine β-casein. Simultaneous addition of ovine or caprine milk and cheese-derived from bovine milk-with the AH4 monoclonal antibody was carried out to the wells of a microtiter plate that had been subjected to sensitization using bovine β-casein. The bovine caseins compete with the bovine β-casein bound to the plate, for the AH4 monoclonal antibody binding sites. Further immunorecognition of AH4 monoclonal antibody bound to the bovine β-casein was achieved using rabbit anti-mouse immunoglobulin conjugated to peroxidase. Subsequently, by enzymatically converting the substrate, the different absorbance values were clearly differentiated during assaying mixtures of ovine and caprine milk and cheese derived from different amounts of bovine milk. The competitive ELISA suggested in this study can be used to quantitatively detect bovine milk in ovine and caprine milk when added at levels ranging from 0.5 to 25% (Anguita et al., 1997). The use of an indirect ELISA was also reported by García et al. (1990). The method aimed at detecting bovine milk in ovine milk. Polyclonal antibodies were raised in rabbits against bovine whey proteins. Anti-bovine whey protein antibodies were obtained from crude antiserum by immunoadsorption and elution from a column containing bovine whey proteins. The antibodies were biotinylated and their specificity to bovine milk was attained by mixing them with lyophilized ovine and caprine whey proteins. ExtrAvidin-peroxidase was employed to detect the biotinylated anti-bovine whey protein antibodies bound to bovine milk proteins. The substrate was then enzymatically converted, resulting in clear differentiation of the optical densities between mixtures of ovine milk, containing different amounts of bovine milk. Moreover, an immunostick ELISA was developed by Anguita et al. (1996) aiming at rapidly detecting cow's milk in ewe's milk or cheese. The assay is relied on a monoclonal antibody (AH4) generated against bovine β-casein for detecting cow's milk or cheese bound to the paddles of immunostick tubes. By using this immunostick ELISA, ewe's milk containing more than 1% cow's milk or cheese containing more than 0.5% cow's cheese can be visually identified.

Haza et al. (1997) reported the production of a stable hybridoma cell line (B2B) that can lead to the generation of a monoclonal antibody specific to goat's milk α_{s2}-casein. The monoclonal antibody B2B was employed in two ELISA formats for detecting and quantifying the presence of goat's milk in ewe's milk. In the indirect

ELISA format the limit of detection was 0.5 to 15% (vol/vol) goat's milk in ewe's milk. Subsequently, a competitive indirect ELISA was generated for detecting 0.25 to 15% (vol/vol) goat's milk in ewe's milk. The method was successful and it was characterised by high specificity. It can give results in less than 5 h, while is not affected by the heat treatment of milk. Similarly, Haza et al. (1999) generated a stable hybridoma cell line (B2B) that secreted a monoclonal antibody specific for the α_{s2}-casein of goats. The monoclonal antibody B2B was applied in two ELISA formats in an effort to detect and quantify goat's cheese in ewe's cheese samples. In the indirect ELISA format the limit of detection was $1 \pm 25\%$ (w/w) ewe's cheese in goat's cheese. Subsequently, a competitive indirect ELISA was successfully generated and applied for detecting 0.5 to 25% (w/w) of goat's cheese in ewe's cheese. This technique can also be accomplished in less than 5h. It is characterised by high sensitivity and is not affected by the ripening process of cheese.

One of the main adulteration issues that milk processing industries and distributors frequently face is the adulteration of liquid milk by the addition of bovine cheese whey. The detection of milk with whey is now frequently conducted by identifying glycomacropeptide. Current non-immunological methods for detecting glycomacropeptide in dairy products require high costs and time and are not very sensitive. Therefore, a novel sandwich ELISA has been developed for detecting and quantifying whey in raw milk, using a polyclonal rabbit anti-glycomacropeptide antibody. The construction of the calibration curves was performed through analysis of raw milk standards, which contained different predetermined amounts of liquid cheese whey (0.02–20%). The method had a detection limit of 0.047% (v/v) and a quantification limit of 0.14% (v/v). The antibody was very sensitive and did not cross react with any milk substances apart from κ-casein. It successfully detected glycomacropeptide in commercial dairy products. The recovery ratio was between 95.62% and 113.88% for all matrices examined. The intra-assay coefficient was found to be <6%, while the interassay coefficient was <7%. Also, it can be stored for a period of three months in the form of a ready-to-use kit, while remains accurate and reproducible (Chávez et al., 2012).

An effort to develop a method for detecting milk adulteration was also made in the study of Hurley et al. (2004a). This study aimed at developing an assay for the detection of adulteration of high premium milk with cheaper milk. An indirect, competitive ELISA was generated to rapidly detect cows' milk in the milk of goat, sheep, and buffalo. The assay was relied on the use of a monoclonal antibody generated against bovine IgG. This antibody recognises a species-specific epitope on the heavy chain of both bovine IgG1 and IgG2. A peroxidase-conjugated anti-mouse IgG antibody was employed for detecting bound monoclonal antibody. Subsequently, the substance was enzymatically converted allowing the clear differentiation of the absorbance when assaying various mixtures of milks adulterated with cows' milk. Following optimization, the ELISA was characterised by high specificity. The detection limits of the assay were 1.0 μg/mL for bovine IgG, or 0.1% (vol/vol) adulteration with cows' milk. The assay had very good reproducibility (CV < 10%) and was also successful when applied for the detection of bovine IgG in mixtures with the three types of milk tested. Also, according to Hurley et al. (2004b), several ELISAs were generated for detecting milk adulteration in dairy products. The antigens of interest were caseins, lactoglobulins, immunoglobulins and other whey proteins. Polyclonal and monoclonal antibodies were applied in various formats including direct, indirect, competitive and sandwich ELISAs. ELISAs were very effective when used for detecting cows' milk adulteration of sheep, goat and buffalo milk. Detection of goat milk adulteration of sheep milk was also successfully performed. Several ELISAs could also be used for examining cheese. It was suggested that the use of ELISA should be combined with PCR to ensure optimum efficiency.

Finally, in the study of Rolland et al. (1993), the comparison of the primary sequences of bovine and ovine milk proteins showed that some short peptide fragments are cow-specific (141–148 fragment of bovine α_{s1} casein). Therefore, chemical synthesis of the 140–149 peptide was performed on a solid phase matrix and it was then used as an immunogen for the production of polyclonal monospecific antibodies in rabbits. These antibodies recognised this fragment both on the peptidyl resin and in the native protein and were monospecific, since no antigen-antibody complex was produced with homologous ovine or caprine proteins. A competitive ELISA was then developed for detecting predetermined amounts of cows' milk in sheep's milk from [0–125 to 64% (v/v)] and in cheese from [0–5 to 25% (v/v)].

6.6 Use of Spectroscopic Methods for Ensuring the Authenticity of Dairies

IR (Infra-Red) spectroscopy is a rapid technique employed to non-destructively authenticate food products. By analyzing a foodstuff using the MIR spectrum (4000–400 cm^{-1}), information related to the molecular bonds present can be revealed. NIR spectroscopy is based on the use of the spectral range of 14,000 to 4000 cm^{-1} providing much more detailed structural information in relation to the vibrational behaviour of food mixtures. These techniques can be effectively used at an industrial level due to their simplicity and their low operational costs of obtaining and using the required equipment (Reid et al., 2006). At present, several physical techniques, based on differences in the freezing point and specific gravity are used for the detection of milk adulteration. However, these methods are quite time consuming and their accuracy may be relatively low. As regards the quantitative analysis, MIR spectroscopy is nowadays commonly used for determining several components in milk. Unfortunately, this technique can only destructively analyze the materials and is also very expensive (Kasensumran et al., 2007). On the other hand, NIR spectroscopy is a nondestructive and rapid technique used increasingly for evaluating the qualitative characteristics of foods. It allows the determination of many characteristics that can be used to assess the quality of foods (Cen and He, 2007). This technique has been relatively recently developed and its main advantage is the short times required for the analysis and the fact that no pretreatment of samples is required (Rodriguez-Otero et al., 1997).

Adulteration of milk and dairy products can have significant effects on both human health and the food industry. Taking into account the vast variety of adulterants that could potentially be found in milk (melamine, urea, tetracycline, sugar/salt, etc.), the development of rapid, widely available, cost-effective and sensitive techniques for detecting each of the components in milk, is required. A method based on FTIR spectroscopy in combination with two-dimensional correlation spectroscopy was used for the discriminative analysis of adulteration in milk. Initially, the peaks of the raw milk were found in the 4000–400 cm^{-1} region by its original spectra. Afterwards, samples containing the adulterant were analyzed using the same method in order to establish a spectral database for subsequent comparison. Two-dimensional correlation spectra of all samples were obtained. They were characterised by high time resolution and provided information about concentration-dependent intensity changes. Also, comparison of the peaks in the synchronous two-dimensional correlation spectra of the suspected samples was performed with those of raw milk. The differences indicated that the suspected milk samples contained adulterants. Melamine, urea, tetracycline and glucose adulterants in milk were effectively identified. It was therefore proved that this method could be applied to non-destructively and accurately determine whether samples of dairies contain adulterants, thus providing a simple and cost-effective alternative for testing these products (He et al., 2010). Also, since the authenticity of milk and milk products is of high importance, having major health, cultural, and financial implications, the FTIR spectroscopy could be applied as a direct biochemical fingerprinting technique for reducing considerably the time required for the analysis. To test the validity of this hypothesis, three types of milk: cow, goat, and sheep milk were examined by Nicolaou et al. (2010). Specifically, four mixtures were prepared. The first three were binary mixtures of sheep and cow milk, goat and cow milk, or sheep and goat milk and amounts of 0 to 100% of each milk in increments of 5% were contained in each mixture. The fourth mixture was a tertiary mixture composed of sheep, cow, and goat milk also in increments of 5%. Analysis using FTIR spectroscopy and multivariate statistical methods, such as PLS regression and nonlinear kernel PLS regression, were applied for quantifying the different levels of milk adulteration. The FTIR spectra demonstrated a very good predictive value for the binary mixtures, with an error level of 6.5 to 8% when analysed using PLS. More accurate predictions were achieved (4–6% error) when KPLS was used. Very accurate predictions were obtained by both PLS and KPLS with errors of 3.4 to 4.9% and 3.9 to 6.4%, respectively, when the tertiary mixtures were analysed. It was therefore shown that FTIR spectroscopy is a promising method and could be successfully used in the dairy industry to rapidly detect and quantify milk adulteration.

A NIR spectroscopic method that can be used for detecting foreign fat adulteration in dairy products was evaluated using fats extracted from butter and margarine mixtures, as well as from milk and soymilk mixtures. Analysis of butter and margarine mixtures was also carried out. The NIR absorptions at 1164, 1660, 2144, and 2176 nm were attributed to cis unsaturation of fatty acid moieties, the ratio of which is intrinsic to each one of the oils. An effort was made to prove that the NIR spectra around these wavelengths could be used as an index for fatty acid profiles in oil. Moreover, the difference of spectral data between the 1164, 1660, 2144, or 2176 nm and 2124 nm was found to correlate very well with the mixing ratios. As regards the second derivative spectra, spectral data at log 1164, 1660, 2144, or 2176 nm were found to be well correlated with the mixing ratios, and further to that, the correlation proved to be better by employing the difference between these values and those at 2124 nm. The use of this NIR technique for the analysis of dairy products showed that detection of adulteration of fat with as little as 3% foreign fat could be achieved simply, rapidly, and non-destructively for butter and margarine mixtures (Sato and Kawano, 1990).

According to Meurens et al. (2005), "the collective term "conjugated linoleic acid" or "CLA" generally refers to a mixture of conjugated positional and geometric isomers of linoleic (*cis*-9,*cis*-12-octodecadienoic) acid". In nature, these isomers are usually generated in the rumen through biohydrogenation of polyunsaturated fatty acids. A study was conducted aiming at determining CLA in cow's milk fat using Raman spectroscopy. Examination of the spectra of pure *cis*-9-oleic, *cis*-9,*cis*-12-linoleic, *cis*-9,*trans*-11-linoleic, and *trans*-10,*cis*-12-linoleic acids was performed in comparison with the spectra of milk-fat samples which contained 0 to 3% CLA. The CLA was determined in cow's milk fat using Raman spectroscopy, reaching to the following conclusions: firstly, by examining the Raman spectra three specific Raman signals of the chemical bonds were identified. These bonds were associated to the cis,trans conjugated C═══C in the rumenic and *trans*-10,*cis*-12-octodecadienoic acids at 1652, 1438, and 3006 cm^{-1}. Furthermore, by calibrating the Raman spectrometer for the CLA determination, it was found that these three specific signals could be used to accurately and reliably determine the CLA content of milk fat.

Tsenkova et al. (1999) made an effort to evaluate the potential use of NIR spectroscopy for measuring the fat, total protein, and lactose concentration of non-homogenised milk. The effects of the spectral region, sample thickness, and spectral data treatment on the accuracy of measurement were all examined. Transmittance spectra of 258 milk samples were obtained during the milking process, using a spectrophotometer in the wavelength range from 400 to 2500 nm with sample thicknesses of 1 mm, 4 mm and 10 mm. It was shown that the spectral region and sample thickness significantly affected the determination of milk fat and total protein, while they had no influence on the determination of lactose. The most accurate results were collected at the 1100 to 2400 nm region and for 1-mm sample thickness and first derivative data transformation. As regards the spectral region from 700 to 1100 nm, accurate results were obtained for fat with a 10-mm sample and for total protein with 1-mm sample thickness. The sample thickness did not significantly affect the accuracy of lactose determination. Different treatments of spectral data had no positive effects on the calibrations of fat and protein. For the region from 700 to 1100 nm, where cost-effective on-line sensors can be employed, the highest positive coefficients for fat were at 930, 968, 990, 1026, 1076 and 1092 nm; for lactose were at 734, 750, 786, 812, 908, 974, 982 and 1064 nm; and for total protein were at 776, 880, 902, 952 and 1034 nm. Another study that used NIR to analyse milk powder was that of Wu et al. (2008b). This study aimed at providing an alternative for milk powder analysis through the use of short-wave NIR spectroscopy. Analysis of the NIR spectra in the 800- to 1,025-nm region of 350 samples was performed for determining the brands and the quality of milk powders. The brands were identified by a least squares support vector machine (LS-SVM) model in combination with fast fixed-point independent component analysis (ICA). The accuracy of the ICA-LS-SVM model was as high as 98%, which is much improved in comparison to the correct answer rate of the LS-SVM (95%). The concentration of fat, proteins and carbohydrates was performed using the LS-SVM and ICA-LS-SVM models. Both processes effectively determined the main components in milk powder based on short-wave NIR spectra. The coefficients of determination for prediction and root mean square error of prediction of ICA-LS-SVM were 0.983, 0.231 and 0.982, and 0.161, 0.980 and 0.410, for fat, protein and carbohydrates, respectively. However, there were less than 10 input variables in the ICA-LS-SVM model compared with 225 in the LS-SVM model. As a result, the processing time required was much shorter and the model was simpler. It was finally demonstrated that the short-wave NIR region could be successfully used to rapidly and reliably determine the brand and the main components in milk powder.

The fat content of milk powder was also successfully determined in the study of Wu et al. (2007). This study used an IR spectroscopy technique to non-destructively measure the fat content in milk powder. Fat is one of the main components of milk powder and is therefore of high importance to develop a method for detecting its content in milk powder both rapidly and non-destructively. For this purpose, NIR and MIR spectroscopy techniques were employed. LS-SVM was employed for the development of the fat-content prediction model using the IR spectral transmission values. The results relied on LS-SVM were better than those of back-propagation artificial neural networks. The determination coefficient for prediction of the results found by the LS-SVM model was 0.9796 and the root mean square error was 0.836708. It was finally proved that this IR spectroscopy technique could be used for the quantification of the fat content in milk powder. The process is rapid and non-destructive. Furthermore, comparison of the prediction results between NIR and MIR spectral data was carried out. It was proved that the model with both MIR and NIR spectral data gave slightly worse results than that of the whole IR spectral data.

Wu et al. (2008a) used the IR spectroscopic technique for measuring the amount of protein in milk powder. LS-SVM was employed to develop the protein prediction model using the spectral transmission rate. The determination coefficient for prediction was 0.981, while the root mean square error for prediction was 0.4115. It was proved that the technique could be used for the quantification of the protein content in milk powder rapidly and non-destructively. The process is quite simple, and the prediction ability of LS-SVM is higher in comparison to that of PLS. In addition, by comparing the prediction results, it was shown that the performance of the model with MIR spectra data was better than the performance of the model that was based on NIR spectra data.

Wu et al. (2009) and Wu et al. (2012) used spectroscopic techniques for detecting the amounts of minerals in dairies. Specifically, in the study of Wu et al. (2009), NIR and MIR spectroscopy were employed for predicting the iron and zinc levels of powdered milk. A hybrid variable selection method (uninformative variable elimination) in combination with a successive projections algorithm was used for selecting the most effective wave number variables from full 2756 NIR and 3727 MIR variables, respectively. Finally, 18 NIR and 18 MIR variables were chosen for predicting the iron content, while 17 NIR and 12 MIR variables were chosen for predicting the zinc content of the product. The wave number variables were input into PLS and LS-SVM, respectively. The selected MIR variables gave much better results in comparison to NIR for the prediction of both iron and zinc contents in the PLS and LS-SVM models. The iron content prediction results based on LS-SVM with 18 MIR spectra gave a coefficient of determination (r^2) of 0.920, residual predictive deviation of 3.321, and root-mean-square error of prediction of 1.444. The zinc content prediction results based on LS-SVM with 12 selected MIR spectra gave r^2 of 0.946, residual predictive deviation of 4.361, and root-mean-square error of prediction of 0.321. It was therefore shown that UVE-SPA is a very effective variable selection tool and the MIR spectroscopy in combination with UVE-SPALS-SVM could be used to rapidly and accurately determine the trace mineral content of powdered milk. In the study of Wu et al. (2012), NIR and MIR spectroscopy techniques were assessed for their ability to determine the calcium content of powdered milk. A hybrid spectral variable selection algorithm in combination with uninformation variable elimination (UVE) and successive projections algorithm, selected 11 NIR and 15 MIR variables from full 2,756 NIR and 3,727 MIR variables, respectively. Predicted results of least-squares support vector machine models for the samples in the prediction set indicated that the 15 MIR variables gave much better results (0.930 for coefficient of determination (r^2), 3.703 for residual predictive deviation, 30.162 for root mean square error of prediction set and 5.22% for relative errors of prediction than the 11 NIR variables (0.636 for r^2, 1.587 for residual predictive deviation, 78.815 for root mean square error of prediction, and 13.40% for relative errors of prediction). It was therefore proved that MIR spectroscopy could be used to rapidly and accurately determine the calcium content of powdered milk.

Cow's milk can be adulterated by adding cheaper materials, such as water or whey. NIRS was used as a non-destructive technique for the detection of milk adulteration. In the study of Kasemsumran et al. (2007) mixtures of cow milk with water and whey were tested. NIR spectra of mixtures and pure milk samples in the region of 1100–2500 nm were obtained. The different types of samples were classified using discriminant PLS and SIMCA methods. The development of PLS calibration models, for determining the water and whey contents in milk adulteration, was also carried out. It was proved that NIR spectroscopy could be employed to effectively detect water or whey adulterants and their levels in milk samples. According to Balabin and Smirnov (2011), melamine (2,4,6-triamino-1,3,5-triazine) is a nitrogen-rich substance connected with some feed and food recalls. Due to the severity of health implications from consuming melamine and the extensive range of products that can potentially be affected, rapid and accurate methods for detecting its presence are of paramount importance. The use of spectroscopic data obtained using NIR and MIR spectroscopies, was suggested for detecting melamine in complex dairy matrixes. It was proved that IR spectroscopy could be successfully used for the detection of melamine in dairy products, such as infant formula, milk powder, or liquid milk. ALOD below 1 ppm (0.76 ± 0.11 ppm) can be obtained subject to the use of the required spectrum pretreatment, a correct multivariate (MDA) algorithm and a spectrum analysis. The MIR/NIR spectra of milk products and melamine content are not linearly related and therefore nonlinear regression methods are required for the correct prediction of the triazine-derivative content of milk products. It was finally shown that both MIR and NIR spectroscopy can rapidly, sensitively and effectively analyse liquid milk, infant formula, and milk powder.

Borin et al. (2006) used LS-SVM as a multivariate calibration method to simultaneously quantify specific adulterants (starch, whey or sucrose) commonly detected in powdered milk, using NIR spectroscopy with direct measurements by diffuse reflectance. The spectral differences of the adulterants lead to the development of a nonlinear behaviour (when all adulterants are in the same data set), thus hampering the use of linear methods such as PLS regression. Optimum models were developed using LS-SVM, characterised by low prediction errors and much better potential in comparison to PLSR. It was therefore shown that effective models could be developed for quantifying a number of adulterants in powdered milk using NIR spectroscopy and LS-SVM as a nonlinear multivariate calibration process.

In another study, the ability of FT-MIR and FT-NIR spectroscopic techniques, in quantifying trace tetracycline levels in milk, was examined. Different amounts of tetracycline were added into milk samples, which were subsequently scanned using FT-MIR and FT-NIR spectroscopy. Selection of the correct spectral wave number regions was carried out for developing PLS regression models. High prediction errors were found when the development of the calibration model was performed using a wide range of tetracycline levels (4 to 2000 ppb) in milk. Maximum correlation coefficient value of about 0.89 was obtained for the validation models built using several concentrations. The prediction errors were high for the FT-NIR method. It was shown that FT-MIR spectroscopy could be employed to rapidly detect tetracycline hydrochloride residues in milk (Sivakesava and Irudayaraj, 2002).

The production of milk is a fundamental part of cow's metabolism and is highly associated with the blood circulation. Therefore, the produced milk can be used to obtain information on the metabolic status of the animal. Measuring milk components on-line during milking could enhance the possibility of detecting systemic and local alterations. A study was carried out to examine the potential use of MIR spectroscopy for measuring the milk composition using micro attenuated total reflection (μATR) and high throughput transmission (HTT). PLS regression was employed for the prediction of fat, crude protein, lactose and urea after pretreating IR data and selecting the most informative wave number variables. The determination of the prediction accuracies was performed individually for raw and homogenized milk samples. As regards fat content, both measurement modes gave excellent prediction for homogenized samples ($R^2 >$ 0.92) but poor prediction for raw samples ($R^2 < 0.70$). Nevertheless, homogenization was not essential for achieving good predictions for crude protein and lactose with both μATR and HTT, or urea with μATR spectroscopy. Excellent predictions of crude protein, lactose and urea content ($R^2 >$ 0.99, 0.98 and 0.86 respectively) were achieved in raw and homogenized milk using μATR IR spectroscopy. These results were considerably better than those obtained when HTT IR spectroscopy was used. Furthermore, the prediction performance of HTT was still good for crude protein and lactose content ($R^2 >$ 0.86 and 0.78 respectively) in raw and homogenized samples. However, the detection of urea in milk with HTT spectroscopy was better ($R^2 = 0.69$ versus 0.16) after homogenizing the samples. It was therefore proved that the μATR technique was the most effective method for rapidly taking on-line measurements of milk composition, although homogenisation is essential for obtaining good prediction of the fat content (Aernouts et al., 2011).

Etzion et al. (2004) suggested that the MIR range could be used for the determination of the protein content in raw cow milk. The protein content can be determined based on the characteristic absorbance of milk proteins, which includes two absorbance bands in the 1500 to 1700 cm^{-1} range, commonly known as the amide I and amide II bands, and absorbance in the 1060 to 1100 cm^{-1} range, related to phosphate groups covalently bound to casein proteins. The minimization of the effects of the strong water band (centered around 1640 cm⁻¹) that overlaps with the amide I and amide II bands, was achieved using an optimized automatic procedure for accurate water subtraction. After subtracting the waster, three methods were employed for analysing the spectra, namely simple band integration, PLS and neural networks. As regards the neural network models, the spectra were initially decomposed by PCA, and the neural network inputs were the spectra principal components scores. Furthermore, the amounts of two constituents expected to interact with the protein (i.e., fat and lactose) were also considered as inputs. 235 spectra of standardised raw milk samples, which corresponded to 26 protein concentrations in the 2.47 to 3.90% (weight per volume) range, were examined. The simple integration method did not gave good results, while PLS led to prediction errors of about 0.22% protein. The neural network approach gave prediction errors of 0.20% protein when based on PCA scores only, and 0.08% protein when lactose and fat concentrations were taken into consideration in the model. It was therefore shown that FTIR/attenuated total reflectance spectroscopy is useful for the rapid determination of protein concentration in raw milk.

Another study examining some of the qualitative characteristics of milk is that of Ferrand-Calmels et al. (2014). Specifically, MIR spectrometry was employed for estimating the fatty acid composition of cow, ewe, and goat milk targeting at comparing different statistical approaches with wavelength selection in order to predict the milk fatty acid composition from MIR spectra, and towards the development of equations for fatty acid determination in cow, goat, and ewe milk. In total, a set of 349 cow milk samples, 200 ewe milk samples, and 332 goat milk samples were subjected to analysis using MIR and gas chromatography (the reference method). The examination of a vast range of fatty acids was ensured, using milk from different breeds and feeding systems. The methods examined were PLS regression, first-derivative pretreatment + PLS, genetic algorithm + PLS, wavelets + PLS, least absolute shrinkage and selection operator method (LASSO), and elastic net. Optimum results were achieved using PLS, genetic algorithm + PLS and first derivative + PLS. The residual standard deviation and the coefficient of determination in external validation were employed for the characterisation of the equations as well as for retaining the best for each FA in each species. It was proved that the predictions were significantly better for fatty acids found at medium to high concentrations (i.e., for saturated fatty acids and a few monounsaturated fatty acids with a coefficient of determination in external validation >0.90). The fatty acids content was very accurately converted from grams per 100 mL milk to grams per 100 g fatty acids. Similar results were found in the study of Soyeurt et al. (2006). According to these authors the fatty acid content of dairy products is of high importance, but measuring fatty acids is commonly performed using gas-LC. Although this method can be effective and give accurate results, it is very time-consuming and expensive, and appropriately trained staff is required for its conduction. Therefore, the MIR spectrometry method could be used as an alternative for the assessment of the fatty acid content of dairies. To test and confirm that, MIR spectrometry was employed for the estimation of the fatty acid contents of milk and milk fat. The measurements of the milk fat were not as reliable as those for the same fatty acids in milk. It was also proved that by increasing the fatty acid concentrations in milk, the prediction efficiency of the IR analysis method increased. Selected prediction equations were characterised high cross-validation coefficient of determination, a high ratio of standard error of cross-validation to standard deviation, and high repeatability of chromatographic data. It was finally shown that the calibration equations predicting 12:0, 14:0, 16:0, 16:1cis -9, 18:1, and saturated and monounsaturated fatty acids in milk could be employed and therefore this IR analysis technique could be used for the assessment and improvement of the qualitative characteristics of milk. It should be mentioned that this method allows the estimation of the fatty acid composition of the milk of each cow and these measurements could play the role of indicator traits for determining the genetic values of underlying fatty acid concentrations. These genetic values could potentially be used for animal selection purposes aiming at improving the nutritional quality of cow milk.

In the study of Sørensen and Jepsen (1998), thirty-two batches of semi-hard cheese with 46% fat in dry matter were produced in a pilot plant while the pH, moisture and microbiological contamination with *Clostridium tyrobutyricum* and propionibacteria factors were taken into account as controlled variables. The samples were analysed after 5, 7, 9 and 11 weeks of ripening using NIR spectroscopy in transmittance mode (850–1050 nm) and in reflectance mode (1110–2490 nm). An evaluation of their sensory characteristics was carried out by a panel of 10 assessors. The development of the calibration equations was performed by PLS regression on consistency and flavour attributes. In general, NIR spectroscopy was more accurate than transmittance spectroscopy. The squared correlation coefficients obtained by NIR reflectance spectroscopy were 0.74–0.88 for consistency attributes and 0.27–0.59 for flavour attributes. Depending on the attribute, the standard error of prediction correlated with the standard error of reproducibility of the average score from 2 to 6 assessors. Another study examining the qualitative characteristics of a cheese product using NIR is that of Downey et al. (2005). Specifically, in this study, production of twenty-four experimental Cheddar cheeses was carried out using five renneting enzymes and the samples were stored at 4°C for up to nine months. At 2, 4, 6 and 9 months, sensory analysis of the samples was performed, evaluating the following attributes: "crumbly", "fragmentability", "firmness", "rubbery", "gritty/grainy", "moist", "chewy", "mouth coating", "greasy/oily", "melting" and "mass forming". Records of NIR (750–2498 nm) reflectance spectra were also taken. Predictive models were developed for the sensory characteristics and age using PLS regression. It was shown that generally, the most reliable models were developed using spectral data in the range of 750–1098 nm after a 2nd derivatisation step. The prediction of age was performed with a root mean square error of cross-validation (RMSECV) equal to 0.61. All sensory attributes were effectively modelled and their respective RMSECV values were "crumbly" (2.3), "rubbery" (3.4), "chewy" (4.0), "mouth coating" (5.0) and "mass forming" (4.1). It is believed that these models could give accurate measurements at an industrial level.

It is therefore obvious that spectroscopic methods offer several advantages over conventional methods for assessing the quality of dairy products and detecting adulterants. However, research is still required to explore and assess the potential of these techniques.

6.7 Determination of the Geographical Origin of Dairies

According to Sacco et al. (2009), the determination of the geographical origin of foods is of high importance for both consumers and manufacturers, since it may be an indicator that could be taken into account in quality, authenticity and typicality certifications. The Protected Designation of Origin (PDO) trademark has been assigned to various products to ensure their geographical origin. This designation can be obtained subject to producing and processing a raw material in the specific region from which the product gets its name (Sacco et al., 2009). As a result, the determination of the authenticity of typical dairy products can only be achieved by determining the geographical origin of milk and other final products derived from it (Brescia et al., 2005).

A technique that employed multivariate data analysis in an effort to combine MIR spectroscopy with front-face fluorescence spectroscopy was used in the study of Karoui et al. (2004a). The aim of this study was the discrimination between Emmental cheeses originating from several European countries: Austria ($n = 12$), Finland ($n = 10$), Germany ($n = 19$), France ($n = 57$), and Switzerland ($n = 65$). In total, 163 Emmental cheeses produced during winter ($n = 91$) and summer ($n = 72$) were examined. The use of Factorial Discriminant Analysis for the IR or fluorescence spectral data did not lead to satisfactory classifications. As a result, the first twenty principal components of the PCA extracted from each data set (MIR and tryptophan fluorescence spectra) were pooled (concatenated) into a single matrix and analysed by Factorial Discriminant Analysis. 89% of the calibration spectra and 76.7% of the validation spectra of the samples were correctly classified. The cheeses from Finland were perfectly discriminated, while good discrimination was also obtained for Austrian, German, French and Swiss cheeses, although some of the samples were not classified correctly. It was shown that concatenation of the data from the two spectroscopic techniques could be used for the authentication of Emmental cheeses. Moreover, MIR and front-face fluorescence spectroscopies, in combination with chemometric techniques, were examined in the study of Karoui et al. (2004b) for their effectiveness in discriminating Emmental cheeses originating from different geographic origins. A total of 74 Emmental cheeses, produced during summer in Denmark ($n = 2$), Finland ($n = 4$), Germany ($n = 6$), Austria ($n = 8$), France ($n = 27$) and Switzerland ($n = 27$), were subjected to analysis. Within the 1500-900 cm^{-1} spectral region, 83.7 and 77% of the calibration and validation data sets, respectively, were correctly classified, while within the 3000-2800 cm^{-1} only 57.4 and 29.7% of the calibration and validation spectra, respectively, were classified correctly. Direct records of protein tryptophan and vitamin A fluorescence spectra were kept. As regards the tryptophan fluorescence spectra, 76.4 and 63.5% of the calibration and the validation samples, respectively, were effectively classified, while a better classification was obtained from the vitamin A fluorescence spectra (correct classification was obtained for 93.9 and 90.5% of the calibration and validation spectra, respectively).

Good separation between cheeses from Finland was achieved using the 1500-900 cm^{-1} spectra and tryptophan or vitamin A fluorescence spectra. As regards the German cheeses, 100% correct classification was only obtained using vitamin A spectra. Cheeses from France and Switzerland were generally discriminated, but a high number of samples was not classified correctly. It was finally proved that vitamin A fluorescence spectra could be used to reliably evaluate the Emmental cheese origin. In the study of Karoui et al. (2005a), spectroscopic techniques were implemented for the determination of the origin of dairy products. Specifically, the potential of NIR, MIR and front-face fluorescence spectroscopies in combination with chemometric methods was evaluated for the discrimination of Emmental cheeses from different European geographic origins. In total, 91 Emmental cheeses produced during winter in Austria ($n=4$), Finland ($n=6$), Germany ($n=13$), France ($n=30$) and Switzerland ($n=38$) were examined. The classification of the samples was performed using factorial discriminant analysis. The use of NIR led to correct classification for 89% and 86.8% of the calibration and validation spectral data sets, respectively. The MIR results were comparable to those observed with NIR. Within the 3000–2800 cm^{-1} region, correct classification of 84.1% and 85.7% of the cheeses was obtained for the calibration and validation data sets, respectively. However, the classification obtained with the tryptophan fluorescence spectra was significantly lower. Specifically, only 67.6% and 41.7% of classification and validation spectra, respectively, were classified correctly. Tryptophan fluorescence spectra, however, allowed a good discrimination of Emmental cheeses made from raw or thermised milk. Different analyses of components and specific weights analysis were carried out on the physico-chemical, IR and tryptophan fluorescence data sets. The results obtained indicated that spectroscopic techniques could be successfully used for identifying Emmental cheeses according to the geographic origin and production conditions.

The usefulness of the spectroscopic techniques in determining the geographic origin of dairies was also demonstrated in the study of Karoui et al. (2005b). Specifically, the effectiveness of MIR and intrinsic fluorescence spectroscopies on determining the geographic origin of hard cheeses was assessed. Records of MIR (1700–1500 cm^{-1} region), fluorescence emission spectra, following excitation at 250 and 290 nm, and fluorescence excitation spectra following emission at 410 nm were taken directly on cheese samples. Twelve hard cheeses were produced using identical and controlled cheese-making conditions with milks obtained from three different regions in Jura (France). The results of factorial discriminant analysis performed on the fluorescence and MIR spectra of the experimental cheeses indicated that the cheeses were well discriminated. Afterwards, 25 Gruyère PDO cheeses produced at different altitudes in Switzerland, i.e., 11 Gruyère PDO cheeses from the lowlands (<800 m), eight Gruyère PDO cheeses from the highlands (1100–1500 m) and six L'Etivaz PDO cheeses (1500–1850 m) were subjected to analysis. Only 80% of the samples were correctly classified by applying factorial discriminant analysis to the MIR spectra of the cheeses examined. However, 100% classification was obtained by relying on the fluorescence spectra. Therefore, front-face fluorescence spectroscopy is a rapid, low cost and effective technique when used for the determination of the geographical origin of Gruyère cheeses. Also, according to Pillonel et al. (2003), four different FTIR techniques (NIR diffuse reflection (NIR/DR), MIR attenuated total reflection (mIR/ATR) using two different instruments and MIR transmission (mIR/ Tr) spectroscopy) combined with multivariate chemometrics were evaluated for their effectiveness in differentiating between Emmental cheeses of different geographic origins. A total of 20 Emmental cheese samples manufactured in winter in Switzerland

($n = 6$), Allgäu (Germany) ($n = 3$), Bretagne (France) ($n = 3$), Savoie (France) ($n = 3$), Vorarlberg (Austria) ($n = 3$) and Finland ($n = 2$) were examined. The normalized spectra or their 2nd derivatives were analysed using principal component analysis and linear discriminant analysis of the PCA-scores. Despite the small number of samples in this preliminary investigation, clear trends were found. The mIR transmission spectra achieved 100% correct classification in linear discriminant analysis when used to differentiate the Swiss Emmental from the other cheeses pooled as one group, while the NIR/DR spectroscopy classified the samples according to the six regions of cheese origin.

In the study of Karoui et al. (2005c), a total of 40 milk samples, eight of which were produced in lowland (430–480 m), 16 in mid-mountain (720–860 m) and 16 in mountain (1070–1150 m) areas of the Haute-Loire department (France), were obtained and analysed using front-face fluorescence spectroscopy. Records of tryptophan, aromatic amino acids and nucleic acid (AAA+NA) and riboflavin fluorescence spectra were taken in triplicate on different aliquots directly on milks with the excitation wavelengths set at 290 nm, 250 nm and 380 nm, respectively. The excitation spectra of vitamin A were also recorded in triplicate on different aliquots with the emission wavelength set at 410 nm. Factorial Discriminant Analysis correctly classified 74.1%, 81.5% and 76.9% of samples for the calibration data sets of tryptophan, AAA+NA and riboflavin spectra, respectively. As regards the validation data sets, correct classification of 69.2%, 76.9% and 70.4% was achieved for tryptophan, AAA+NA and riboflavin spectra, respectively. Afterwards, the first five principal components of the principal component analysis extracted from each data set (tryptophan fluorescence, AAA+NA fluorescence, vitamin A fluorescence and riboflavin fluorescence spectra) were gathered together into one matrix and subjected to analysis using factorial discriminant analysis. 100 and 69.2% of samples were correctly classified for the calibration and the validation groups, respectively. Good discrimination between the milk samples from the lowlands and the others was obtained. It is proved that fluorescence spectra of milks could provide data in relation to the geographic origin. Nevertheless, the size of the data set should be increased in order to improve the robustness of the algorithm.

According to Sacco et al. (2009), in Apulia, Italy, the production of local cow milk is of high importance for the local economy. Cow Apulian mozzarella is a type of cheese produced from this local milk. The determination of the geographical origin of this product is essential for the protection of its authenticity. Thus classical techniques, high performance ion chromatography (HPIC), inductively coupled plasma atomic emission spectroscopy (ICP-AES), nuclear magnetic resonance (NMR) and isotope ratio mass spectrometry (IRMS) were employed combined with chemometric methods to determine several compounds and the geographical origin of Apulian and other cow milk samples. By statistically comparing the constituents of milk obtained from several regions, the milk samples could be geographically characterised. The most significant factors contributing to discrimination were fatty acids, identified by chemically analysing the samples. The fatty acids were affected by the feeding diet. The existence and amounts of sugars, amino acids and organic acids found using NMR spectra, could give further information for the discrimination. d13C results, obtained from IRMS measurements, were the most important discrimination parameters, also affected by the cows' feeding regime. This method could also be employed for the examination of other dairy products awarded European Community Protected Designation of Origin. Finally, in the study of Pillonel et al. (2003), the volatile compounds of Emmental cheese obtained from several geographic regions were examined to assess their suitability as markers of geographic origin. A total of 20 Emmental cheese samples derived from Switzerland, Allgäu (D), Bretagne (F), Savoie (F), Vorarlberg (A) and Finland (produced in winter) were subjected to analysis using dynamic headspace gas chromatography followed by flame ionisation and mass spectrometry. Successful differentiation between all regions was achieved by using compounds found only in the corresponding region or by combining a few compounds by PCA. The samples were further analysed using a mass spectrometry-based electronic nose. PCA achieved 90 and 91% of correct classifications for the Emmental from Switzerland and other regions, respectively.

The stable isotope ratios ($\delta^{13}C$, $\delta^{15}N$ and $\delta^{34}S$ of casein and $\delta^{13}C$ and $\delta^{18}O$ of glycerol) measured by IRMS of French, Italian, and Spanish cheeses were examined in the study of Camin et al. (2004). Variability parameters such as animal-feeding regimen, geographical origin, and climatic and seasonal conditions were taken into account in an effort to examine all possibilities of cheese characterisation obtained by each isotopic parameter. $\delta^{13}C$ values of both casein and glycerol were well correlated with the amount of maize in the animal diet. $\delta^{15}N$ and $\delta^{34}S$ of casein were the isotopes mostly affected by the geoclimatic conditions of the area (aridity, closeness to the sea, altitude). $\delta^{18}O$ of glycerol highly depended on the geographical origin of the cheeses and on climatic/ seasonal parameters. The use of a multivariate stepwise canonical discriminant analysis offered a good discrimination potential for the different European cheeses. The results were confirmed by the classification analysis, when correct reclassification of >90% of the samples was achieved. A pilot study was conducted by Crittenden et al. (2007) to assess the suitability of multi-element isotope ratio analysis for determination of the origin of cows' milk produced within Australasia. A milk sample from pasture-fed cows was obtained from seven regions in Australia and New Zealand and subjected to analyses for the ratios of $^{13}C/^{12}C$, $^{15}N/^{14}N$, $^{18}O/^{16}O$, $^{34}S/^{32}S$, and $^{87}Sr/^{86}Sr$. Each milk sample played the role of a different isotopic fingerprint. Specifically, isotope ratios for $^{18}O/^{16}O$ and $^{13}C/^{12}C$ conformed to predicted isotope fractionation patterns based on the latitude and climate of each region. As regards the skim milk, casein was enriched in ^{13}C and ^{15}N, and lacks ^{34}S, whereas $^{87}Sr/^{86}Sr$ ratios were the same in skim milk and casein. The milk samples from Australasia had a higher content of ^{18}O and ^{34}S in comparison to the results obtained for the majority of the European dairy products. Generally, multi-element isotopic analysis could be suitable for the determination of the geographic origin of dairy products manufactured within Australasia. Also, in the study of Kornexl et al. (1997), relative carbon and nitrogen stable isotope levels in total milk offer an indication of the isotopic composition of the diet fed to the dairy cows. This diet and its δ-values are highly dependent on geographical and climatic factors. Milk originated from regions dominated by grassland usually has negative $\delta^{13}C$-values, while milk obtained from regions dominated by crop cultivation has usually positive $\delta^{13}C$-values. The $\delta^{15}N$-values were affected by different factors such as soil conditions, intensity of agricultural use and climate. Casein in authentic milk samples was quite rich in both ^{13}C and ^{15}N in comparison to total milk, while the whey fraction was slightly enriched in ^{13}C and lacks ^{15}N. The isotopic level of milk, casein and whey from one location was examined during a period of longer than one year, indicating that any variations are usually lower than 1‰. In milk water, the ^{18}O content was increased by between 2 and 6‰ in comparison to ground water. Similarly, in the study of Manca et al. (2001), the stable isotope ratios ($^{13}C/^{12}C$ and $^{15}N/^{14}N$) of casein determined using IRMS, and some free amino acid ratios (His/Pro, Ile/Pro, Met/Pro, and Thr/Pro) measured using HPLC in samples of ewes' milk cheese from Sardinia, Sicily, and Apulia, were not found to be correlated with ripening time. Multivariate data treatments were carried out through application of both unsupervised (principal component analysis and cluster analysis) and supervised (LDA) methods and showed good potential for discrimination of the geographical origin of the cheeses. In relation to that, the variables Ile/Pro, Thr/Pro, $^{13}C/^{12}C$, and $^{15}N/^{14}N$ ratios were of high importance and their use led to 100% discrimination and classification of the samples by LDA.

According to Manca et al. (2006), characterisation of the isotopic composition of "Peretta" cows' milk cheese was carried out aiming at protecting the origin of this Sardinian product against other cheeses offered under the same name, but produced of raw materials from northern Europe. Three types of cheeses were examined: those produced in local dairies from milk derived from free-grazing or pasture-grazing cows in Sardinia, cheeses industrially produced from milk in Sardinia (intensive farming), and cheeses produced with raw materials obtained from other countries. To differentiate between local cheeses and cheeses made of raw materials from other countries, the stable isotope ratios $^{13}C/^{12}C$, $^{15}N/^{14}N$, D/H, $^{34}S/^{32}S$ and $^{18}O/^{16}O$ were used. The isotopic data $\delta^{13}C$, $\delta^{15}N$, δ^2H and $\delta^{34}S$ were determined in the casein fraction, whereas determination of $\delta^{18}O$ and $\delta^{13}C$ was performed in the glycerol fraction. Measurements were taken using IRMS. By comparing the mean values of the isotope ratios using statistical analysis (ANOVA and Tukey's test) it was shown that the greatest differences between the three types of cheese were in the $^{13}C/^{12}C$, $^{34}S/^{32}S$ and $^{18}O/^{16}O$ isotope ratios. As regards the other parameters, either no differences ($\delta^{15}N$) or slight differences (δ^2H) were detected. The data were evaluated by multivariate statistical analysis (PCA and hierarchical cluster analysis) revealing that the isotope characteristics of the factory products were similar to those of the cheeses manufactured from imported raw materials, whereas a difference was found between the local cheeses and the other types of cheeses. Also, Pillonel et al. (2003) analyzed twenty Emmental cheeses from six European regions (Allgäu (D), Bretagne (F), Finland, Savoie (F), Switzerland and Vorarlberg (A)) to detect stable isotope ratios such as $^{13}C/^{12}C$, $^{15}N/^{14}N$, $^{18}O/^{16}O$, D/H and $^{87}Sr/^{86}Sr$, as well as major (Ca, Mg, Na, K), trace (Cu, Mn, Mo, I) and radioactive elements (^{90}Sr, ^{234}U, ^{238}U). The discrimination potential of these factors was assessed using difference tests of mean values and PCA. Good separation was achieved for "Finland", "Bretagne" and "Savoie" cheeses using $\delta^{13}C$-, $\delta^{15}N$-, δ^2H- and $\delta^{87}Sr$-values. Concentrations of molybdenum and sodium effectively separated the groups "Switzerland", "Vorarlberg" and "Allgäu". ^{90}Sr activity presented a good correlation with the altitude of the production area. This factor can be used for the separation of "Finland" and "Bretagne" from "Vorarlberg" cheese.

6.8 Conclusions

Although some reliable, rapid and effective techniques are already being used for the detection of the adulteration of dairy products, it is believed that the continuously increasing interest of both consumers and the industry towards the authenticity of dairy products will lead to the development of even more specific and sensitive methods and their established application at an industrial level. However, the complexity of the global market is expected to hinder the development of commonly established methods. Moreover, it has been shown that people committing fraudulent activities usually tend to be a step forward and thus the development of effective techniques will probably be continuously required. The development of new methods and the improvement of currently-used ones should aim not only to the melioration of the techniques, but also to the establishment of methods for quicker and more cost-effective preparation of any samples, as well as to the reduction of the analyses time required.

References

Abernethy, G. and K. Higgs. 2013. Rapid detection of economic adulterants in fresh milk by liquid chromatography–tandem mass spectrometry. Journal of Chromatography A. 1288: 10–20.

Aernouts, B., E. Polshin, W. Saeys and J. Lammertyn. 2011. Analytica Chimica Acta. 705: 88–97.

Amigo, L., M. Ramos and R.J. Martin-Alvarez. 1991. Effect of Technological Parameters on Electrophoretic Detection of Cow's Milk in Ewe's Milk Cheeses. Journal of Dairy Science. 74: 1482–1490.

Anema, S.G. 2009. The use of "lab-on-a-chip" microfluidic SDS electrophoresis technology for the separation and quantification of milk proteins. International Dairy Journal. 19: 198–204.

Anguita, G., R. Martín, T. García, P. Morales, A.I. Haza, I. González, B. Sanz and P.E. Hernández. 1995. Indirect ELISA for detection of cows' milk in ewes' and goats' milks using a monoclonal antibody against bovine β-casein. Journal of Dairy Research. 62: 655–659.

Anguita, G., R. Martín, T. García, P. Morales, A.I. Haza, I. González, B. Sanz and P.E. Hernández. 1996. Immunostick ELISA for Detection of Cow's Milk in Ewe's Milk and Cheese Using a Monoclonal Antibody against Bovine β-Casein. Journal of Food Protection. 4: 341–440.

Anguita, G., R. Martín, T. García, P. Morales, A.I. Haza, I. González, B. Sanz and P.E. Hernández. 1997. A Competitive Enzyme-Linked Immunosorbent Assay for Detection of Bovine Milk in Ovine and Caprine Milk and Cheese Using a Monoclonal Antibody against Bovine β-Casein. Journal of Food Protection. 1: 66–87.

Balabin, R.M. and S.V. Smirnov. 2011. Melamine detection by mid- and near-infrared (MIR/NIR) spectroscopy: A quick and sensitive method for dairy products analysis including liquid milk, infant formula, and milk powder. Talanta. 85: 562–568.

Beer, M., I. Krause, M. Stapf, C. Schwarzer and H. Klostermeyer. 1996. Indirect competitive enzyme-linked immunosorbent assay for the detection of native and heat-denatured bovine β-lactoglobulin in ewes' and goats' milk cheese. Z Lebensm Unters Forsch. 203: 21–26.

Borin, A., M.F. Ferrão, C. Mello, D.A. Maretto and R.J. Poppi. 2006. Least-squares support vector machines and near infrared spectroscopy for quantification of common adulterants in powdered milk. Analytica Chimica Acta. 579: 25–32.

Branciari, R., I.J. Nijman, M.E. Plas, E. Di Antonio and J.A. Lenstra. 2000. Species origin of milk in Italian mozzarella and Greek feta cheese. Journal of Food Protection. 63: 408–411.

Brescia, M.A., M. Monfreda, A. Buccolieri and C. Carrino. 2005. Characterisation of the geographical origin of buffalo milk and mozzarella cheese by means of analytical and spectroscopic determinations. Food Chemistry. 89: 139–147.

Camin, F., K. Wietzerbin, A.B. Cortes, G. Haberhauer, M. Lees and G. Versini. 2004. Application of Multielement Stable Isotope Ratio Analysis to the Characterization of French, Italian, and Spanish Cheeses. Journal of Agricultural and Food Chemistry. 52: 6592–6601.

Cen, H. and Y. He. 2007. Theory and application of near infrared reflectance spectroscopy in determination of food quality. Trends in Food Science & Technology. 18: 72–83.

Chávez, N.A., J. Jauregui, L.A. Palomares, K.E. Macías, M. Jiménez and E. Salinas. 2012. A highly sensitive sandwich ELISA for the determination of glycomacropeptide to detect liquid whey in raw milk. Dairy Science & Technology. 92: 121–132.

Cheng, Y.-H., S.-D. Chen and C.-F. Weng. 2006. Investigation of Goats' Milk Adulteration with Cows' Milk by PCR. Asian-Australasian Journal of Animal Sciences. 19(10): 1503–1507.

Chmilenko, F.A., N.P. Minaeva and L.P. Sidorova. 2011. Complex Chromatographic Determination of the Adulteration of Dairy Products: A New Approach. Journal of Analytical Chemistry. 66: 572–581.

Crittenden, R.G., A.S. Andrew, M. LeFournour, M.D. Young, H. Middleton and R. Stockmann. 2007. Determining the geographic origin of milk in Australasia using multi-element stable isotope ratio analysis. International Dairy Journal. 17: 421–428.

Czerwenka, C., L. Müller and W. Lindner. 2010. Detection of the adulteration of water buffalo milk and mozzarella with cow's milk by liquid chromatography–mass spectrometry analysis of β-lactoglobulin variants. Food Chemistry. 122: 901–908.

Dąbrowska, A., E. Wałecka, J. Bania, M. Żelazko, M. Szołtysik and J. Chrzanowska. 2010. Quality of UHT goat's milk in Poland evaluated by real-time PCR. Small Ruminant Research. 94: 32–37.

Darwish, S.F., H.A. Allam and A.S. Amin. 2009. Evaluation of PCR Assay for Detection of Cow's Milk in Water Buffalo's Milk. World Applied Sciences Journal. 7: 461–467.

Deshmukh, A.R. and A.J. Donker. 1989. Cellulose Acetate and Polyacrylamide Gel Electrophoresis for Quantification of Milk Protein Fractions. Journal of Dairy Science. 72: 12–17.

Díaz, I.L.-C., I.G. Alonso, V. Fajardo, I. Martín, P. Hernández, T.G. Lacarra and R. Martín de Santos. 2007. Application of a polymerase chain reaction to detect adulteration of ovine cheeses with caprine milk. European Food Research and Technology A. 225: 345–349.

Domingo, E., A.A. Tirelli, C.A. Nunes, M.C. Guerreiro and S.M. Pinto. 2014. Melamine detection in milk.

Downey, G., E. Sheehan, C. Delahunty, D. O'Callaghan, T. Guinee and V. Howard. 2005. Prediction of maturity and sensory attributes of Cheddar cheese using near-infrared spectroscopy. International Dairy Journal. 15: 701–709.

Etzion, Y., R. Linker, U. Cogan and I. Shmulevich. 2004. Determination of Protein Concentration in Raw Milk by Mid-Infrared Fourier Transform Infrared/Attenuated Total Reflectance Spectroscopy. Journal of Dairy Science. 87: 2779–2788.

Feligini, M., I. Bonizzi, V.C. Curik, P. Parma, G.F. Greppi and G. Enne. 2005. Detection of Adulteration in Italian Mozzarella Cheese Using Mitochondrial DNA Templates as Biomarkers. Food Technology and Biotechnology. 43: 91–95.

Ferrand-Calmels, M., I. Palhière, M. Brochard, O. Leray, J.M. Astruc, M.R. Aurel, S. Barbey, F. Bouvier, P. Brunschwig, H. Caillat, M. Douguet, F. Faucon-Lahalle, M. Gelé, G. Thomas, J.M. Trommenschlager and H. Larroque. 2014. Prediction of fatty acid profiles in cow, ewe, and goat milk by mid-infrared spectrometry. Journal of Dairy Science. 97: 17–35.

Fuente, M.A. and M. Juárez. 2005. Authenticity Assessment of Dairy Products. Critical Reviews in Food Science and Nutrition. 45: 563–585.

García, T., R. Martín, E. Rodríguez, P. Morales, P.F. Hernández and B. Sanz. 1990. Detection of Bovine Milk in Ovine Milk by an Indirect Enzyme-Linked Immunosorbent Assay. Journal of Dairy Science. 73: 1489–1493.

Haza, A.I., P. Morales, R. Martin, T. Garcia, G. Anguita, I. Gonzalez, B. Sanz and P.E. Hernandez. 1996. Development of monoclonal antibodies against caprine alphaS2-casein and their potential for detecting the substitution of ovine milk by caprine milk by an indirect ELISA. Journal of Agricultural and Food chemistry. 44: 1756–1761.

Haza, A.I., P. Morales, R. Martín, T. García, G. Anguita, I. González, B. Sanz and P.E. Hernandez. 1997. Use of a Monoclonal Antibody and Two Enzyme-Linked Immunosorbent Assay Formats for Detection and Quantification of the Substitution of Caprine Milk for Ovine Milk. Journal of Food Protection. 8: 883–1012.

Haza, A.I., P. Morales, R. Martín, T. García, G. Anguita, B. Sanz and P.E. Hernández. 1999. Detection and quantification of goat's cheese in ewe's cheese using a monoclonal antibody and two ELISA formats. Journal of the Science of Food and Agriculture. 79: 1043–1047.

He, B., R. Liu, R. Yang and K. Xu. 2010. Adulteration detection in milk using infrared spectroscopy combined with two-dimensional correlation analysis. Optical Diagnostics and Sensing X: Toward Point-of-Care Diagnostics, doi: 10.1117/12.841580.

Hurley, I.P., H.E. Ireland, R.C. Coleman and J.H.H. Williams. 2004b. Application of immunological methods for the detection of species adulteration in dairy products. International Journal of Food Science and Technology. 39: 873–878.

Hurley, I.P., H.E. Ireland, R.C. Coleman and J.H.H. Williams. 2006. Use of sandwich IgG ELISA for the detection and quantification of adulteration of milk and soft cheese. International Dairy Journal. 16: 805–812.

Hurley, I.P., R.C. Coleman, H.E. Ireland and J.H.H. Williams. 2004a. Measurement of Bovine IgG by Indirect Competitive ELISA as a Means of Detecting Milk Adulteration. Journal of Dairy Science. 87: 543–549.

Izco, J.M., A.I. Ordóñez, P. Torre and Y. Barcina. 1999. Validation of capillary electrophoresis in the analysis of ewe's milk casein. Journal of Chromatography A. 832: 239–246.

Kaminarides, S.E. and P. Koukiassab. 2002. Detection of bovine milk in ovine yoghurt by electrophoresis of para-κ-casein. Food Chemistry. 78: 53–55.

Karoui, R. and J. De Baerdemaeker. 2007. A review of the analytical methods coupled with chemometric tools for the determination of the quality and identity of dairy products. Food Chemistry. 102: 621–640.

Karoui, R., J.-O. Bosset, G. Mazerolles, A. Kulmyrzaev and É. Dufour. 2005b. Monitoring the geographic origin of both experimental French Jura hard cheeses and Swiss Gruyère and L'Etivaz PDO cheeses using mid-infrared and fluorescence spectroscopies: a preliminary investigation. International Dairy Journal. 15: 275–286.

Karoui, R., É. Dufour, L. Pillonel, D. Picque, T. Cattenoz and J.-O. Bosset. 2004a. Determining the geographic origin of Emmental cheeses produced during winter and summer using a technique based on the concatenation of MIR and fluorescence spectroscopic data. European Food Research and Technology. 219: 184–189.

Karoui, R., É. Dufour, L. Pillonel, D. Picque, T. Cattenoz and J.-O. Bosset. 2004b. Fluorescence and infrared spectroscopies: a tool for the determination of the geographic origin of Emmental cheeses manufactured during summer. Dairy Science and Technology. 84: 359–374.

Karoui, R., É. Dufour, L. Pillonel, E. Schaller, D. Picque, T. Cattenoz and J.-O. Bosset. 2005a. The potential of combined infrared and fluorescence spectroscopies as a method of determination of the geographic origin of Emmental cheeses. International Dairy Journal. 15: 287–298.

Karoui, R., B. Martin and É. Dufour. 2005c. Potentiality of front-face fluorescence spectroscopy to determine the geographic origin of milks from the Haute Loire department (France). Dairy Science and Technology. 85: 223–236.

Kasemsumran, S., W. Thanapase and A. Kiatsoonthon. 2007. Feasibility of Near-Infrared Spectroscopy to Detect and to Quantify Adulterants in Cow Milk. Analytical Sciences. 23: 907–910.

Kirk, R.S and R. Sawyer. 1991. Introduction. Legislation, Standards and Nutrition. pp. 1–7. In: Pearson's Composition and Analysis of Foods. Longman Scientific & Technical, Essex, UK.

Kornexl, B.E., T. Werner, A. Roßmann and H.-L. Schmidt. 1997. Measurement of stable isotope abundances in milk and milk ingredients—a possible tool for origin assignment and quality control. Zeitschrift für Lebensmitteluntersuchung und -Forschung A. 205: 19–24.

Kotowicz, M., E. Adamczyk and J. Bania. 2007. Application of a Duplex-PCR for Detection of Cows' Milk in Goats' Milk. Annals of Agricultural and Environmental Medicine. 14: 215–218.

Levieux, D. and A. Venien. 1994. Rapid, sensitive two-site ELISA for detection of cows' milk in goats' or ewes' milk using monoclonal antibodies. Journal of Dairy Research. 61: 91–99.

Lin, S., J. Sun, D. Cao, J. Cao and W. Jiang. 2010. Distinction of different heat-treated bovine milks by native-PAGE fingerprinting of their whey proteins. Food Chemistry. 121: 803–808.

López-Calleja, I., I. González, I. Fajardo, I. Martín, P.E. Hernández, T. García and R. Martín. 2007. Quantitative detection of goats' milk in sheep's milk by real-time PCR. Food Control. 18: 1466–1473.

López-Calleja, I., I. González, V. Fajardo, I. Martín, P.E. Hernández, T. García and R. Martín. 2005. Application of Polymerase Chain Reaction to Detect Adulteration of Sheep's Milk with Goats' Milk. Journal of Dairy Science. 88: 3115–3120.

Manca, G., F. Camin, G.C. Goloru, A. Del Caro, D. Depentori, M.A. Franco and G. Versini. 2001. Characterization of the Geographical Origin of Pecorino Sardo Cheese by Casein Stable Isotope ($^{13}C/^{12}C$ and $^{15}N/^{14}N$) Ratios and Free Amino Acid Ratios. Journal of Agricultural and Food Chemistry. 49: 1404–1409.

Manca, G., M.A. Franco, G. Versini, F. Camin, A. Rossmann and A. Tola. 2006. Correlation Between Multielement Stable Isotope Ratio and Geographical Origin in Peretta Cows' Milk Cheese. Journal of Dairy Science. 89: 831–839.

Mašková, E. and I. Paulíčková. 2006. PCR-Based Detection of Cow's Milk in Goat and Sheep Cheeses Marketed in the Czech Republic. Czech Journal of Food Sciences. 24: 127–132.

Meurens, M., V. Baeten, S.H. Yan, E. Mignolet and Y. Larondelle. 2005. Determination of the Conjugated Linoleic Acids in Cow's Milk Fat by Fourier Transform Raman Spectroscopy. Journal of Agricultural and Food Chemistry. 53: 5831–5835.

Moatsou, G. and E. Anifantakis. 2003. Recent developments in antibody-based analytical methods for the differentiation of milk from different species. International Journal of Dairy Technology. 56: 133–138.

Motta, T.M.C., R.B. Hoff, F. Barreto, R.B.S. Andrade, D.M. Lorenzini, L.Z. Meneghini and T.M. Pizzolato. 2014. Detection and confirmation of milk adulteration with cheese whey using proteomic-like sample preparation and liquid chromatography—electrospray-tandem mass spectrometry analysis. Talanta. 120: 498–505.

Nicolaou, N., Y. Xu and R. Goodacre. 2010. Fourier transform infrared spectroscopy and multivariate analysis for the detection and quantification of different milk species. Journal of Dairy Science. 93: 5651–5660.

Pesic, M., M. Barac, M. Vrvic, N. Ristic, O. Macej and S. Stanojevic. 2011. Qualitative and quantitative analysis of bovine milk adulteration in caprine and ovine milks using native-PAGE. Food Chemistry. 125: 1443–1449.

Pillonel, L., S. Ampuero, R. Tabacchi and J. Bosset. 2003a. Analytical methods for the determination of the geographic origin of Emmental cheese: volatile compounds by GC/MS-FID and electronic nose. European Food Research and Technology. 216: 179–183.

Pillonel, L., R. Badertscher, P. Froidevaux, G. Haberhauer, S. Hölzl, P. Horn, A. Jakob, E. Pfammatter, U. Piantini, A. Rossmann, R. Tabacchi and J.O. Bosset. 2003b. Stable isotope ratios, major, trace and radioactive elements in emmental cheeses of different origins. Lebensmittel-Wissenschaft & Technologie. 36: 615–623.

Pillonel, L., W. Luginbühl, D. Picque, E. Schaller, R. Tabacchi and J. Bosset. 2003c. Analytical methods for the determination of the geographic origin of Emmental cheese: mid- and near-infrared spectroscopy. European Food Research and Technology. 216: 174–178.

Reid, L.M., C.P. O'Donnell and G. Downey. 2006. Recent technological advances for the determination of food authenticity. Trends in Food Science & Technology. 17: 344–353.

Richter, W., I. Krause, C. Graf, I. Sperrer, C. Schwarzer and H. Klostermeyer. 1997. An indirect competitive ELISA for the detection of cows' milk and caseinate in goats' and ewes' milk and cheese using polyclonal antibodies against bovine γ-caseins. Z Lebensm Unters Forsch A. 204: 21–26.

Rodrigues, N.P.A., P.E.N. Givisiez, R.C.R.E. Queiroga, P.S. Azevedo, W.A. Gebreyes and C.J.B. Oliveira. 2012. Milk adulteration: Detection of bovine milk in bulk goat milk produced by small holders in northeastern Brazil by a duplex PCR assay. Journal of Dairy Science. 95: 2749–2752.

Rodriguez-Otero, J.L., M. Hermida and J. Centeno. 1997. Analysis of Dairy Products by Near-Infrared Spectroscopy: A Review. Journal of Agricultural and Food Chemistry. 45: 2815–2819.

Rodríguez, E., R. Martín, T. García, J.I. Azcona, B. Sanz and P.E. Hernández. 1991. Indirect ELISA for detection of goats' milk in ewes'milk and cheese. International Journal of Food Science and Technology. 26: 457–465.

Rodríguez, E., R. Martín, T. García, P.E. Hernández and B. Sanz. 1990. Detection of cows' milk in ewes' milk and cheese by an indirect enzyme-linked immunosorbent assay (ELISA). Journal of Dairy Research. 57: 197–205.

Rolland, M.-P., L. Bitri and P. Besançon. 1993. Polyclonal antibodies with predetermined specificity against bovine a_{s1}-casein: application to the detection of bovine milk in ovine milk and cheese. Journal of Dairy Research. 60: 431–420.

Romero, C., O. Perez-Andújar, A. Olmedo and S. Jiménez. 1996. Detection of Cow's Milk in Ewe's or Goat's Milk by HPLC. Chromatographia. 42: 181–184.

Sacco, D., M.A. Brescia, A. Sgaramella, G. Casiello, A. Buccolieri, N. Ogrinc and A. Sacco. 2009. Discrimination between Southern Italy and foreign milk samples using spectroscopic and analytical data. Food Chemistry. 144: 1559–1563.

Santos, P.M., E.R. Pereira-Filho and L.E. Rodriguez-Saona. 2013. Rapid detection and quantification of milk adulteration using infrared microspectroscopy and chemometrics analysis. Food Chemistry. 138: 19–24.

Sato, T. and S. Kawano. 1990. Detection of Foreign Fat Adulteration of Milk Fat by Near Infrared Spectroscopic Method. Journal of Dairy Science. 73: 3408–3413.

Sivakesava, S. and J. Irudayaraj. 2002. Rapid Determination of Tetracycline in Milk by FT-MIR and FT-NIR Spectroscopy. Journal of Dairy Science. 85: 487–493.

Song, H., H. Xue and Y. Han. 2011. Detection of cow's milk in Shaanxi goat's milk with an ELISA assay. Food Control. 22: 883–887.

Sørensen, L.K. and R. Jepsen. 1998. Assessment of Sensory Properties of Cheese by Near-infrared Spectroscopy. International Dairy Journal. 8: 863–871.

Soyeurt, H., P. Dardenne, F. Dehareng, G. Lognay, D. Veselko, M. Marlier, C. Bertozzi, P. Mayeres and M. Gengler. 2006. Estimating Fatty Acid Content in Cow Milk Using Mid-Infrared Spectrometry. Journal of Dairy Science. 89: 3690–3695.

Tay, M., G. Fang, P.L. Chia and S.F.Y. Li. 2013. Rapid Screening for detection and differentiation of detergent powder adulteration in infant milk formula by LC-MS. Forensic Science International. 232: 32–39.

Tsenkova, R., S. Atanassova, K. Toyoda, Y. Ozaki, K. Itoh and T. Fearn. 1999. Near-Infrared Spectroscopy for Dairy Management: Measurement of Unhomogenized Milk Composition. Journal of Dairy Science. 82: 2344–2351.

Veloso, A.C.A., N. Teixeira, A.M. Peres, Á. Mendonça and I.M.P.L.V.O. Ferreira. 2004. Evaluation of cheese authenticity and proteolysis by HPLC and urea–polyacrylamide gel electrophoresis. Food Chemistry. 87: 289–295.

Wu, D., S. Feng and Y. He. 2007. Infrared Spectroscopy Technique for the Nondestructive Measurement of Fat Content in Milk Powder. Journal of Dairy Science. 90: 3613–3619.

Wu, D., S. Feng and Y. He. 2008b. Short-Wave Near-Infrared Spectroscopy of Milk Powder for Brand Identification and Component Analysis. Journal of Dairy Science. 91: 939–949.

Wu, D., Y. He, S. Feng and D.-W. Sun. 2008a. Study on infrared spectroscopy technique for fast measurement of protein content in milk powder based on LS-SVM. Journal of Food Engineering. 84: 124–131.

Wu, D., Y. He, J. Shi and S. Feng. 2009. Exploring Near and Midinfrared Spectroscopy to Predict Trace Iron and Zinc Contents in Powdered Milk. Journal of Agricultural and Food Chemistry. 57: 1697–1704.

Wu, D., P. Nie, Y. He and Y. Bao. 2012. Determination of Calcium Content in Powdered Milk Using Near and Mid-Infrared Spectroscopy with Variable Selection and Chemometrics. Food and Bioprocess Technology. 5: 1402–1410.

7

Honey Authenticity

Ioannis S. Arvanitoyannis* and Konstantinos V. Kotsanopoulos

7.1 Introduction

The detection of adulteration and the determination of the geographical and botanical origin of bee-products are highly important at a worldwide level and directly related to the international market of food commodities. Chemometrics are new analytical complements, which, in recent years, have been effectively applied to determine the geographical and botanical origin of food raw materials or for detecting possible adulterations. The classification can be performed by employing different analysis criteria, such as minor and trace elements, as well as sugars or other parameters (Camina et al., 2012). Honey has been a valuable food product since ancient times and has been adopted by many cultures as an important part of the diet. It can act both as a sweetener and a condiment (Devine, 2005). It is a naturally sweet material, which is produced through collection of nectar of living plant organs by *Apis mellifera* bee species or insects. Delivery of the material to the workers in the hive is followed by enzymatic alteration, storing, ripening and dehydration performed by the insects (Sass-Kiss, 2008). The significance of honey has been recently upgraded due to its nutrient and therapeutic effect but the adulteration of honey has increased exponentially as regards both its geographic and/or botanical origin. As a result, a need has arisen for the development of more effective quality control methods, which can effectively detect adulteration. Numerous novel, fast, and accurate methods like Atomic Absorption Spectrophotometry (AAS), High Performance Liquid Chromatography (HPLC), Gas Chromatography-Mass Spectrometry (GC-MS), High Performance Anion-Exchange chromatography and Pulsed Amperometric Detection (HPAEC-PAD), Nuclear Magnetic Resonance (NMR), Fourier-Transform Raman Spectroscopy (FT-Raman), and Near Infrared Spectroscopy (NIR) can contribute to the accomplishment of this

School of Agricultural Sciences, Department of Agriculture, Ichthyology and Aquatic Environment, University of Thessaly, Fytoko St., 38446 Nea Ionia Magnesias, Volos, Hellas, Greece.
* Corresponding author

target. Also, apart from these novel methods, the use of multivariate analysis and, specifically, Principal Component Analysis (PCA) and Cluster Analysis (CA), can be extremely useful for the categorisation or detection of honey of various origins. Mineral and trace element analysis have repeatedly proven to be very accurate for classifying honey of various origins (geographical and botanical) (Arvanitoyannis et al., 2005).

Honey is a relatively expensive commodity. Its dry weight consists mainly of the carbohydrates, glucose and fructose in approximately equal amounts. Therefore, it can potentially be adulterated with cheaper sources of sugar, such as high fructose corn syrup (HFCS), which can be sourced at only a fraction of the cost, while it very closely resembles the carbohydrate composition of honey. Since the nectar collected by bees is mainly derived from propionate flora, the potential of detecting HFCS derived from the butyrate corn plant has been extensively examined (Kelly, 2003). According to White et al. (1998), stable carbon isotope ratio analysis (SCIRA) of honey has been used for several years for detecting adulteration with cane or corn sugars. Six years of data from the internal standard carbon isotope ratio analysis (ISCIRA) method verify its worldwide applicability for honey analysis. The ISCIRA database of pure honeys has been extended from 64 U.S. samples to 224, through addition of data from Germany, United Kingdom, Mexico, Italy and Spain. ISCIRA analysis of 131 commercial honeys from the United States, Mexico and Spain indicated that 17 of them were adulterated. Analyses of 303 Chinese honeys proved that their carbon isotope values should be similar to honeys from other areas, which contradicts with the theory that the observed differences are intrinsic because of the variability of environmental conditions and plants used in honey production in China. Addition of corn or cane (C_4) sugars to honeys at levels that do not lead to the production of a $\delta^{13}C$ value higher than $-23.5‰$ for the mixture, cannot be detected by the original 1978 SCIRA procedure. Detection of such adulteration, however, can be performed by the ISCIRA procedure using the $\delta^{13}C$ value of the protein contained in the product, which offers an indication of the isotopic composition of the honey before addition of C_4 sugars. 43% of 98 honeys received in the United States in 1994–1997 with $\delta^{13}C$ <$-23.5‰$ were suspected and found to be adulterated.

Retail honey has always been subjected to some form of processing during production. Nowadays, the production of most, if not of all, commercial honey is carried out through centrifugation. An appellation such as "harvested in the cold" is a mislabelling, since honey is harvested naturally at temperatures between 25–32°C, which is the same as the temperature in the beehive (35°C). Filtering of honey is of high significance. In the regulations of many European beekeeping associations, the use of honey filters is prescribed. Harvesting of honey should be performed using filters with a mesh size not smaller than 0.2 mm aiming at preventing pollen removal. However, some packers, mainly in North America, will use smaller filters in order to filter out certain contaminants. According to the international honey legislation, such honey should be labelled as "filtered" (Bogdanov and Martin, 2002). The International Honey Commission (IHC) is a network created in 1990, under the umbrella of Apimondia, to enhance the knowledge on honey quality and research. The main work of the IHC is the improvement of honey analysis methods and the proposal of new quality criteria (Oddo and Bogdanov, 2004). Due to the variation of botanical origin, honeys may be very different in terms of appearance, sensory perception and composition. The most important nutritional and health relevant components are carbohydrates, mainly fructose and glucose, as well as 25 different oligosaccharides. Although honey contains high amounts of carbohydrates, its glycemic index can vary within a wide range from 32 to 85, depending on the botanical source. It contains low levels of proteins, enzymes, amino acids, minerals, trace elements, vitamins, aroma compounds and polyphenols (Bogdanov et al., 2008). Determination of the botanical origin of honey can be performed using classical methods such as by determining physicochemical parameters (electrical conductivity, sugars, enzyme activity, colour, pH, etc.) as well as other methods such as HPLC, spectroscopic techniques, etc. (Bogdanov et al., 2004).

Numerous analytical methods have been applied aiming at ensuring the authenticity of honey, but special emphasis has been put on the development of suitable methods for determining the geographical and botanical origin of the product. Although the determination of certain parameters, such as 5-hydroxymethylfurfural (HMF), moisture, enzyme activity, nitrogen, mono- and disaccharides, and residues from medicinal treatment or pesticides in honey does not offer any information about the botanical and geographical origin, some effective methods have been developed, which are based on the analysis of specific components or on multi-component analysis. In most cases, such methods can indicate the botanical origin, through investigation of flavonoids› patterns, distribution of pollen, aroma compounds and special marker compounds. There are some other profiles of substances, which could potentially be employed for detecting the geographical origin of honey (e.g., oligosaccharides, amino acids, trace elements). A combination of methods could be a promising approach to prove authenticity, especially when modern statistical data evaluation techniques are used (Anklam, 1998). Table 7.1 summarises some representative examples of techniques reported in different studies for the authentication of honey, both in terms of biological/geographical origin and addition of sugars from cheaper sources. According to Ruoff (2006), "the botanical origin of honey is currently determined by experts through evaluation of results derived from several analytical methods, such as pollen analysis, electrical conductivity and sugar composition. Although the composition of unifloral honeys has been described in various studies, internationally accepted criteria and the measurements to be considered for their authentication have not been defined yet. Pollen analysis has been considered to be the most important technique to classify different honey types. However, changes in legislation have recently allowed the removal of pollen with filtration. The altered pollen content does no more allow reliable conclusions to be drawn on the botanical and geographical origin of honey, thereby facilitating honey fraud. Moreover, various factors influencing the presence of pollen in honey lead to uncertainties in the interpretation of pollen analytical results".

Volatile compound determination aiming at evaluating the botanical origin of honey can be extremely useful. Nonetheless, further studies are required for developing a simple, rapid and objective method. Current methods that use solvents or headspace analysis (static and dynamic) have shown good approximations, but are rather complicated and demand additional equipment. Among them, dichloromethane extraction under an inert atmosphere followed by simultaneous steam distillation-dichloromethane extraction was shown to be very effective for honey aroma characterisation. On the other hand, the relatively recent Solid Phase Microextraction-Gas Chromatography/Mass Spectrometry (SPME-GC/MS) method is highly reliable,

Table 7.1. Representative techniques that can be used for the authentication of honey and their effectiveness

Type of honey	Type of adulteration to be detected	QC method	Efficiency	References
Clover honey	Addition of high-fructose corn syrup	HPLC, determination of compositional properties and statistical analysis	Simple tests/good indicators of adulteration	Abdel-Aal et al., 1993
Honeys of different botanical and geographical origin	False description of botanical origin	Analysis of amino acids with HPLC	Free amino acids proved to be good indicators of the botanical origin	Qamer et al., 2007
Acacia, chestnut, dandelion, lime, fir and rape honeys		Electronic nose based on mass spectrometry, PCA and DFA models, SPME, INDEX	Fast, reliable and powerful method/ best classification using SPME	Ampuero et al., 2004
Honey marketed in Southwestern Nigeria	Heat treatment and adulteration with sugar syrup	Analysis of physicochemical parameters, sucrose and Hydroxymethylfurfural	Accurate techniques when used for this purpose	Ayansola and Banjo, 2011
Honey produced in Córdoba (Argentina)	False description of geographical origin	Chemometrics	99% correct classification	Baroni et al., 2009
Multifloral honey from different countries		Saccharides analysis by NMR	Correct discrimination of samples	Consonni et al., 2013
Brazilian honey	False description of geographic origin, production location and authenticity assurance	Multi-element determination by ICP-MS and MLP, SVM and RF	Rapid way of generation of multi-element fingerprints	Batista et al., 2012
Multifloral and unifloral honeys (black locust, chestnut, lime, indigobush, rapeseed)	False description of botanical origin	Determination of Cd, Pb, Hg, As and Cu by AAS	Successful discrimination	Bilandžić et al., 2012
Acacia, Jujube, Vitex, Rape and Linden honey types		Near infrared spectroscopy and multivariate analysis	Very high percentages of correct classification (80–97.1%)	Chen et al., 2012
Acacia, Vitex, Rape and Linden honey types		Determination of mineral elements by ICP-MS and use of PCA, BP-ANN and PLS-DA	Powerful tool for biological origin classification	Chen et al., 2014
Lavandula, Robiniae and Fir honeys		DSC	The varieties could be easily differentiated	Cordella et al., 2003

Table 7.1. contd....

Table 7.1. contd.

Type of honey	Type of adulteration to be detected	QC method	Efficiency	References
Acacia, chestnut and lavender honeys	Adulteration by addition of sugar syrups	Analysis of sugars by gas chromatography and liquid chromatography	Effective detection of fraud resulting from 5 to 10% addition of foreign syrups	Cotte et al., 2003
Citrus, heather, Eucalyptus, rosemary, Echium and Rosaceae honeys	False description of botanical origin	GC method for analysis of carbohydrates and multivariate statistical analysis	Unambiguous classification cannot be obtained	de la Fuente et al., 2011
Heather (*Calluna vulgaris* and *Erica arborea*) honeys		Determination of floral origin markers by GC-MS	Markers were effectively identified	Guyot et al., 1999
		Identification, quantification and carbon stable isotopes determinations of organic acids by ion chromatography	Effective when used for this purpose	Gaëlle et al., 2012
Monofloral honeys	Detection of foreign sugars	Deuterium NMR Spectroscopy and SCIRA/MS	Can be used to detect added cane or corn sugars, but not beet sugar. Some useful information is given by Deuterium NMR	Giraudon et al., 2000
Different honeys produced in west Spain	False description of geographical origin	Employment of sugar patterns and common chemical quality parameters (use of linear stepwise discriminant analysis)	85.00–100% of samples were classified correctly	Gómez Bárez et al., 2000
"Mel de Galicia" honey		NIR-based method and pattern recognition techniques	Fast and cost-effective	Latorre et al., 2013
Lavandula stoechas, Lavandula angustifolia, and *Lavandula angustifolia x latifolia* honeys	False description of botanical origin	Determination of the floral markers by GC-FID	Markers were effectively detected	Guyot-Declerck et al., 2002
Brazilian honeys of different floral types		Determination of oligosaccharides by chemical method	Successful differentiation	Leite et al., 2000
Orange honey		Study of the volatile fraction of orange honey and by SPME-GC-MS	Rapid and easy method for floral origin authenticity	Verzera et al., 2014

reproducible and sensitive when used for extracting and identifying honey volatile compounds without the need of complex traditional methods. Furthermore, SPME is very flexible, simple and relatively economic extraction technique. To date, composite SPME fibre coatings, such as CAR/PDMS fibre (75 μm), PDMS/DVB (65 μm) and DVD/CAR/PDMS (50/30 l m), have shown to be effective for honey volatile extraction. The effectiveness of these methods depends on both the fibre characteristics and extraction conditions used for the analysis. Headspace solid phase dynamic extraction (SPDE) is another promising extraction technique with the advantage of the SPME coatings. However, further studies are demanded before it is established as a widely accepted technique for honey aroma extraction (Cuevas-Glory et al., 2007). According to Kaškonienė and Venskutonis (2010), "in view of the expanding global market, authentication and characterisation of botanical and geographic origins of honey has become a more important task than ever. Many studies have been performed with the aim of evaluating the possibilities to characterise honey samples of various origins by using specific chemical marker compounds. These have been identified and quantified for numerous honey samples. It has been demonstrated that currently it is rather difficult to find reliable chemical markers for the discrimination of honey collected from different floral sources because the chemical composition of honey also depends on several other factors, such as geographic origin, collection season, mode of storage, bee species, and even interactions between chemical compounds and enzymes in the honey". Therefore, according to the same authors, characterisation of honey can be more reliably performed through the determination of more than a single class of compounds and preferably in conjunction with modern data management of results, such as PCA or CA.

7.2 Use of High Performance Liquid Chromatography for the Authentication of Honey

HPLC was employed in the study of Qamer et al. (2007) for the identification of the amino acids content of honey. Specifically, analysis of Pakistani honey samples of different botanical and geographical origin was carried out for the development of a free amino acids profile. Detection of a total of seventeen free amino acids was performed in sunflower honey samples from Honeybee Research Farms of University of the Punjab, Phulai honey samples from Islamabad and sidder/ber honey samples from Bannu, Karak and Chunian. Proline was found to be the most dominant amino acid followed by aspartic acid and glutamic acid. Methionine was only found in three sidder honey samples obtained from different localities of Pakistan. Lysine and tryptophan were exclusively found in sunflower honey samples. Arginine was the only essential amino acid found in "Phulai" honey. Free amino acids were proved to be useful indicators of the botanical origin of honey. In addition, in the study of Cotte et al. (2004), amino acid analysis of honey using HPLC was employed for discriminating different botanical origins and detecting adulteration. Pure honeys of seven selected floral varieties were subjected to analysis. A PCA was also performed to select the most discriminating parameters. Lavender honeys were thus perfectly characterised, but no complete characterisation was obtained with the six other varieties. This method (analysis by HPLC and statistical processing by PCA) allowed the detection of added sugar syrup into rape and fir honeys.

In another study, deliberate contamination of pure honey with HFCS at levels of 10%, 20%, 30%, 40% and 50% (w/w) was performed. The sugar composition was determined by HPLC for all samples. Determination of the following compounds was carried out for pure and adulterated honey: moisture, total soluble solids, nitrogen, apparent viscosity, HMF, ash, sodium, calcium, potassium, proline, refractive index and diastatic activity. Statistical analysis proved that the following compositional parameters were negatively correlated with sugar composition: dry matter, apparent viscosity, sodium, potassium, proline and nitrogen. On the other hand, ash, calcium, HMF and moisture were positively correlated with sugar composition for both pure and adulterated honey. As a result, it was shown that such simple tests could be used as good indicators for the detection of the adulteration of honey with HFCS, at adulteration levels ranging from 10% to 50% (Abdel-Aal et al., 1993).

An HPLC method that can be used to determine organic acids in honey after sample purification by solid-phase extraction was described in the study of Cherchi et al. (1994). The chromatographic separation was achieved with two Spherisorb ODS-1 SS columns connected in series and sulphuric acid (pH 2.45) as the mobile phase. This method can separate a large number of organic acids. Simultaneous determination of many of them can be performed with the mobile phase at pH 2.45. On varying the mobile phase pH, separation of some acids whose peaks partially or completely overlap at pH 2.45 can be performed. In fact, the elution rate of organic acids changes significantly by changing the mobile phase acidity. It was shown that polycarboxylic acids, such as citric and fumaric acid, are characterised by higher sensitivity than others to any change in the mobile phase pH. The average recoveries of the acids ranged from 89% to 104% and the detection limits from 0.002 to 3 ppm (w/w).

Also, analysis of sugar profiles of fifty honey samples from different regions of Algeria was carried out using HPLC with PAD. 25 multifloral and 25 unifloral honeys were analyzed. Quantification of eleven sugars (two monosaccharides and nine oligosaccharides) was performed. The mean values of fructose and glucose ranged from 35.99 to 42.57% and from 24.63 to 35.06%, respectively. These monosaccharides were the most abundant sugars in all honey samples. The sucrose, maltose, isomaltose, turanose and erlose were found in almost all samples, while raffinose and melezitose were found in only few samples. Moreover, trehalose was detected in only two samples and none of the samples was found to contain melibiose. Low levels of melezitose, raffinose and erlose were detected in the range of 0.03–2.14%, 0.03–0.35% and 0.01–2.35%, respectively. PCA indicated that the cumulative variance was approximately 40% and correct classification of Apiaceae honeys was achieved using FDA (Factorial Discriminant Analysis) (Ouchemoukh et al., 2010).

The levels of ten oligosaccharides in 70 genuine Brazilian honeys of different floral types were determined in the study of Leite et al. (2000). Specifically, the oligosaccharides were determined by employing HPLC with a pump (Knauer, Germany), an injection valve of 20 m l loop (Reodhyne, USA) and a refractive index monitor (Waters, USA). The sucrose and isomaltose levels were 0.07 ± 0.77 and 0.18 ± 0.71%, respectively. The mean values for maltose ranged from 1.58 to 3.77%. The level of turanose (0.78 ± 2.03%) was similar to that of nigerose (1 11 +

2.81%). Low levels of melibiose (0.05 ± 0.15%) and panose (0.03 ± 0.08%) were detected in Brazilian honeys. Maltotriose, melezitose and raffinose were also found (0.24 ± 1.03, 0.21 ± 0.37 and 0.10 ± 0.25%, respectively). Between Brazilian states, honeys from São Paulo had mean values for melibiose significantly (P < 0.05) lower in comparison to those from Minas Gerais and Rio de Janeiro. The mean values for maltose and nigerose detected in the Mato Grosso do Sul and Goiás states were higher in comparison to those from the Mato Grosso state. Honey samples from the Paraná state had a mean value of maltotriose significantly higher than the samples from the Rio Grande do Sul state. The levels of maltose, nigerose, turanose and maltotriose proved to be useful for differentiating between honey samples from different geographical regions and could also be effective for testing the authenticity of Brazilian honeys. It is also important to mention the study of Kuhn et al. (2014), according to which it has been demonstrated that the chromatographic separation of mixtures of saccharides may be improved by using activated charcoal, a low cost material employed for separating sugars such as fructo-oligosaccharides. This study focused on developing a method for the separation of fructo-oligosaccharides from glucose, fructose and sucrose, using a fixed bed column packed with activated charcoal. The sugars were quantified with ion exchange chromatography with a PADE valuation of the influence of temperature, eluant concentration and step gradients were performed in an effort to enhance the separation efficiency and fructo-oligosaccharide purity. The final degrees of fructo-oligosaccharide purification and separation efficiency were 94% and 3.03 respectively, using ethanol gradient concentration ranging from 3.5% to 15% (v/v) at 40°C. The fixed bed column packed with the activated charcoal was proved to be very effective for separating sugars and especially fructo-oligosaccharides, giving highly pure solutions.

In another study, an optimization of an o-phthaldialdehyde method was successfully used to separate and quantify 23 amino acids. HPLC was employed as follows: the sample extracts (0.5 ml for honey and 1.0 ml for bee-pollen) were submitted to an automatic pre-column reaction using 100 µl of derivatising reagent. The chromatographic conditions were as follows: flow 0.1 ml/min until minute 3 and then 1.5 ml/min; volume of injection 10 µl for honey and 20 µl for bee-pollen; and solvents, A, sodium phosphate buffer (10 mM, pH 7.3) : methanol : tetrahydrofurane (80 : 19 : 1) and B, sodium phosphate buffer (10 mM, pH 7.3) : methanol (20 : 80). Fluorimetric detection was performed using excitation and emission wavelengths of 340 and 426 nm, respectively. Detection limits ranged from 0.24 to 10.1 pmol in honey and from 29.1 to 0.42 pmol in bee-pollen, the reproducibility ranged from 5.3% to 20.4%, while the recoveries were above 78.8%. Analysis of forty mono-varietal honey samples from ilex, oak, heather and chestnut-tree was carried out for their free amino acid profiles. α-aminoadipic acid and homoserine were reported for the first time in honeys. Thirty-two samples of Spanish bee-pollen, made of *Cistus Ladanifer* (67.1%) and *Echium plantagineum* (8.9%), were examined in order to determine their free and total amino acid profiles. Free γ-aminobutyric acid was detected (0.53 mg/g), while Hser and Orn were not frequent. Manually separated monofloral pellets from *C. ladanifer* and *E. plantagineum* were subjected to analysis in an effort to determine their free amino acid contents and were found to contain 32.46 and 21.87 mg/g for the former and 22.18 and 12.23 mg/g for the latter. In contrast, the total amino acid percentage (on a dry weight basis) was 13.95% for *C. ladanifer* and 32.22% for *E. plantagineum* (Paramás et al., 2006).

The HPLC phenolic profiles of 52 selected unifloral honey of European origin were subjected to analysis for detecting possible markers that could be used to determine the floral origin of the products. Lime-tree (five markers), chestnut (five markers), rapeseed (one marker), eucalyptus (six markers) and heather (three markers) honeys gave specific markers with distinctive UV spectra. Furthermore, it was shown that the flavanone hesperetin could be used as a marker of citrus honey and kaempferol and quercetin could be effectively used as markers for rosemary honey and sunflower honey, respectively. Abscisic acid, which had been reported to be a potential marker for heather honey, was also found in rapeseed, lime-tree and acacia honeys. Detection of ellagic acid in heather honey and the hydroxycinnamates caffeic, p-coumaric and ferulic acids in chestnut, sunflower, lavender and acacia honeys was also achieved. The characteristic propolis-derived flavonoids pinocembrin, pinobanksin and chrysin were detected in the majority of samples at various levels (Tomás-Barberán et al., 2001). Furthermore, Estevinho et al. (2008) examined the antioxidant and antimicrobial potential of phenolic compounds derived from dark and clear honeys from Trás-os-Montes of Portugal. The extraction was performed using Amberlite XAD-2 and the antioxidant effect was assessed by an *in vitro* test capacity of scavenge the 2,2 diphenyl-1-picryhydrazyl (DPPH) free radical as well as using the reduction of power of iron (III)/ferricyanide complex. Screening of the antimicrobial activity was carried out using three Gram-positive bacteria (*Bacillus subtilis, Staphylococcus aureus* and *Staphylococcus lentus*) and three Gram-negative bacteria (*Pseudomonas aeruginosa, Klebsiella pneumoniae* and *Escherichia coli*). The results obtained by partially identifying honey phenolic compounds using HPLC with a diode array detector indicated that p-hydroxybenzoic acid, cinnamic acid, naringenin, pinocembrin and chrysin were the phenolic compounds present in the majority of the samples examined. The antioxidant activity depended on honey extract concentration and it was shown that dark honey phenolic compounds were characterised by higher activity than those obtained from clear honey. It was also proved that *S. aureus* was the most sensitive microrganism and *B. subtilis, S. lentus, K. pneumoniae* and *E. coli* presented moderate sensitivity to the antimicrobial activity of honey extracts. No antimicrobial activity was observed in the case of *P. aeruginosa*.

Finally, the development of a simple, rapid, and effective HPLC–DAD method was carried out for detecting honey adulteration by rice syrup, based on the use of 2-acetylfuran-3-glucopyranoside (AFGP) which is a compound of rice syrup. The HPLC analyses indicated that the average concentration of AFGP was 92 ± 60 mg/kg in rice syrup. Since no AFGP was detected in any of the natural honey samples, it could be applied as a marker for detecting honey adulteration by rice syrup. The developed method allowed an accurate detection of honey samples adulterated with 10% rice syrup. Using this method, 16 out of 186 honey samples obtained from the market were found to be adulterated with rice syrup (Xue et al., 2013).

7.3 Use of Gas Chromatography for the Authentication of Honey

Portuguese lavender honeys are produced from the nectar of *Lavandula stoechas*, whereas French lavender honeys are exclusively derived from *Lavandula angustifolia*, *Lavandula latifolia*, or hybrids of these two species. To authenticate these types of honeys, volatile compounds from *L. stoechas*, *L. angustifolia*, and *L. angustifolia* × *latifolia* unifloral honeys were analysed. A Hewlett Packard Model 5890 gas chromatograph, equipped with a Hewlett Packard Model 7673 automatic sampler, a cold on-column injector, a flame ionisation detector, and a Shimadzu CR4A integrator was employed. The volatile compounds were analyzed on a 50 m × 0.32 mm i.d. wall-coated open tubular CP-SIL5 CB (Chrompack, Antwerp, Belgium) capillary column (film thickness, 1.2 μm), preceded by a 1 m × 0.53 mm i.d. capillary column, coated with a thin film of methyl silicone phase (Hewlett Packard, Brussels, Belgium). The aromatic profiles of French and Portuguese lavender honey samples differed significantly both in terms of quality and quantity, but no volatile compound characteristic of *L. stoechas* honeys only was found. It was confirmed that n-hexanal, n-heptanal, phenylacetaldehyde, and n-hexanol, previously proposed for the authentication of French lavender honeys, were contained at levels far above the published discrimination thresholds. Coumarin, previously proposed for the characterisation of French lavender honeys was found to be a better indicator of the freshness of lavender honey, being mainly released from glycosides during storage. Finally, *L. angustifolia* honeys were easily discriminated from hybrid-derived samples due to their lower phenylacetaldehyde and higher heptanoic acid content (Guyot-Declerck et al., 2002). According to de la Fuente et al. (2011), "characterisation of the most important Spanish floral honeys was carried out by analysing the carbohydrate content of 109 honey samples. The main unifloral sources, detected with pollen analysis, were citrus, heather, eucalyptus, rosemary, echium and rosaceae. A high number of multifloral samples were also examined. A GC method was employed for the analysis, based on the use of two different stationary phases. A quantitative procedure, taking into consideration the possible errors caused by unidentified overlapping compounds, was used and quantitative determination of two monosaccharides, 14 disaccharides and 21 trisaccharide peaks was carried out. Although similar qualitative results were reported for all samples, the quantitative results were highly variable, even within the same source. Multivariate statistical techniques were applied to the carbohydrate concentration data in an effort to detect possible relationships amongst the floral sources and sugar composition. Numerous carbohydrates were identified as characteristic of the most important honey types, although their levels in the samples did not allow an unambiguous classification of the main unifloral sources.

The chemical analysis of honey is mainly focused on parameters related to its state of preservation, such as HMF, diastase activity, and water content. Generally, numerous "minor" components of honey, such as flavours, di- and tri-saccharides, and free amino acids, have been employed for certifying the botanical origin of the product. To examine the potential of developing a method towards the authentication of honey, six kinds of honey from different botanical sources (acacia, citrus fruit, chestnut-tree, rhododendron, rosemary, and lime-tree) were subjected to analysis using capillary GC, and the data extracted were statistically assessed for determining whether the amino acid profile could be employed for verifying the botanical source of the material. It was found that the presence of amino acids such as arginine, tryptophan, and cystine is characteristic of a particular kind of honey, and other amino acids, such as proline, asparagine, lysine and methionine, could be used provided that quantitative data in relation to the levels of the compounds present can be provided (Pirini et al., 1992).

7.4 Use of Other Chromatographic Methods and GC-MS for the Authentication of Honey

In the study of Cotte et al. (2003), simultaneous use of GC and liquid chromatography was employed for the analysis of sugars in honey. After statistical processing by PCA, it was shown that the detection of added exogenous sugars could be performed by the appropriate fingerprints of adulteration. The limits of detection were very good (between 5 and 10%) for acacia, chestnut and lavender honeys. This method has the advantage of being universal with respect to a number of syrup types (C3 and C4) and therefore shows a better potential than carbon 13 isotopic analysis. Through analysing commercial samples, it was shown that some of the products were adulterated and in some cases the type of syrups used to adulterate the products was identified. This method can also be effectively used to authenticate other honey varieties (Cotte et al., 2003).

A method based on SPME using a 100 μm poly(dimethylsiloxane) (PDMS) fibre, followed by GC-MS was employed for analysing some monoterpenoids in honey. The extraction was carried out through direct immersion of the fibre using a sampling period of 15 min with constant magnetic stirring (1100 rpm) at an extraction temperature of 20°C. A 7 mL sample volume of an aqueous solution of honey with 25% of NaCl was placed in 15 mL glass vial fitted with screw cap and PTFE/silicone septum. Desorption was carried out in the gas chromatograph injector port during 5 min at 250°C using the splitless mode. The method is characterised by high sensitivity with detection limits between 11 and 25 μg L^{-1}, and is also very accurate with coefficients of variation ranging from 1.28 to 3.71%, and linear over more than one order of magnitude. Recoveries of 71.8 to 90.9% were observed. SPME remains an effective alternative technique due to the fact that it is rapid and solvent-free extraction method (Peña et al., 2004). GC-MS was also employed in the study of Radovic et al. (2001) to obtain the volatile profiles of 43 authentic honey samples of different botanical and geographical origins. A qualitative analysis of the volatile compounds found was carried out for assessing any potential marker compounds for both botanical and geographical origin. It was shown that a number of marker compounds could be used for the floral origins examined (e.g., acacia, chestnut, eucalyptus, heather, lavender, lime, rape, rosemary and sunflower).

Characterisation and classification of Greek thyme honeys (*Thymus capitatus* L.) according to their geographical origin was performed by Karabagias et al. (2014) through determination of volatile compounds and physicochemical parameters using MANOVA and Linear Discriminant Analysis. Forty-two thyme honey samples originated from five different regions of Greece were obtained and analysis of their volatile compounds was carried out using Headspace SPME-GC/MS. Identification and semi-quantification of forty-seven volatile compounds was achieved. Physicochemical analysis included the determination of pH, free, lactonic and total acidity, electrical conductivity, moisture, ash, lactonic/free acidity ratio and the colour parameters: L*, a* and b*. Nine volatile compounds and 11 physicochemical parameters were used to accurately classify the samples according to their geographical origin. The use of volatile compounds led to 64.3% correct prediction, while the use of physicochemical parameters, and the combination of both volatiles and physicochemical parameters led to 92.7% and 92.9% correct prediction, respectively.

According to Guler et al. (2014), one hundred pure and adulterated honey samples collected from feeding honeybee colonies and characterised by different levels (5, 20 and 100 L/colony) of different added commercial sugar syrups such as HFCS-85, HFCS-55, Bee Feeding Syrup (BFS), Glucose Monohydrate Sugar (GMS) and Sucrose Sugar (SS) were examined in terms of $\delta^{13}C$ and its protein, difference between the $\delta^{13}C$ value of protein and honey ($D\delta^{13}C$), and $C_4\%$ sugar ratio. Determination of the $\delta^{13}C$ values was carried out with Elemental Analyser–Isotope Ratio Mass Spectrometry after complete sample combustion to carbon dioxide. Sugar type, sugar level and the sugar type × sugar level interaction were found to be significant ($P < 0.001$) in relation to the evaluated characteristics. Adulteration could be detected in the 5 L/colony syrup level of all sugar types when the $\delta^{13}C$ value of honey, $D\delta^{13}C$ (protein–honey), and $C_4\%$ sugar ratio were employed as criteria according to the AOAC standards. On the other hand, detection of the adulteration was possible by using the same criteria in the honeys taken from the 20 and 100 L/colony of HFCS-85 and the 100 L/colony of HFCS-55. Detection of adulteration at low syrup level (20 L/colony) was more easily performed by increasing the fructose content of HFCS syrup. Therefore, the indirect adulteration of honey obtained by feeding the bee colonies with the syrups produced from C_3 plants such as sugar beet (*Beta vulgaris*) and wheat (*Triticium vulgare*) cannot be performed using the current methods and thus the development of novel methods and standards for detecting the presence and the level of indirect adulterations are urgently required.

Verzera et al. (2014) developed a new method for assessing the botanical origin of honeys, which relied on the enantiomeric ratio investigation of chiral volatile constituents derived from the plants being used by the bees. The method was used for the examination of orange honeys. The volatile fraction of orange honey and flowers was examined by SPME-GC-MS. Many substances were identified in orange honeys and linalool was the most abundant among orange flower volatiles. Determination of the enantiomeric ratios of linalool and its oxides was achieved and analogous values resulted between honey and flowers. Although the typical volatile constituents of orange honeys varied greatly, the enantiomeric ratios of linalool and its oxides were stable and therefore less affected by factors such as the production period, conditioning, packaging and storage. It was thereby suggested that enantiomeric distribution of honey volatile constituents that directly come from flowers could be used to rapidly and easily authenticate the floral origin of honeys.

In another study, fourteen organic acids were quantitatively and qualitatively analysed using ion chromatography with an electrochemical detector. Determination of the $^{13}C/^{12}C$ isotopic ratios of the honeys, and the organic acids, extracted from them with an anion exchange resin, was performed by IRMS. Gluconic acid was the most abundant organic acid in honey, occurring at levels of 1.8 to 12.7 g/kg. As regards fir honey, the predominant acid was the galacturonic acid (approximately 4.6 g/kg). The isotopic ratios of honeys and of their acids are highly correlated. The $\delta^{13}C$ values of the honey and the acids were significantly correlated. This study described a method that can be used for the differentiation of honeys from seven botanical origins, based on organic acid analysis. The combination of various organic acid levels and isotopic ratio values using PCA can differentiate honey samples according to their botanical origin (Gaëlle et al., 2012).

7.5 Use of Inductively Coupled Plasma Spectrometry-based Methods for the Authentication of Honey

According to Akbari et al. (2012), various elements are present in honey at different concentrations. These elements can offer several nutritional advantages at these concentrations, but can cause health problems when present at higher levels. Since food safety can be assured through regular monitoring of food quality, an attempt was made to establish a method for determining heavy metals and trace element levels using inductively coupled plasma atomic emission spectrometer (ICP-AES). Ten different honey brands from Iranian markets were examined. A Varian 720-ES ICP-AES (Agilent Technologies, Inc., Santa Clara, CA, USA) was employed for determining the elements. A Speed Wave 4 microwave digestion system (Berghof Products. Instruments GmbH, Eningen, Germany), maximum pressure 40 bar, maximum temperature 230°C, with Teflon reaction vessels was used for all digestion procedures. All heavy metal concentrations determined in the samples were within the ranges reported in literature, except for Hg, Al and As. Comparison with recommended daily intakes indicated that heavy metals or trace elements intoxication due to honey consumption in Iran is unlikely. Moreover, in the study of Batista et al. (2012), multi-element analysis of honey samples was performed in order to develop an accurate method that could be used to trace the origin of honey. Determination of 42 chemical elements (Al, Cu, Pb, Zn, Mn, Cd, Tl, Co, Ni, Rb, Ba, Be, Bi, U, V, Fe, Pt, Pd, Te, Hf, Mo, Sn, Sb, P, La, Mg, I, Sm, Tb, Dy, Sd, Th, Pr, Nd, Tm, Yb, Lu, Gd, Ho, Er, Ce and Cr) was carried out by inductively coupled plasma mass spectrometry (ICP-MS). Afterwards, three machine learning tools for classification and two for attribute selection were used aiming at proving data that can be used to determine the origin of honey. It was shown that Support Vector Machine (SVM), Multilayer Perceptron (MLP) and Random Forest (RF) chemometric tools could be very effective when used for honey origin authentication. Furthermore, the selection tools allowed a reduction from 42 trace element concentrations to only five.

Determination of the levels of 23 chemical elements (Al, As, Ba, Ca, Cd, Co, Cr, Cu, Fe, Hg, K, Mg, Mn, Na, Ni, Pb, Sb, Se, Sr, Th, Tl, U and Zn) in 51 honey samples of different botanical origin produced in Siena County (Italy) was performed by Pisani et al. (2008). The analyses were performed by inductively coupled plasma optical emission spectroscopy (ICP-OES), using a Perkin-Elmer Optima

2000 spectrophotometer, and by ICP-MS, using a Perkin-Elmer Sciex Elan 6100 spectrometer. The elements determined were Al, As, Ba, Ca, Cd, Co, Cr, Cu, Fe, Hg, K, Mg, Mn, Na, Ni, Pb, Sb, Se, Sr, Th, Tl, U and Zn, while K (1195 mg/kg), Ca (257 mg/kg), Na (96.6 mg/kg) and Mg (56.7 mg/kg) were the most abundant elements. The Fe, Zn and Sr concentrations ranged from 1 to 5 mg/kg. With the exception of Ba, Cu, Mn and Ni, the trace elements contents were below 100 μg/kg. It was therefore proved that there is a significant influence of the botanical origin on the element composition while a number of local geological and geochemical factors also affected the chemistry of the honey.

In another study, mineral elements and chemometric methods were applied for the classification of Chinese honeys according to their botanical origin. Determination of twelve mineral elements (Na^{23}, Mg^{24}, P^{31}, K^{39}, Ca^{43}, Mn^{55}, Fe^{56}, Cu^{63}, Zn^{66}, Rb^{85}, Sr^{88} and Ba^{137}) of 163 Chinese honey samples, including linden, vitex, rape, and acacia from Heilongjiang, Beijing, Hebei, and Shaanxi, China, was carried out by a ICP-MS method. PCA limited the 10 variables to four principal components and could explain 93.06% of the total variance. Partial least-squares discriminant analysis (PLS-DA) and back-propagation artificial neural network (BP-ANN) were used for the development of a classification model. Using PLS-DA, the total correct classification rates for model training and cross-validation were 90.9 and 88.4%, respectively, while the use of BP-ANN led to total correct classification rates for model training and cross-validation of 100 and 92.6%, respectively. The performance of BP-ANN was better than that of PLS-DA. After validation of the method, it was confirmed that the profiles of mineral elements created by ICP-MS with chemometric methods could be successfully used for classifying Chinese honey samples of different botanical origins (Chen et al., 2014).

7.6 Use of Atomic Absorption Spectrometry for the Authentication of Honey

According to Bilandžić et al. (2012), multifloral and unifloral honeys [black locust (*Robinia pseudoacacia* L.), chestnut (*Castanea sativa* Mill.), lime (*Tilia* spp.), indigobush (*Amorfa fruticoza* L.) and rapeseed (*Brassica napus*)] were collected from Koprivnica-Križevci County in northwestern Croatia during 2010 and 2011. Determination of their Cd, Pb, Hg, As and Cu contents was carried out and the mean levels of elements (mg/kg) in honey samples were as follows: in multifloral 1.26 for Cd, 163 for Pb, 135 for As, 1.35 for Hg and 11.7 for Cu; in black locust 1.52 for Cd, 182 for Pb, 23.2 for As, 0.46 for Hg and 7,697 for Cu; and in lime 2.92 for Cd, 340 for Pb, 116 for As, 0.74 for Hg and 7,798 for Cu. AAS was employed for measuring As, Cd, Cu and Pb concentrations. Argon was used for graphite furnace measurements. Pyrolytic coated graphite tubes with a platform were employed. Measurement of the atomic absorption signal was performed in peak area mode against a calibration curve. Quantification of mercury in honey samples was performed using the AMA-254 (Advanced Mercury Analyser, Leco, Poland) without acid digestion by direct combustion of the sample in an oxygen-rich atmosphere. Standards were used for determining the accuracy of the applied methods. It was found that Hg and Cu concentrations differed significantly between honey types. The average Cu levels detected in lime and black locust honey types were much higher than the levels reported for other countries around Europe. The elements contained in the highest amounts were: Cd 4.0 mg/kg and As 502 mg/kg in rapeseed, Hg 6.11 mg/kg in chestnut, Pb 2,159 mg/kg in black locust and Cu 79,167 mg/kg in indigobush. Lead content determined in all honey samples was much higher than that determined in Italy, Slovenia, Poland, Romania and Turkey. It was therefore suggested that it is of high importance to ensure positions of hive in zones of bee forage that are more distant from highways and railways. It was finally proved that the trace element content of honeys of different botanical origins collected at the same area was different.

Slurry sampling electrothermal AAS was employed in the study of de Andrade et al. (2014) to directly determine Cr, Pb and Cd in honey without the need of any sample pretreatment. Preparation of the slurries was carried out in aqueous solution containing hydrogen peroxide and nitric acid. The slurries were directly introduced in the pyrolytic graphite tubes. Pd-Mg was employed as a chemical modifier only for determining Cd. Analytical curves were developed with aqueous standards for Pb and Cr and by adding fructose for Cd. The quantification limits for Cd, Pb and Cr were determined at 2.0, 5.4 and 9.4 ng g^{-1}, respectively. The method proved to be quite precise and the recoveries were high (94–101%) for all three elements. The method was used for the examination of honey from Paraná (Brazil) and the levels of Pb, Cd and Cr ranged from 141 to 228 ng g^{-1}, <2.0 to 8 ng g^{-1} and 83 to 94 ng g^{-1}, respectively. According to Tuzen et al. (2007), "a survey of 25 honey samples from different botanical origin, collected all over Turkey was conducted to assess their trace element contents. The aim of this study was to determine the levels of cadmium (Cd), lead (Pb), iron (Fe), manganese (Mn), copper (Cu), nickel (Ni), chromium (Cr), zinc (Zn), aluminium (Al) and selenium (Se) in honey samples from different regions of Turkey. Trace element contents were determined by a flame and graphite furnace AAS technique after dry-ashing, microwave digestion and wet-digestion. The accuracy of the method was corrected by the standard reference material, NIST-SRM 1515 Apple leaves. The contents of trace elements in honey samples were in the range of 0.23–2.41 µg g^{-1}, 0.32–4.56 µg g^{-1}, 1.1–12.7 µg g^{-1}, 1.8–10.2 µg g^{-1}, 8.4–105.8 µg kg^{-1}, 2.6–29.9 µg kg^{-1}, 2.4–37.9 µg kg^{-1}, 0.9–17.9 µg kg^{-1}, 83–325 µg kg^{-1} and 38–113 µg kg^{-1} for Cu, Mn, Zn, Fe, Pb, Ni, Cr, Cd, Al and Se, respectively. Iron was the most abundant element while cadmium was the lowest element in Turkish honeys surveyed. The results showed that trace element concentrations in the honeys from different regions were generally correlated with the degree of trace element contamination of the environment".

7.7 Use of Near Infrared Spectroscopy and Chemometrics for the Authentication of Honey

According to Chen et al. (2012), NIR spectroscopy and multivariate analysis could be used for the classification of Chinese honey samples according to their different floral origins. Five different kinds of honey—acacia, linden, rape, vitex and jujube —were subjected to analysis using an NIR spectrophotometer with a fibre optic probe. Development of classification models relying on the NIR spectra was carried out using Mahalanobis-distance discriminant analysis (MD-DA) and a BP-ANN. Using the MD-DA model, a total of 87.4% and 85.3% of calibration and validation samples, respectively, were correctly classified. The ANN model correctly classified 90.9% and 89.3% of the calibration and validation sets, respectively. Using ANN, the corresponding correct classification rates of linden, acacia, vitex, rape and jujube were 97.1%, 94.3%, 80.0%, 97.1% and 85.7% in calibration, and 100%, 93.3%, 80.0%, 100% and 73.3% in validation. It was therefore proved that NIR in combination with a classification technique could be an effective way of classifying Chinese honeys from different botanical origins.

Authenticity is considered to be a very significant food quality criterion. Rapid methods that can confirm the authenticity status of foods, and detect adulteration are widely required by food producers, processors, consumers and regulatory authorities. Latorre et al. (2000) developed a model that could confirm the authenticity status of Galician-labelled honeys. Determination of nine metals was performed in 42 honey samples, which were divided into "Galician" and "non-Galician" honeys. Multivariate chemometric techniques such as CA, PCA, Bayesian methodology, partial least-squares regression and neural networks were used to modelling classes as per the chemical data. It was proved that a very accurate classification and prediction potential existed for both the neural networks and partial least-squares methods. The mineral profiles proved to be valuable tools that could be used to develop classification rules for the identification of honeys in accordance with their geographical origin. Furthermore, Latorre et al. (2013) used the NIR spectra of honeys with protected geographical indication (PGI) "Mel de Galicia" for developing an authentication system for this product. Different chemometric techniques were applied. Honey spectra were obtained in a rapid and single way, and pre-treatment was carried out in terms of standard normal variate transformation to remove the influence of particle size, scattering and other factors, prior to their use as input data. Initially, display techniques such as PCA and CA were used to confirm that the NIR data contained useful information for developing a pattern recognition classification system that could be used towards the authentication of honeys with PGI. Afterwards, application of different pattern recognition techniques (such as D-PLS: Discriminant partial least squares regression; SIMCA: Soft independent modelling of class analogy; KNN: K-nearest neighbours; and MLF-NN: Multilayer feedforward neural networks) was used to derive diverse models for PGI-honeyclass in order to detect possible falsification of these high-quality honeys. Amongst all the classification chemometric procedures, SIMCA proved to be the best PGI-model, being highly sensitive (sensitivity of 93.3%) and specific (100% specificity). As a result, combination of NIR information data with SIMCA led to the development of a single and fast method that could be effectively used for differentiating between genuine PGI-Galician honey samples and other commercial honey samples of different origin.

With the objective of further investigating the applicability of NIR spectroscopy on the rapid detection of honey adulteration, NIR spectroscopy combined with chemometric methods was used to qualitatively and quantitatively detect beet syrup adulteration of honey. Total prediction accuracy of testing set was 90.2% when PLS-DA was used for authentic and adulterated honey samples. Total prediction accuracy of testing sets was lower than 33.3% by different discriminant methods for classes of adulteration level. The quantitative analysis of adulteration level by PLS regression was satisfactory when adulterated honey samples were obtained from the same authentic honey sample, but it was not effective when used for examining adulterated samples derived from different botanical origins or different samples of the same botanical origin. It was reported that NIRS could be used to rapidly detect qualitatively authentic and adulterated honey samples, but not for detecting the adulteration level or for quantifying adulteration levels with beet syrup (Li et al., 2013).

7.8 Use of NMR-based Methods for the Authentication of Honey

Quantitative deuterium NMR spectroscopy was applied in combination with SCIRA/ mass spectrometry to detect sugars added to mono-floral honeys. The ^{13}C content of sugars can be used to indicate the type of photosynthetic metabolism of the plant that synthesized them, while the deuterium content is characteristic of the secondary metabolism and of environmental factors. Therefore, determination of the ^{13}C content of honeys and proteins extracted from the honeys can be used for the detection of added C_4 plant sugars (cane or corn), but cannot be used to reveal the addition of C_3 plant sugars such as beet sugar. Deuterium NMR gives useful information for some mono-floral honeys. NMR analysis is carried out on ethanol obtained from fermented honey after extraction by distillation. The isotopic composition of the ethanol gives an indication of the nature of the sugars it originates from. Different types of mono-floral honeys were examined, and the results obtained for commercial honeys proved that isotopic analysis is a valuable tool. The development of a database of authentic honeys to validate or affirm certain results could also be extremely useful (Giraudon et al., 2000). Consonni et al. (2013) also examined the saccharide content of honey samples from three different floral sources aiming at detecting a pattern that could be used for their geographical discrimination. Specifically, multi-floral honey from different countries, "high mountain multifloral" and rhododendron honeys from different regions of the northern part of Italy were examined using ^1H NMR spectroscopy and modelled with OPLS-DA in order to accurately classify the samples. NMR data allowed an accurate geographical discrimination of honeys highlighting the discriminating saccharides. It was therefore concluded that NMR and chemometrics are very effective in addressing quality requirements, and can thus be used for the quality assessment of honey in terms of geographical determination.

7.9 Use of Other Spectroscopic Methods for the Authentication of Honey

A chemometric analysis was employed to examine the adulteration of Mexican honey by sugar syrups such as corn syrup and cane sugar syrup. Fourier transform infrared spectroscopy (FTIR) was employed for measuring the absorption of a group of bee honey samples from Mexico. PCA was applied to process FTIR spectra aiming at determining the adulteration of bee honey. Furthermore, the concentrations of individual sugars from honey samples (glucose, fructose, sucrose and monosaccharides) were measured using PLS-FTIR analysis and the results were validated by HPLC measurements.

This analytical methodology based on infrared spectroscopy and chemometrics can accurately determine the purity and authenticity of honey (Rios-Corripio et al., 2011). Moreover, in a study carried out by Siobhán et al. (2008), FTIR and chemometrics were employed towards the verification of the origin of honey samples ($n = 150$) from Europe and South America. Authentic honey samples were obtained from five sources and included unfiltered samples from Mexico in 2004, commercially filtered samples from Ireland and Argentina in 2004, commercially filtered samples from the Czech Republic in 2005 and 2006, and commercially filtered samples from Hungary in 2006. Dilution of samples with distilled water was carried out to a standard solids content (70° Brix) and their spectra (2500–12,500 nm) were recorded at room temperature using an FTIR spectrometer equipped with a germanium attenuated total reflection accessory. First- and second-derivative and standard normal variate data pretreatments were used followed by PLS regression analysis, FDA and SIMCA. Examination of an attenuated wavelength range (6800–11,500 nm) instead of the whole spectrum (2500–12,500 nm) gave more accurate classification. It was finally shown that 93.3% of the samples were correctly classified with PLS discriminant analysis, while FDA techniques classified correctly 94.7% of honey samples. All samples were accurately classified using SIMCA, but models describing some classes were characterised by very high false positive rates.

Direct application of front-face fluorescence spectroscopy, on honey samples, was employed for authenticating 11 uni-floral and poly-floral honey types ($n = 371$ samples) which had previously been classified using traditional methods such as chemical, pollen, and sensory analysis. Records of excitation spectra (220–400 nm) were taken using the emission measured at 420 nm. Moreover, emission spectra were recorded between 290 and 500 nm (excitation at 270 nm) as well as between 330 and 550 nm (excitation at 310 nm). A total of four different spectral data sets were taken into account for data analysis. Chemometric evaluation of the spectra was performed using PCA and linear discriminant analysis; the error rates of the discriminant models were calculated by using Bayes' theorem and ranged from <0.1% (poly-floral and chestnut honeys) to 9.9% (fir honeydew honey) by using single spectral data sets and from <0.1% (metcalfa honeydew, poly-floral, and chestnut honeys) to 7.5% (lime honey) by combining two data sets. It was shown that front-face fluorescence spectroscopy can be successfully used for the authentication of the botanical origin of honey and may also be applied for determining the geographical origin of honeys of the same uni-floral type (Ruoff et al., 2006).

In the study of Latorre et al. (1999), a method was developed to confirm the geographical authenticity of Galician-labelled honeys. Determination of eleven metals was carried out in 42 honey samples, which were categorised into natural Galician honeys and processed non-Galician honeys. Multivariate chemometric techniques such as PCA, linear discriminant analysis, KNN and SIMCA were applied for classifying honeys in accordance with their type and origin on the basis of the chemical data. Measurements of eleven selected metals, Li, Rb, Na, K, Mg, Zn, Cu, Fe, Mn, Ni and Co were taken using a Varian AA 10 Plus atomic spectrometer. Li, Rb, Na and K were determined by atomic emission spectroscopy and determination of the remaining elements was performed using AAS. The use of only three attributes, Cu, Mn and Li, was sufficient for an almost correct classification.

Lachman et al. (2010) determined the antioxidant activity and total polyphenol content of Czech honey samples originated from the region North Moravia. Forty honey samples (multifloral, lime, rape, raspberry, mixture and honeydew honeys) were analyzed. The determination of antioxidant activity and total phenolics were performed using a UV-VIS spectrophotometer Helios g (Spectronic Unicam, Great Britain), centrifuge Janetzki T30, ultrasonic bath, magnetic stirrer, analytical weigh, pH-meter, SPE columns and plastic cuvettes. Determination of the total phenolics was performed with the modified Folin-Ciocalteau method. The antioxidant activity was evaluated using three different methods namely the ferric reducing antioxidant power (FRAP) assay, the 1,1-diphenyl-2-picrylhydrazyl (DPPH) assay and the 2,20-azinobis (3-ethylbenzothiazolin)-6-sulphonate (ABTS) assay. It was shown that the total phenolics and the antioxidant activity of the samples varied greatly between the honey kinds, location and time of harvest. The average total phenolic values ranged from 89.9 mg GA eq. kg^{-1} in lime honey up to 215.2 mg GA eq. kg^{-1} in honeydew honey. The antioxidant activity determined with the DPPH, ABTS and FRAP methods was lowest in floral honeys. The highest values were found in the cases of honeydew and mixture honeys. ABTS and FRAP assays proved to be the optimal methods for antioxidant activity determination in honey. It was also shown that the antioxidant activity and total phenolics were positively correlated. This is a proof that phenolics are one of the main substances adding antioxidant activity to honey. In this instance, it is also important to mention the study of Roshan et al. (2013) in which a model was described for authenticating mono-floral Yemeni Sidr honey using UV spectroscopy and chemometric techniques of hierarchical cluster analysis (HCA), PCA and SIMCA. The development of the model was performed using 13 genuine Sidr honey samples and validated using 25 honey samples of different botanical origins. HCA and PCA presented a successful preliminary clustering pattern that could be used for the segregation of the genuine Sidr samples from the lower priced local poly-floral and non-Sidr samples. The SIMCA model presented a clear demarcation of the samples and was applied for the identification of genuine Sidr honey samples as well as the detection of admixture with lower priced poly-floral honey with detection limits of >10%. This model is a simple and effective method of analysis and can also be effectively used for authenticating other honey types.

7.10 Use of Electronic Nose and Differential Scanning Calorimetry towards the Authentication of Honey

Analysis of 15 samples of honey, and specifically 14 samples from two Lombard provinces (four from Brescia and 10 from Sondrio) and a sample purchased in China was carried out in the study of Ghidini et al. (2008). Sensory, chemical and melissopalynological analysis were carried out aiming at characterising the samples. Furthermore, the obtained samples were also analysed using an electronic nose. pH, humidity, potassium, calcium, manganese and selenium were the most effective parameters for honey characterisation. Mineralisation of aliquots (0.5 g) of the samples was performed through acid digestion assisted with microwave oven and the levels of elements contained in the samples were determined by means of ICP-AES (Jobin

Yvon Ultima 2). Analysis was also performed by employing an artificial olfactory system equipped with 12 MOS sensors (MOS-AOS system), to obtain an aroma-characteristic fingerprint of the substance. An ISE Nose 2000 electronic nose equipped with a 16-position semi-automatic sampling unit was employed for the analyses. The electronic nose demonstrated its effectiveness since it allowed the discrimination of the Chinese sample from the Italian ones and the chestnut honey from the rest of the honey samples.

An electronic nose relied on mass spectrometry was used to verify the authenticity of the botanical origin of honey. PCA and DFA models were built using groups of samples declared as typical uni-floral honey by classical methods such as combined sensory, pollen and physicochemical analysis. Analysis of Swiss uni-floral honeys of acacia, chestnut, dandelion, lime, fir and rape types was performed. Three different sampling modes were tested: static headspace, SPME and inside-needle dynamic extraction (INDEX). SPME and INDEX proved to be effective in extracting volatile components at a higher concentration as well as heavier compounds. The best classification was achieved using the SPME sampling mode. This method was fast, reliable and very effective. A good correlation of results was found between the present approach and the classical method (Ampuero et al., 2004).

The thermal behaviour of authentic honeys and sugar syrups, both industrially produced and homemade, was investigated using Differential Scanning Calorimetry (DSC). The effect of adulteration on the thermal behaviour of authentic honeys was evaluated through examination of 30 honey samples (Robinia, Lavender, Chestnut and Fir). The T_g of samples was measured following an appropriate experimental protocol. It was shown that this parameter could be successfully used for distinguishing and characterising these varieties between them. When used to examine honey samples artificially adulterated with different industrial syrups, DSC had a detection level of 5–10% depending on the type of syrup (Cordella et al., 2003).

7.11 Physicochemical and Microscopic Analysis and Alternative Methods Proposed for Ensuring the Authenticity of Honey

In the study of Ayansola and Banjo (2011), the authenticity of honey from six states of southwestern Nigeria (Lagos, Ogun, Oyo, Osun, Ondo and Ekiti) was assessed to determine the quality status and give an indication of the extent of adulteration. The following physicochemical parameters were examined: Moisture content, Ash content, Total solids, Total titratable acidity, Glucose content, Fructose content, Sucrose content, and HMF content. According to these authors, "honey samples from the six states give the values of most of their physicochemical parameters within the acceptable range of the IHC except for the sucrose and HMF content. The HMF contents for all the six states were above the acceptable range. Lagos, Ogun, Ondo and Oyo states had their sucrose content above the permissible maximum value of the IHC. The high HMF values indicate heat-treatment of honey being sold in local markets in southwestern Nigeria thus rendering the honey nutritionally and medicinally valueless which is a form of adulteration. The high sucrose content is an indication of honey adulteration with sugar syrup. This confirms the insinuations about adulteration of marketed honey in southwestern Nigeria".

Analysis of sixty honey samples derived from six different production zones of the provinces of Salamanca, Zamora and Cáceres was performed for determining 13 common legal physicochemical parameters and 17 sugars in an effort to classify them as per their geographical origin. The use of linear stepwise discriminant analysis (LSDA) for several variables made of a selection of analytical results and simple mathematical functions of them allowed the differentiation between honeys of different geographical origins, as well as discrimination between honeys originated from the three zones of the province of Slamanca. The eight most discriminant variables selected for the six zones were a combination of nine physicochemical parameters and sugars, giving correct classification in 85.00% of samples. As regards the three zones of the province of Salamanca, seven variables were selected (consisting of seven single physicochemical parameters and sugars) and all samples (100.00%) were classified correctly (Gómez Bárez et al., 2000). Moreover, in the study of Sancho et al. (1991), the maxminf, direct and Wilks' lambda methods of discriminant analysis from the SPSSX statistical package were applied to the 61 pollen taxa found in 115 honeys from three provinces (Vizcaya, Guipuzcoa and Alava) in the Basque Country of northern Spain. By employing the direct method (use of 58 taxa), the highest discrimination degree was achieved, correctly classifying 93% of the honeys to their province of origin. When seven taxa [Compositae type *Carduus* sp., Ericaceae type *Arbutus unedo*, *C. sativa*, Leguminosae type *Genista* sp., *Eucalyptus* sp., *Rubus* sp. and fruit-bearing Rosaceae (*Prunus* sp.)] were considered, the discrimination was 81% for the maxminf method with an Fto-enter of 4.0. Moreover, in the study of Baroni et al. (2009) characterisation of honey samples originated from Córdoba (Argentina) and their classification by geographical provenance (North/South) was carried out using chemometrics. Analysis of 22 variables was performed taking into account both chemical properties and mineral profile. The samples met the international specifications for the parameters of interest. Classification of honey according to its geographical provenance (North/South) was performed using pattern recognition techniques that were applied to 15 out of 22 variables. Selection of glucose, pH, free acidity, free amino acids, calcium and zinc was performed by stepwise discriminant analysis, classifying the samples in accordance with their geographical origin. By applying k-nearest-neighbour classification procedure to these six selected variables, a successful assignation (99% correct) of honey to its provenance was observed. However, only 83% right assignation was achieved, when 15 variables were used, thus indicating that the use of all variables is not necessary for achieving an accurate geographical discrimination.

In the study of Huidobro et al. (1993), determination of the glycerol content of 33 honey samples of Galicia (northwestern Spain) was carried out using the Boehringer-Mannheim enzymatic method modified for this purpose. Volumes of 0.5 mL of potassium hexacyanoferrate(II)-trihydrate solution (Carrez I) and zinc acetate-dihydrate solution (Carrez II) were used and, following clarification, about 4 mL of 0.1 N NaOH. The enzymatic determination was carried out spectrophotometrically at 365 nm using pyruvate kinase, lactate dehydrogenase and glycerokinase in double the quantities recommended by the supplier. The method proved to be very precise (CV%

less than 1.1%), sensitive (30 mg/kg), simple and cost-effective, while good recovery (102.2%) was observed as well. The glycerol content of the samples ranged from 50.0 to 366.2 mg/kg (mean 137 mg/kg), which is in agreement with values obtained in other studies that used alternative methods (HPLC, GC).

Using a routine microscopic analysis of some honey samples, parenchyma cells, single rings of ring vessels and epidermal cells can be detected. These cells originate from the sugar cane stem. The potential relation between these plant fragments and the $\delta^{13}C$ value of honey was examined. Analysis of 17 honey samples and six cane sugar samples was carried out. Microscopic analysis of the samples was performed quantitatively by counting the parenchyma cells, rings, and epidermal cells present in 10 g of the sample using polarized light microscopy. Determination of the repeatability of the microscopic analysis was achieved by calculating the standard deviation of the values and examination of eight sub-samples from one honey sample. It was found that the presence of more than 150 parenchyma cells and/or 10 rings in 10 g indicated that the samples had been adulterated with C_4 sugars (from sugar cane or corn) according to the $\delta^{13}C$ method. Lower microscopic counts indicated honey with suspected adulteration below 7%, which was the limit of detection of the $\delta^{13}C$ method. It was finally shown that this microscopic procedure was an effective screening method for detecting adulteration of honey with cane sugar products (Kerkvliet and Meijer, 2000).

In another study, nectar and honey samples from alfalfa, alsike, canola, red clover, sweet clover and trefoil were analysed for monosaccharides (glucose and fructose) and oligosaccharides. Sucrose was found in all the nectars except from canola nectar, while a very small level of maltose was found in alsike nectar. No other oligosaccharides were found in the nectar samples. The oligosaccharides detected in the honey samples were present in similar amounts and seemed to originate from the enzyme activity of a- and b-glucosidase. Many of the oligosaccharides present in honey could be formed through the addition of isolated honey a- and b-glucosidase to a nectar-like carbohydrate solution. Analysis of honey samples from *Apis dorsata, Apis cerana* and *Apis florae* revealed that the oligosaccharides present were very similar to those detected in the *A. mellifera* honey samples (Low et al., 1988).

According to Guyot et al. (1999), 'Heather' is commonly used to qualify honeys issuing from the Ericaceae family. As mellissopalynology and sensory assessments alone were not sufficient to authenticate their floral origin, volatile compounds from *Erica arborea* and *Calluna vulgaris* unifloral honeys were investigated. Flavours were isolated via dichloromethane solubilisation, followed by a Likens-Nickerson simultaneous steam-distillation/solvent extraction. The extracts exhibited an intense honey aroma, representative of their floral origin. Four hundred compounds were separated by GC and MS. Among them, and in comparison with 11 other honey types, four proved to be markers of the Ericaceae family. Moreover, three were specific for *C. vulgaris* species and three others discriminant for the *E. arborea* samples".

In the study of Isengard et al. (2001), a modified Karl Fischer titration method (measurement at 50°C) was employed to accurately determine the water content in honey. Furthermore, the adjustment of a drying method was carried out using an infrared technique. The results obtained with the infrared method at temperatures of 98–100°C were found to be very similar to those obtained by the Karl Fischer technique. Afterwards, the determination of the water content was performed for 39 authentic honey samples from various geographical and botanical origins using Karl Fischer titration, refractive index measurements, official oven-drying method and a special oven method. The results obtained with different methods differed significantly (maximum differences ranged from 0.9 to 5.1%). The results obtained with Karl Fischer titration are characterised by the lowest standard deviations. With the exception of a few honey types, this method gave the highest values.

Paramás et al. (2000) used sixty honey samples originated from six different production zones of the provinces of Salamanca, Zamora and Cáceres (western Spain) to carry out analyses for 13 common legal physicochemical parameters and 17 mineral elements (13 cations and four anions) in an effort to achieve the geographical classification of the samples. As in the study of Bárez et al. (2000), application of linear stepwise discriminant analysis to a number of variables made of a selection of analytical results and their simple mathematical functions allowed the clear discrimination between honeys from all six zones as well as discrimination between honeys from the three zones of the province of Salamanca. Ten variables were the most discriminant and were selected for the six zones. Specifically, a combination of three physicochemical parameters and nine elements correctly classified 91.38% of samples. As regards the three zones of the province of Salamanca, six variables were selected (made of eight elements) leading to 97.07% of correctly classified samples.

In another study, analysis of forty-eight honeys from the La Rioja region of Spain was carried out to determine their quality status. Measurement of fourteen parameters required by legal authorities for quality control purposes was performed. Samples were obtained from Valley and Sierra. These areas are characterised by distinct agroclimatic conditions and, therefore, different flora types. The samples were classified according to their geographic origin by using multivariate statistical analysis to the chemical and physical data. Acidity (free acidity and pH), mineral content (electrical conductivity and ash) and factors connected with the degree of freshness (HMF and diastatic activity), were found to be the most important factors for the classification. The origin of samples of La Rioja honey (Valley or Sierra) was accurately determined in 83% of samples using only the legally required quality control parameters (Sanz et al., 1995).

The potential applicability of physical and chemical measurements for determining the botanical origin of honey using both the classical profiling approach and chemometrics was examined. An effort to authenticate nine uni-floral (acacia, rhododendron, chestnut, dandelion, heather, lime, rape, fir honeydew, metcalfa honeydew) and poly-floral honey types (in total $n = 693$ samples) was made. The classical approach that employs a profile for determining the botanical origin of honey indicated that the physical and chemical measurements alone are not sufficient for achieving a reliable determination. Pollen analysis is thus essential for discriminating between uni- and poly-floral honeys. However, chemometric evaluation of the physical and chemical data by linear discriminant analysis (LDA) proved to be sufficient for reliably authenticating the samples with no need to specialised expertise, pollen or sensory analysis. The error rates calculated by Bayes' theorem ranged from 1.1% (rape and lime honeys) to 9.9 % (acacia honey) (Ruoff et al., 2007).

In another study, the antioxidant activities and phenolic concentrations of five different types of Yemeni honey, namely *Acacia ehrenbergina* (Salam-Tehamah), *Acacia edgeworhi* (Somar-Hadramout), *Ziziphus spinachristi* L. (Sidr-Hadramout),

Ziziphus spinachristi L. (Sidr-Taiz), and Tropical blossom (Marbai- Hadramout), and four types of imported honeys (American-Tropical blossom, American-Orange source, Swiss-blossom, and an Iranian-Tropical blossom) were assessed. The total phenolic levels of diluted samples ranged from 56.32 to 246.21 mg/100 g honey as Catechin equivalent by the Folin-Ciocalteu method. Four of the five Yemeni honey samples were characterised by significantly higher total phenolic content in comparison to the imported honeys. Percentage antioxidant levels of diluted honey samples were assayed *in vitro* by the inhibition of liver homogenate oxidation mediated through a $FeSO_4$/ascorbate system. The antioxidant activity increased with increasing the levels of honey samples. The total antioxidant activities of diluted samples ranged from –6.48% (prooxidant activity) to 65.44% inhibition. It was shown that the *A. ehrenbergina* (Salam-Tehamah) presented the highest antioxidant activity and total phenolic content. It was also found that the percentage antioxidant and total phenolics were positively correlated (Al-Mamary et al., 2002). Moreover, analysis of mono-floral Cuban honeys was performed in the study of Alvarez-Suarez et al. (2010) for determining their total phenolic, flavonoid, ascorbic acid, amino acid, protein and carotenoid levels as well as their radical-scavenging activity and antimicrobial potential. The total phenolic, flavonoid and carotenoid levels presented high variations. The highest values were detected in Linen vine (*Govania polygama* (Jack) Urb) honey, which is classified as an amber honey. The highest amino acid content was determined in Morning glory (*Ipomoea triloba* L.) while the highest protein levels were detected in Linen vine. Linen vine honey had also the highest antioxidant potential while the corresponding lowest value was detected in Christmas vine (*Turbina corymbosa* (L.) Raf). The presence of ascorbic acid was not detected. Hydroxyl radical formation was analysed with EPR and spin trapping, and detected in all samples. Screening of the antimicrobial activity was carried out using two Gram-positive and Gram-negative bacteria. *S. aureus* was the most sensitive microorganism while *P. aeruginosa* had higher minimum active dilution values. *B. subtilis* and *E. coli* presented moderate sensitivity to honey antimicrobial activity. It was shown that radical-scavenging activity and total phenolic content were correlated. A correlation was also found between colour and phenolics levels, colour and flavonoid content, and between phenolic and flavonoid levels. Finally, Ferreira et al. (2009) evaluated the antioxidant potential of Portuguese honeys by taking into account their phenolic extracts. Numerous chemical and biochemical assays (namely reducing power, DPPH radical-scavenging capacity, and inhibition of lipid peroxidation using the β-carotene linoleate model system and the thiobarbituric acid reactive substances assay) were employed for screening the antioxidant properties of honeys with different colour intensity and phenolic extracts. Determination of the levels of phenols, flavonoids, ascorbic acid, β-carotene, lycopene and sugars in the samples was also carried out. The highest antioxidant levels and the lowest EC_{50} values for antioxidant activity were detected in the dark honey.

7.12 Conclusions

Honey is a highly nutritious food used by human for centuries but unfortunately it can be very easily adulterated. Addition of lower cost sources of honey as well as declaration of false geographical or biological origins are the most common ways of adulteration. Several techniques have been employed to detect adulteration or perform routine analyses (for quality control purposes) in honey, including HPLC, GC-MS, AAS, NMR, NIR, etc. The use of chemometrics is also commonly reported in relevant studies, while multivariate analysis (PCA, CA, LDA, etc.) has proved to be very useful when large numbers of data are analysed and particularly when the authentication of the origin of honey is examined. Future research should focus on the development of more accurate, sensitive and cost-effective techniques with the potential of rapidly authenticating honey. Due to the nature of the production/collection process of the product, particular interest should also be paid on the development of more user-friendly authentication and quality control measures in order to extend the implementation of controls towards the beginning of the supply chain.

References

Adbel-Aal, E.-S.M., H.M. Ziena and M.M. Youssef. 1993. Adulteration of honey with high-fructose corn syrup: Detection by different methods. Food Chemistry. 48: 209–212.

Akbari, B., F. Gharanfoli, M.H. Khayyat, Rezaee R. Khashyarmanesh and G. Karimi. 2012. Determination of heavy metals in different honey brands from Iranian markets. Food Additives and Contaminants: Part B: Surveillance. 5: 1–7.

Al-Mamary, M., A. Al-Meeri and M. Al-Habori. 2002. Antioxidant activities and total phenolics of different types of honey. Nutrition Research. 22: 1041–1047.

Alvarez-Suarez, J.M., S. Tulipani, D. Díaz, Y. Estevez, S. Romandi, F. Giampieri, E. Damiani, P. Astolfi, S. Bompadre and M. Battino. 2010. Antioxidant and antimicrobial capacity of several monofloral Cuban honeys and their correlation with color, polyphenol content and other chemical compounds. Food and Chemical Toxicology. 48: 2490–2499.

Ampuero, S., S. Bogdanov and J.-O. Bosset. 2004. Classification of unifloral honeys with an MS-based electronic nose using different sampling modes: SHS, SPME and INDEX. European Food Research and Technology A. 218(2): 198–207.

De Andrade, C.K., V.E. Anjos, M.L. Felsner, Y.R. Torres and S.P. Quináia. 2014. Direct determination of Cd, Pb and Cr in honey by slurry sampling electrothermal atomic absorption spectrometry. Food Chemistry. 146: 166–173.

Anklam, E. 1998. A review of the analytical methods to determine the geographical and botanical origin of honey. Food Chemistry. 63(4): 549–562.

Arvanitoyannis, I.S., C. Chalhoub, P. Gotsiou, N. Lydakis-Simantiris and P. Kefalas. 2005. Novel Quality Control Methods in Conjunction with Chemometrics (Multivariate Analysis) for Detecting Honey Authenticity. Critical Reviews in Food Science and Nutrition. 45: 193–203.

Ayansola, A.A. and A.D. Banjo. 2011. Physico-chemical Evaluation of the Authenticity of Honey Marketed in Southwestern Nigeria. Journal of Basic and Applied Scientific Research. 1(12): 3339–3344.

Bárez, J.A.G., R.J. Garcia-Villanova, S.E. Garcia, T.R. Palá, A.M.G. Paramás and J.S. Sánchez. 2000. Geographical discrimination of honeys through the employment of sugar patterns and common chemical quality parameters. European Food Research and Technology. 210(6): 437–444.

Baroni, M.V., C. Arrua, M.L. Nores, P. Fayé, M.P. Díaz, G.A. Chiabrando and D.A. Wunderlin. 2009. Composition of honey from Córdoba (Argentina): Assessment of North/South provenance by chemometrics. Food Chemistry. 114: 727–733.

Batista, B.L., L.R.S. Silva, B.A. Rocha, J.L. Rodrigues, A.A. Berretta-Silva, T.O. Bonates, V.S.D. Gomes, R.M. Barbosa and F. Barbosa. 2012. Multi-element determination in Brazilian honey samples by inductively coupled plasma mass spectrometry and estimation of geographic origin with data mining techniques. Food Research International. 49: 209–215.

Bilandžić, N., M. Đokić, M. Sedak, I. Varenina, B.S. Kolanović, A. Končurat, B. Šimić and N. Rudan. 2012. Content of five trace elements in different honey types from Koprivnica-Križevci county. Slovenia Veterinary Research. 49(4): 167–175.

Bogdanov, S. and P. Martin. 2002. Honey Authenticity: A review. Mitteilungen aus dem Gebiete der Lebensmitteluntersuchung und Hygiene. 93: 232–254.

Bogdanov, S., T. Jurendic, R. Sieber and P. Gallmann. 2008. Honey for Nutrition and Health: A review. American Journal of the College of Nutrition. 27: 677–689.

Bogdanov, S., K. Ruoff and Oddo L. Perdano. 2004. Physico-chemical methods for the characterisation of unifloral honeys: A review. Apidologie. 35: S4–S17.

Camina, M.J., G.P. Roberto and J.M. Eduardo. 2012. Geographical and Botanical Classification of Honeys and Apicultural Products by Chemometric Methods. A review. Current Analytical Chemistry. 8(3): 408–425.

Chen, H., C. Fan, Q. Chang, G. Pang, X. Hu, M. Lu and W. Wang. 2014. Chemometric Determination of the Botanical Origin for Chinese Honeys on the Basis of Mineral Elements Determined by ICP-MS. Journal of Agricultural and Food Chemistry. 62(11): 2443–2448.

Chen, L., J. Wang, Z. Ye, J. Zhao, X. Xue, Y.V. Heyden and Q. Sun. 2012. Classification of Chinese honeys according to their floral origin by near infrared spectroscopy. Food Chemistry. 135: 338–342.

Cherchi, A., L. Spanedda, C. Tuberoso and P. Cabras. 1994. Solid-phase extraction and high-performance liquid chromatographic determination of organic acids in honey. Journal of Chromatography A. 669: 59–64.

Consonni, R., L.R. Cagliani and C. Cogliati. 2013. Geographical discrimination of honeys by saccharides analysis. Food Control. 32: 543–548.

Cordella, C., J.P. Faucon, D. Cabrol-Bass and N. Sbirrazzuoli. 2003. Application of DSC as a tool for honey floral species characterization and adulteration detection. Journal of Thermal Analysis and Calorimetry. 71: 275–286.

Cotte, J., H. Cadabianca, B. Giroud, M. Albert, J. Lheritier and M. Grenier-Loustalot. 2004. Characterization of honey amino acid profiles using high-pressure liquid chromatography to control authenticity. Analytical and Bioanalytical Chemistry. 378(5): 1342–1350.

Cotte, J.F., H. Casabianca, S. Chardon, J. Lheritier and M.F. Grenier-Loustalot. 2003. Application of carbohydrate analysis to verify honey authenticity. Journal of Chromatography A. 1021: 145–155.

Cuevas-Glory, L.F., J.A. Pino, L.S. Santiago and E. Sauri-Duch. 2007. A review of volatile analytical methods for determining the botanical origin of honey. Food Chemistry. 103: 1023–1043.

Devine, B. 2005. Feasibility of the Spectroscopic Authentication of Honey Yearbook 2005, DIT School of Physics. 7–8.

Estevinho, L., A.P. Pereira, L. Moreira, L.G. Dias and E. Pereira. 2008. Antioxidant and antimicrobial effects of phenolic compounds extracts of Northeast Portugal honey. Food and Chemical Toxicology. 46: 3774–3779.

Ferreira, I.C.F.R., E. Aires, J.C.M. Barreira and L.M. Estevinho. 2009. Antioxidant activity of Portuguese honey samples: Different contributions of the entire honey and phenolic extract. Food Chemistry. 114: 1438–1443.

de la Fuente, E., A.I. Ruiz-Matute, R.M. Valencia-Barrera, J. Sanz and I.M. Castro. 2011. Carbohydrate composition of Spanish unifloral honeys. Food Chemistry. 129: 1483–1489.

Gaëlle, D., M. Dany and C. Hervé. 2012. Identification, quantification and carbon stable isotopes determinations of organic acids in monofloral honeys. A powerful tool for botanical and authenticity control. Rapid Communication in Mass Spectrometry. 26(17): 1993–1998.

Ghidini, S., C. Mercanti, E. Dalcanale, R. Pinalli and P.G. Bracchi. 2008. Italian Honey Authentication. Ann. Fac. Medic. Vet. di Parma. XXVIII: 113–120.

Giraudon, S., M. Danzart and M.H. Merle. 2000. Deuterium nuclear magnetic resonance spectroscopy and stable carbon isotope ratio analysis/mass spectrometry of certain monofloral honeys. Journal of AOAC International. 83(6): 1401–1409.

Guler, A., H. Kocaokutgen, A.V. Garipoglu, H. Onder, D. Ekinci and S. Biyik. 2014. Detection of adulterated honey produced by honeybee (*Apis mellifera* L.) colonies fed with different levels of commercial industrial sugar (C3 and C4 plants) syrups by the carbon isotope ratio analysis. Food Chemistry. 155: 155–160.

Guyot-Declerck, C., S. Renson, A. Bouseta and S. Collin. 2002. Floral quality and discrimination of *Lavandula stoechas*, *Lavandula angustifolia*, and *Lavandula angustifolia* × *latifolia* honeys. Food Chemistry. 79: 453–359.

Guyot, C., V. Scheirman and S. Collin. 1999. Floral origin markers of heather honeys: *Calluna vulgaris* and *Erica arborea*. Food Chemistry. 64: 3–11.

Hennessey, S., G. Downey and C. O'Donnell. 2008. Multivariate Analysis of Attenuated Total Reflection–Fourier Transform Infrared Spectroscopic Data to Confirm the Origin of Honeys. Applied Spectroscopy. 62(10): 1049–1171.

Huidobro, J.F., M.E. Rea, P.C.B. Andrade, M.T. Sancho, S. Muniategui and J. Simal-Lozano. 1993. Enzymic determination of glycerol in honey. Journal of Agriculture and Food Chemistry. 41(4): 557–559.

Isengard, H.-D., D. Schultheiss, B. Radović and E. Anklam. 2001. Alternatives to official analytical methods used for the water determination in honey. Food Control. 12: 459–466.

Karabagias, I.K., A. Badeka, S. Kontakos, S. Karabournioti and M.G. Kontominas. 2014. Characterization and classification of *Thymus capitatus* (L.) honey according to geographical origin based on volatile compounds, physicochemical parameters and chemometrics. Food Research International. 55: 363–372.

Kaškonienė, V. and P.R. Venskutonis. 2010. Floral Markers in Honey of Various Botanical and Geographic Origins: A Review. Comprehensive Reviews in Food Science and Food Safety. 9(6): 620–634.

Kelly, S.D. 2003. Using stable isotope ratio mass spectrometry (IRMS) in food authentication and traceability. pp. 156–183. *In:* M. Lees (ed.). Food Authenticity and Traceability. Woodhead Publishing Ltd, England.

Kerkvliet, J.D. and H.A.J. Meijer. 2000. Adulteration of honey: relation between microscopic analysis and $\delta^{13}C$ measurements. Apidologie. 31: 717–726.

Kuhn, R.C., M.A. Mazutti, L.B. Albertini and F.M. Filho. 2014. Evaluation of fructooligosaccharides separation using a fixed-bed column packed with activated charcoal. New Biotechnology. 31(3): 237–241.

Lachman, J., M. Orsák, Hejtmánková and E. Kovářová. 2010. Evaluation of antioxidant activity and total phenolics of selected Czech honeys. LWT—Food Science and Technology. 43: 52–58.

Latorre, C.H., R.M.P. Crecente, S.G. Martin and J.B. Garcia. 2013. A fast chemometric procedure based on NIR data for authentication of honey with protected geographical indication. Food Chemistry. 141: 3559–3565.

Latorre, M.J., R. Pena, S. Garcia and C. Herrero. 2000. Authentication of Galician (N.W. Spain) honeys by multivariate techniques based on metal content data. Analyst. 125: 307–312.

Latorre, M.J., R. Pena, C. Pita, A. Botana, S. Garca and C. Herrero. 1999. Chemometric classification of honeys according to their type. II. Metal content data. Food Chemistry. 66(2): 263–268.

Leite, J.M.D.C., L.C. Trugo, L.S.M. Costa, L.M.C. Quinteiro, O.M. Barth, V.M.L. Dutra and C.A.B. Maria. 2000. Determination of oligosaccharides in Brazilian honeys of different botanical origin. Food Chemistry. 70: 93–98.

Li, S.-F., R.-Z. Wen, Y. Yin, Z. Zhou and Y. Shan. 2013. Qualitative and Quantitative Detection of Beet Syrup Adulteration of Honey by Near-Infrared Spectroscopy: A Feasibility Study. Spectroscopy and Spectral Analysis. 33(10): 2637–2641.

Low, N.H., D.L. Nelson and P. Sporns. 1988. Carbohydrate analysis of Western Canadian honeys and their nectar sources to determine the origin of honey oligosaccharides. Journal of Apiculture. 27(4): 245–251.

Oddo, L.P. and S. Bogdanov. 2004. Determination of honey botanical origin: problems and issues. Apidologie. 35: S2–S3.

Ouchemoukh, S., P. Schweitzer, M.B. Bey, H. Djoudad-Kadji and H. Louaileche. 2010. HPLC sugar profiles of Algerian honeys. Food Chemistry. 121: 561–568.

Paramás, A.M.G., J.A.G. Bárez, R.J. García-Villanova, T.R. Palá, R.A. Albajar and J.S. Sánchez. 2000. Geographical discrimination of honeys by using mineral composition and common chemical quality parameters. Journal of the Science of Food and Agriculture. 80(1): 157–165.

Paramás, A.M.G., J.A.G. Bárez, C.C. Marcos, R.J. García-Villanova and J.S. Sánchez. 2006. HPLC-fluorimetric method for analysis of amino acids in products of the hive (honey and bee-pollen). Food Chemistry. 95: 148–156.

Peña, R.M., J. Barciela, C. Herrero and S. García-Martín. 2004. Solid-phase microextraction gas chromatography-mass spectrometry determination of monoterpenes in honey. Journal of Separation Science. 27: 1540–1544.

Pirini, A., L.S. Conte, O. Francioso and G. Lercker. 1992. Capillary gas chromatographic determination of free amino acids in honey as a means of discrimination between different botanical sources. Journal of High Resolution Chromatography. 15(3): 165–170.

Pisani, A., G. Protano and F. Riccobono. 2008. Minor and trace elements in different honey types produced in Siena County (Italy). Food Chemistry. 107: 1553–1560.

Qamer, S., M. Ehsan, S. Nadeem and A.R. Shakoori. 2007. Free Amino Acids Content of Pakistani Unifloral Honey Produced by *Apis mellifera*. Pakistan Journal of Zoology. 39(2): 99–102.

Radovic, B.S., M. Careri, A. Mangia, M. Musci, M. Gerboles and E. Anklam. 2001. Contribution of dynamic headspace GC ± MS analysis of aroma compounds to authenticity testing of honey. Food Chemistry. 72: 511–520.

Rios-Corripio, M.A., E. Rios-Leal, M. Rojas-López and R. Delgado-Mecuil. 2011. FTIR characterization of Mexican honey and its adulteration with sugar syrups by using chemometric methods. Journal of Physics: Conference Series. 274(1): 12098–12102.

Roshan, A.-R.A., H.A. Gad, S.H. El-Ahmady, M.S. Khanbash, M.I. Abou-Shoer and M.M. Al-Azizi. 2013. Authentication of Monofloral Yemeni Sidr Honey Using Ultraviolet Spectroscopy and Chemometric Analysis. Journal of Agriculture and Food Chemistry. 61(32): 7722–7729.

Ruoff, K. 2006. Authentication of the Botanical Origin of Honey. Diss. ETH No. 16857.

Ruoff, K., W. Luginbühl, V. Kilchenmann, J.O. Bosset, K. Von der Ohe, W. Von der Ohe and R. Amadó. 2007. Authentication of the botanical origin of honey using profiles of classical measurands and discriminant analysis. Apidologie. 38: 438–452.

Ruoff, K., W. Luginbühl, R. Künzli, S. Bogdanov, J.O. Bosset, K. Von Der Ohe, W. Von Der Ohe and R. Amadó. 2006. Authentication of the Botanical and Geographical Origin of Honey by Front-Face Fluorescence Spectroscopy. Journal of Agriculture and Food Chemistry. 54: 6858–6866.

Sass-Kiss, A. 2008. Chromatographic Technique: High Performance Liquid Chromatography (HPLC). pp. 361–410. *In*: D.-W. Sun (ed.). Modern Techniques for Food Authentication. Academic Press/ Elsevier, California, USA.

Sancho, M.T., S. Muniategui, J.F. Huidobro and J. Simal-Lozano. 1991. Discriminant analysis of pollen spectra of Basque Country (northern Spain) honeys. Journal of Apicultural Research. 30(3-4): 162–167.

Sanz, S., C. Perez, A. Herrera, M. Sanz and T. Juan. 1995. Application of a statistical approach to the classification of honey by geographic origin. Journal of the Science of Food and Agriculture. 69: 135–140.

Tomás-Barberán, F.A., I. Martos, F. Ferreres, B.S. Radovic and E. Anklam. 2001. HPLC flavonoid profiles as markers for the botanical origin of European unifloral honeys. Journal of the Science of Food and Agriculture. 81: 485–496.

Tuzen, M., S. Silici, D. Mendil and M. Soylak. 2007. Trace element levels in honeys from different regions of Turkey. Food Chemistry. 103: 325–330.

Verzera, A., G. Tripodi, C. Condurso, G. Dima and A. Marra. 2014. Chiral volatile compounds for the determination of orange honey authenticity. Food Control. 39: 237–243.

White, J.W., K. Winters, P. Martin and A. Rossmann. 1998. Stable carbon isotope ratio analysis of honey: validation of internal standard procedure for worldwide application. Journal of AOAC International. 81(3): 610–619.

Xue, X., Q. Wang, Y. Li, I. Wu, L. Chen, J. Zhao and F. Liu. 2013. 2-Acetylfuran-3-Glucopyranoside as a Novel Marker for the Detection of Honey Adulterated with Rice Syrup. Journal of Agriculture and Food Chemistry. 61(31): 7488–7493.

PART C

Legislation for Traceability and Authenticity of Foods

8

Food Traceability and Authentication Legislation in EU, USA, Canada and Australia-New Zealand

Ioannis S. Arvanitoyannis* and Persefoni Tserkezou

8.1 Introduction

As a result of recent food crises and outbreaks (dioxins, bovine spongiform encephalopathy (BSE), *Escherichia coli* O157:H7, etc.), several countries have developed and implemented legal requirements on traceability, and defined methods and control authorities to monitor foodstuffs that have to be swiftly removed from the market by recall procedures (Dabbenea et al., 2014).

Traceability is a theory applicable to all products and all types of supply chain. At the present time, in an economic system in which businesses emulate against each other in an environment mainly based on customer satisfaction, traceability is a crucial tool in gaining the market sense. Main profits are product safety, supply chain optimization and market advantages (Regattieri et al., 2007). According to European Union (EU) legislation, "traceability is defined the ability to track any food, feed, food-producing animal or substance that will be used for consumption, through all stages of production, processing and distribution" (European Communities, 2007).

McEntire et al. (2010) described the level of traceability. It can be based on four quantities: (i) breadth (amount of characteristics linked with each traceable unit), (ii) depth (how far upstream or downstream in the food supply chain the traceability

University of Thessaly, School of Agricultural Sciences, Department of Agriculture Ichthyology and Aquatic Environment, Fytokou Str., Nea Ionia Magnessias, 38446 Volos, Hellas (Greece).
* Corresponding author: parmenion@uth.gr

system traces the lot/unit correctly), (iii) precision (the degree of guarantee with which the system can identify a particular foodstuff movement or attributes) and (iv) access (the rapidity with which tracking and tracing information can be transferred to supply chain members and the rapidity with which the information can be disseminated to public health professionals during food-related crises).

Traceability is a way of reacting on potential risks that can come up in food and feed. It is also a way of certifying that all foodstuffs in the EU are safe for consumers (Arvanitoyannis et al., 2008). It is important that when national authorities or food operators identify a risk they can trace it back to its source in order to quickly face and solve the problem and not to allow contaminated foodstuffs to getting in consumers. Moreover, traceability permits focused removals and the provision of accurate information to the public, thus minimising distribution to trade (European Communities, 2007).

In 1985, a United Nations General Assembly resolution developed the "Guidelines for consumer protection" published in 1986. In 2002, The United States Department of Agriculture published "Traceability for Food Marketing and Food Safety: What's the Next Step". This publication presented the case for voluntary traceability in food industry, and suggested that government should ensure that the private companies meet performance targets for food safety. In 2002, EU General Food Law stated that a broad nondescriptive traceability requirement would be introduced from 1st January 2005 (European Parliament, 2002). At the same time, several countries introduced their own traceability legislation.

The legislative framework could not help companies to succeed product traceability. Nevertheless, in recent years several significant conduciveness to trace foodstuffs was made. The first important involvement dates back to the 1970s (Regattieri et al., 2007). The major principles of product traceability were established by Pugh (1973). The impact and principles of a traceability system were presented by Borst et al. (1997) and Gordijn and Akkermans (2001). Abbott (1991) noticed that the traceability helps product recall.

However, food traceability system is a kind of procedure based on the creditability of originality. It means that the information stored in the system is dependent on the willingness and credit of the enterprise. It is very easy for companies to knowingly provide incorrect or altered information if there is not sufficient supervision during the operation. International and national research has revealed several food adulteration cases (Al-Jowder et al., 1999; Marcos Lorenzo et al., 2002; Zhang et al., 2011). Adulteration means to make impure by adding foreign, extraneous, poisonous, insanitary, or inferior substances/ingredients to a food product; a food is adulterated if it fails to meet standards (Whitsitt et al., 2013).

The aim of this chapter is to analyse legal framework related to food traceability and authenticity or adulteration in EU, USA, Canada and Australia-New Zealand. A comprehensive summary of the relevant legislation is given in Table 8.1.

Table 8.1. Legislation related to traceability and adulteration of animal origin foodstuffs

Law	Entry into force	Main points	Comments
		European Union legislation	
Regulation (EC) No 178/2002 laying down the general principles and requirements of food law, establishing the European Food Safety Authority and laying down procedures in matters of food safety	1 January 2002 (except for article 11, 12 and 14–20 that applied from 1 January 2005)	• Definition of traceability • The traceability of food, feed, food-producing animals, and any other substance intended to be, or expected to be, incorporated into a food or feed shall be established at all stages of production, processing and distribution • Establishment of rapid alert system (article 50)	Amendments • Regulation (EC) No 1642/2003 of the European Parliament and of the Council of 22 July 2003 • Commission Regulation (EC) No 575/2006 of 7 April 2006 • Commission Regulation (EC) No 202/2008 of 4 March 2008 • Regulation (EC) No 596/2009 of the European Parliament and of the Council of 18 June 2009
Regulation (EU) No 1169/2011 of the European Parliament and of the Council of 25 October 2011 on the provision of food information to consumers	12 December 2011 (except for Article 9(1), applied from 13 December 2016 and Annex VI applied from 1 January 2014)	• General principles of food information and list of mandatory information are given • Specific requirement for meat, poultry, fish, meat-based products and fishery products are given in Annexes of this Regulation	This Regulation amended: • Regulation (EC) No 1924/2006 • Regulation (EC) No 1925/2006 This Regulation repeals • Commission Directive 87/250/EEC • Council Directive 90/496/EEC • Commission Directive 1999/10/EC • Directive 2000/13/EC • Commission Directive 2002/67/EC • Commission Directive 2008/5/EC • Commission Regulation (EC) No 608/2004
Regulation (EC) No 1760/2000 of the European Parliament and of the Council of 17 July 2000 establishing a system for the identification and registration of bovine animals and regarding the labelling of beef and beef products	13 August 2000 (applicable to beef from animals slaughtered on or after 1 September 2000)	• General principles of the system for the identification and registration of bovine animals • Establishment of Compulsory Community beef labelling system	Amendments • Council Regulation (EC) No 1791/2006 of 20 November 2006 • Council Regulation (EU) No 517/2013 of 13 May 2013 Repeal • Regulation (EC) No 820/97

Regulation (EC) No 2065/2001of 22 October 2001 laying down detailed rules for the application of Council Regulation (EC) No 104/2000 as regards informing consumers about fishery and aquaculture products	8 July 2013	• Changes to lists of commercial designations and requirements governing consumer information • Define the requirements for traceability and control	Amendments • Commission Regulation (EC) No 1792/2006 • Commission Regulation (EU) No 519/2013
Regulation (EC) No 589/2008 of 23 June 2008 laying down detailed rules for implementing Council Regulation (EC) No 1234/2007 as regards marketing standards for eggs	1 July 2008 (except for Article 33 applied until 30 June 2009)	• Marking of eggs for cross-border delivery • Requirements regarding industrial eggs	Amendment • Regulation (EU) No 548/2013 • Regulation (EU) No 342/2013 Repeal • Regulation (EC) No 557/2007
USA legislation			
Food Safety Modernization Act	January 2011	• Introduction to product tracing system • Requirements of traceability of food articles • Importance of food registration and traceback during inspection	–
US Code, Paragraph 342—Adulterated food	July 1954	• List the cases that a food could be deemed adulterated	Last update on 13 August 2013
Poultry Products Inspection Act—Chapter 10	August 1968	• Definition of adulterated poultry and poultry products • Procedures followed when adulterated poultry or poultry products are identified during an inspection	Last update on 20 May 2009
Federal Meat Inspection Act—Subchapter I: Inspection requirements, adulteration and misbranding	July 1948	• Sanitary inspection and regulation of slaughtering and packing establishments; rejection of adulterated meat or meat food products • Inspectors of meat food products; marks of inspection; destruction of condemned products; products for export	Last update on 25 May 2009

Table 8.1. contd....

Table 8.1. contd.

Law	Entry into force	Main points	Comments
		Canadian legislation	
Food and Drug Regulations	1 June 2009	• Define the adulteration of food • Determine when fish or fishery products, poultry and poultry products could be considered adulterated	Last amended on 19 June 2013
Meat Inspection Regulations	1990	• Define when meat is considered adulterated • No adulterated meat product shall be identified as edible	Last amended on 26 April 2013
		Australia—New Zealand legislation	
Standard 3.2.2 regarding Food Safety Practices and General Requirements	2005	• Cover the "one step back and one step forward" elements of traceability • Define the requirements related to food receipt and food recall	Updates: 2006, 2011, 2012
Food Standards Code	2003	• Define the requirements related to traceability on poultry, dairy, egg and seafood • Define the requirements regarding labelling and recalls	

8.2 EU Legislation

In European Union, Regulation (EC) No 178/2002, applied since 2002 and followed by further amendments concerning specific matters, requires the establishment of a traceability system for all foodstuffs. This Regulation clearly states that the detail of traceability is to be prolonged also to each ingredient of the food (Arvanitoyannis et al., 2008). However, the General Food Law does not state any specific method or technique that food operators have to follow (Folinas et al., 2006; Asioli et al., 2011).

According to Regulation (EC) No 178/2002, food law shall aim at the protection of the interests of consumers and shall provide a basis for consumers to make informed choices in relation to the foods they consume. It shall aim at the prevention of: (a) fraudulent or deceptive practices; (b) the adulteration of food; and (c) any other practices which may mislead the consumer. The traceability of food, feed, food-producing animals, and any other substance intended to be, or expected to be, incorporated into a food or feed shall be established at all stages of production, processing and distribution. Food and feed business operators shall be able to identify any person from whom they have been supplied with a food, a feed, a food-producing animal, or any substance intended to be, or expected to be, incorporated into a food or feed. To this end, such operators shall have in place systems and procedures which allow for this information to be made available to the competent authorities on demand. Food and feed business operators shall have in place systems and procedures to identify the other businesses to which their products have been supplied. This information shall be made available to the competent authorities on demand. Food or feed which is placed on the market or is likely to be placed on the market in the Community shall be adequately labelled or identified to facilitate its traceability, through relevant documentation or information in accordance with the relevant requirements of more specific provisions. Article 33 summarises that the Authority shall search for, collect, collate, analyse and summarise relevant scientific and technical data in the fields within its mission. This shall involve in particular the collection of data relating to: (a) food consumption and the exposure of individuals to risks related to the consumption of food; (b) incidence and prevalence of biological risk; (c) contaminants in food and feed and (d) residues.

According to Article 50, a rapid alert system for the notification of a direct or indirect risk to human health deriving from food or feed is hereby established as a network. It shall involve the Member States, the Commission and the Authority. The Member States, the Commission and the Authority shall each designate a contact point, which shall be a member of the network. The Commission shall be responsible for managing the network. Where a member of the network has any information relating to the existence of a serious direct or indirect risk to human health deriving from food or feed, this information shall be immediately notified to the Commission under the rapid alert system. The Commission shall transmit this information immediately to the members of the network. The Authority may supplement the notification with any scientific or technical information, which will facilitate rapid, appropriate risk management action by the Member States. Without prejudice to other Community legislation, the Member States shall immediately notify the Commission under the rapid alert system of: (a) any measure they adopt which is aimed at restricting the placing on the market or forcing the withdrawal from the market or the recall of food or feed in order to protect human health and requiring rapid action; (b) any recommendation or agreement with professional operators which is aimed, on a voluntary or obligatory basis, at preventing, limiting or imposing specific conditions on the placing on the market or the eventual use of food or feed on account of a serious risk to human health requiring rapid action; and (c) any rejection, related to a direct or indirect risk to human health, of a batch, container or cargo of food or feed by a competent authority at a border post within the European Union. The notification shall be accompanied by a detailed explanation of the reasons for the action taken by the competent authorities of the Member State in which the notification was issued. It shall be followed, in good time, by supplementary information, in particular where the measures on which the notification is based are modified or withdrawn. The Commission shall immediately transmit to members of the network the notification and supplementary information received under the first and second subparagraphs. Where a batch, container or cargo is rejected by a competent authority at a border post within the European Union, the Commission shall immediately notify all the border posts within the European Union, as well as the third country of origin. Where a food or feed which has been the subject of a notification under the rapid alert system has been dispatched to a third country, the Commission shall provide the latter with the appropriate information. The Member States shall immediately inform the Commission of the action implemented or measures taken following receipt of the notifications and supplementary information transmitted under the rapid alert system. The Commission shall immediately transmit this information to the members of the network. Participation in the rapid alert system may be opened up to applicant countries, third countries or international organisations, on the basis of agreements between the Community and those countries or international organisations, in accordance with the procedures defined in those agreements. The latter shall be based on reciprocity and shall include confidentiality measures equivalent to those applicable in the Community.

Regulation (EU) No 1169/2011 provides the basis for the assurance of a high level of consumer protection in relation to food information, taking into account the differences in the perception of consumers and their information needs whilst ensuring the smooth functioning of the internal market. This Regulation establishes the general principles, requirements and responsibilities governing food information, and in particular food labelling. It lays down the means to guarantee the right of consumers to information and procedures for the provision of food information, taking into account the need to provide sufficient flexibility to respond to future developments and new information requirements. This Regulation shall apply to food business operators at all stages of the food chain, where their activities concern the provision of food information to consumers. It shall apply to all foods intended for the final consumer, including foods delivered by mass caterers, and foods intended for supply to mass caterers. The food business operator responsible for the food information shall be the operator under whose name or business name the food is marketed or, if that operator is not established in the Union, the importer into the Union market. The food business operator responsible for the food information shall ensure the presence and accuracy of the food information in accordance with the applicable food information law and requirements of relevant national provisions. Food business operators which do not affect food information shall not supply food which they know or presume, on the basis of the information in their possession as professionals, to be non-compliant with the applicable food information law and requirements of relevant national provisions. Food business operators, within the businesses under their control, shall not modify the information accompanying a food if such modification would mislead the final consumer or otherwise reduce the level of consumer protection and the possibilities for the final consumer to make informed choices. Food business operators are responsible for any changes they make to food information accompanying a food. Indication of the following particulars shall be mandatory: (a) the name of the food; (b) the list of ingredients; (c) any ingredient or processing aid listed in Annex II or derived from a substance or product listed in Annex II causing allergies or intolerances used in the manufacture or preparation of a food and still present in the finished product, even if in an altered form; (d) the quantity of certain ingredients or categories of ingredients; (e) the net quantity of the food; (f) the date of minimum durability or the 'use by' date; (g) any special storage conditions and/or conditions of use; (h) the name or business name and address of the food business operator; (i) the country of origin or place of provenance; (j) instructions for use where it would be difficult to make appropriate use of the food in the absence of such instructions; (k) with respect to beverages containing more than 1.2% by volume of alcohol, the actual alcoholic strength by volume and (l) a nutrition declaration.

According to Regulation (EC) No. 1760/2000 regarding the labelling of beef and beef products, the system for the identification and registration of bovine animals shall comprise the following elements: (a) ear tags to identify animals individually; (b) computerised databases; and (c) animal passports; (d) individual registers kept on each holding. The Commission and the competent authority of the Member State concerned shall have access to all the information covered by this title. The Member States and the Commission shall take the measures necessary to ensure access to these data for all parties concerned, including consumer organisations having an interest which are recognised by the Member State, provided that the data confidentiality and protection prescribed by national law are ensured. The label shall contain the following indications: (a) a reference number or reference code ensuring the link between the meat and the animal or animals. This number may be the identification number of the individual animal from which the beef was derived or the identification number relating to a group of animals; (b) the approval number of the slaughter house at which the animal or group of animals was slaughtered and the Member State or third country in which the slaughter house is established. The indication shall read: 'Slaughtered in (name of the Member State or third country) (approval number)'; and (c) the approval number of the cutting hall which performed the cutting operation on the carcass or group of carcasses and the Member State or third country in which the hall is established. The indication shall read: 'Cutting in: (name of the Member State or third country) (approval number)'.

Regulation (EC) No 2065/2001 refers to labelling of fishery and aquaculture products. Any species not included on the list of commercial designations accepted by a Member State may be marketed under a provisional commercial designation laid down by the competent authority of the Member State. A definitive commercial designation included on the list of accepted designations shall be laid down by the Member State within five months of the date on which the species in question is given the provisional commercial designation. Any changes to the list of commercial designations accepted by a Member State shall be notified forth with to the Commission, which shall inform the other Member States thereof. According to Article 8 related to traceability, the information required concerning the commercial designation, the production method and the catch area shall be available at each stage of marketing of the species concerned. This information together with the scientific name of the species concerned shall be provided by means of the labelling or packaging of the product, or by means of a commercial document accompanying the goods, including the invoice.

Regulation (EC) No 589/2008 regards marketing standards for eggs determines that eggs delivered from a production site to a collector, a packing centre or non-food industry situated in another Member State shall be marked with the producer code before leaving the production site. A Member State on whose territory the production site is situated may grant an exemption from the requirement provided for in the previous paragraph, where a producer has signed a delivery contract with a packing centre in another Member State requiring the marking in accordance with this Regulation. Such an exemption may be granted only at the request of both operators concerned and with the prior written agreement of the Member State where the packing centre is situated. In such cases, a copy of the delivery contract shall accompany the consignment. According to Article 9 the producer code shall consist of the codes and letters which shall be easily visible and clearly legible and be at least 2 mm high. Industrial eggs shall be marketed in packaging containers with a red band or label. Those bands and labels shall show: (a) the name and address of the operator for whom the eggs are intended; (b) the name and address of the operator who has dispatched the eggs; and (c) the words 'industrial eggs' in capital letters 2 cm high, and the words 'unsuitable for human consumption' in letters atleast 8 mm high.

8.3 USA Legislation

In the US, obligatory traceability was only recently introduced for the food operations (Smith et al., 2005; Donnelly and Thakur, 2010). Bioterrorism was one of the reasons that the traceability first became compulsory (United States, 2002). The Food Safety Modernization Act, signed on January 2011 by the US President, introduces a system of preventive controls, inspections and compliance authorities, as a response to violations (recalls) on domestic as well as on foreign US food (United States, 2011).

According to Food Safety Modernization Act, the Secretary shall include in the report developed an analysis of the Food and Drug Administration's performance in foodborne illness outbreaks during the five-year period preceding the date of enactment of this Act involving fruits and vegetables that are raw agricultural commodities and recommendations for enhanced surveillance, outbreak response, and traceability. Such findings and recommendations shall address communication and coordination with the public, industry, and State and local governments, as such communication and coordination relates to outbreak identification and traceback. The Secretary, in consultation with the Secretary of Agriculture, shall, as appropriate, establish within the Food and Drug Administration a product tracing system to receive information that improves the capacity of the Secretary to effectively and rapidly track and trace food that is in the United States or offered for import into the United States. Prior to the establishment of such product tracing system, the Secretary shall examine the results of applicable pilot projects and shall ensure that the activities of such system are adequately supported by the results of such pilot projects.

Title 21 (Paragraph 342) of US Code refers to adulterated foods. A food shall be deemed to be adulterated: (a) Poisonous, insanitary, etc., ingredients (e.g., if it bears or contains any poisonous or deleterious substance which may render it injurious to health; but in case the substance is not an added substance such food shall not be considered adulterated under this clause if the quantity of such substance in such food does not ordinarily render it injurious to health); (b) absence, substitution, or addition of constituents (e.g., if any valuable constituent has been in whole or in part omitted or abstracted therefrom; or if any substance has been substituted wholly or in part therefore; or if damage or inferiority has been concealed in any manner; or if any substance has been added thereto or mixed or packed therewith so as to increase its bulk or weight, or reduce its quality or strength, or make it appear better or of greater value than it is); (c) colour additives—if it is, or it bears or contains, a colour additive which is unsafe within the meaning of section 379e of this Title; (d) confectionery containing alcohol or nonnutritive substance; (e) oleomargarine containing filthy, putrid, etc., matter—if it is oleomargarine or margarine or butter and any of the raw material used therein consisted in whole or in part of any filthy, putrid, or decomposed substance, or such oleomargarine or margarine or butter is otherwise unfit for food; (f) dietary supplement or ingredient: safety; (g) dietary supplement: manufacturing practices; (h) reoffer of food previously denied admission (if it is an article of food imported or offered for import into the United States and the article of food has previously been refused admission under section 381(a) of this title, unless the person reoffering the article affirmatively establishes, at the expense of the owner or consignee of the article, that the article complies with the applicable requirements of this chapter, as determined by the Secretary); and (i) noncompliance with sanitary transportation practices (if it is transported or offered for transport by a shipper, carrier by motor vehicle or rail vehicle, receiver, or any other person engaged in the transportation of food under conditions that are not in compliance with regulations promulgated under section 350e of this title).

The Poultry Products Inspection Act in Chapter 10 regarding poultry and poultry products inspections refers to adulterated foodstuffs. Poultry and poultry products are an important source of the Nation's total supply of food. They are consumed throughout the Nation and the major portion thereof moves in interstate or foreign commerce. It is essential in the public interest that the health and welfare of consumers be protected by assuring that poultry products distributed to them are wholesome, not adulterated, and properly marked, labelled, and packaged. Unwholesome, adulterated, or misbranded poultry products impair the effective regulation of poultry products in interstate or foreign commerce, are injurious to the public welfare, destroy markets for wholesome, not adulterated, and properly labelled and packaged poultry products, and result in sundry losses to poultry producers and processors of poultry and poultry products, as well as injury to consumers. It is hereby found that all articles and poultry which are regulated under this chapter are either in interstate or foreign commerce or substantially affect such commerce, and that regulation by the Secretary of Agriculture and cooperation by the States and other jurisdictions as contemplated by this chapter are appropriate to prevent and eliminate burdens upon such commerce, to effectively regulate such commerce, and to protect the health and welfare of consumers. The term "adulterated" shall apply to any poultry product under one or more of the following circumstances: (1) if it bears or contains any poisonous or deleterious substance which may render it injurious to health; but in case the substance is not an added substance, such article shall not be considered adulterated under this clause if the quantity of such substance in or on such article does not ordinarily render it injurious to health; (2) (a) if it bears or contains (by reason of administration of any substance to the live poultry or otherwise) any added poisonous or added deleterious substance; (b) if it is, in whole or in part, a raw agricultural commodity and such commodity bears or contains a pesticide chemical which is unsafe within the meaning of section 346a of this title; (c) if it bears or contains any food additive which is unsafe within the meaning of section 348 of this title; (d) if it bears or contains any colour additive which is unsafe within the meaning of section 379e of this title: Provided, That an article which is not otherwise deemed adulterated under clause (b), (c), or (d) shall nevertheless be deemed adulterated if use of the pesticide, chemical, food additive, or color additive in or on such article is prohibited by regulations of the Secretary in official establishments; (3) if it consists in whole or in part of any filthy, putrid, or decomposed substance or is for any other reason unsound, unhealthful, unwholesome, or otherwise unfit for human food; (4) if it has been prepared, packed, or held under insanitary conditions whereby it may have become contaminated with filth, or whereby it may have been rendered injurious to health; (5) if it is, in whole or in part, the product of any poultry which has died otherwise than by slaughter; (6) if its container is composed, in whole or in part, of any poisonous or deleterious substance which may render the contents injurious to health; (7) if it has been intentionally subjected to radiation, unless the use of the radiation was in conformity with a regulation or exemption in effect pursuant to section 348 of this title; and (8) if any valuable constituent has been in whole or in part omitted or abstracted therefrom; or if any substance has been substituted, wholly or in part therefore; or if damage or inferiority has been concealed in any manner; or if any substance has been added thereto or mixed or packed therewith so as to increase its bulk or weight, or reduce its quality or strength, or make it appear better or of greater value than it is. All poultry carcasses and parts thereof and other poultry products found to be adulterated shall be condemned and shall, if no appeal be taken from such determination of condemnation, be destroyed for human food purposes under the supervision of an inspector: Provided, That carcasses, parts, and products, which may by reprocessing be made not adulterated, need not be so condemned and destroyed if so reprocessed under the supervision of an inspector and thereafter found to be not adulterated. If an appeal be taken from such determination, the carcasses, parts, or products shall be appropriately marked and segregated pending completion of an appeal inspection, which appeal shall be at the cost of the appellant if the Secretary determines that the appeal is frivolous. If the determination of condemnation is sustained the carcasses, parts, and products shall be destroyed for human food purposes under the supervision of an inspector.

The Federal Meat Inspection Act in Subchapter I refers to inspection requirements, adulteration and misbranding. For the purpose of preventing the use in commerce of meat and meat food products which are adulterated, the Secretary shall cause to be made, by inspectors appointed for that purpose, an examination and inspection of all cattle, sheep, swine, goats, horses, mules, and other equines before they shall be allowed to enter into any slaughtering, packing, meat-canning, rendering, or similar establishment, in which they are to be slaughtered and the meat and meat food products thereof are to be used in commerce; and all cattle, sheep, swine, goats, horses, mules, and other equines found on such inspection to show symptoms of disease shall be set apart and slaughtered separately from all other cattle, sheep, swine, goats, horses, mules, or other equines, and when so slaughtered the carcasses of said cattle, sheep, swine, goats, horses, mules, or other equines shall be subject to a careful examination and inspection, all as provided by the rules and regulations to be prescribed by the Secretary, as provided for in this subchapter. The Secretary shall cause to be made, by experts in sanitation or by other competent inspectors, such inspection of all slaughtering, meat canning, salting, packing, rendering, or similar establishments in which cattle, sheep, swine, goats, horses, mules and other equines are slaughtered and the meat and meat food products thereof are prepared for commerce as may be necessary to inform himself concerning the sanitary conditions of the same, and to prescribe the rules and regulations of sanitation under which such establishments shall be maintained; and where the sanitary conditions of any such establishment are such that the meat or meat food products are rendered adulterated, he shall refuse to allow said meat or meat food products to be labelled, marked, stamped or tagged as "inspected and passed". For the purposes herein before set forth the Secretary shall cause to be made by inspectors appointed for that purpose a post mortem examination and inspection of the carcasses and parts thereof of all cattle, sheep, swine, goats, horses, mules, and other equines to be prepared at any slaughtering, meat-canning, salting, packing, rendering, or similar establishment in any State, Territory, or the District of Columbia as articles of commerce which are capable of use as human food; and the carcasses and parts thereof of all such animals found to be not adulterated shall be marked, stamped, tagged, or labelled as "Inspected and passed"; and said inspectors shall label, mark, stamp, or tag as "Inspected and condemned" all carcasses and parts thereof of animals found to be adulterated; and all carcasses and parts thereof thus inspected and condemned shall be destroyed for food purposes by the said establishment in the presence of an inspector, and the Secretary may remove inspectors from any such establishment which fails to so destroy any such condemned carcass or part thereof, and said inspectors, after said first inspection, shall, when they deem it necessary, re-inspect said carcasses or parts thereof to determine whether since the first inspection the same have become adulterated, and if any carcass or any part thereof shall, upon examination and inspection subsequent to the first examination and inspection, be found to be adulterated, it shall be destroyed for food purposes by the said establishment in the presence of an inspector, and the Secretary may remove inspectors from any establishment which fails to so destroy any such condemned carcass or part thereof.

8.4 Canadian Legislation

In 2006, industry leadership built the foundation for animal traceability, National Agriculture and Food Traceability System (NAFTS). The Industry-Government Advisory Committee (IGAC) was also established to lead the increase and implementation of the livestock and poultry constituents of a NAFTS. The IGAC consisted of 22 members from industry and 15 representatives from federal, provincial and territorial governments (Canadian Federation of Agriculture, 2013).

According to Consolidation on Safe Food for Canadians Act (2013), food traceability will be included in regulations respecting the traceability of any food commodity, including regulations requiring persons to establish systems to (i) identify the food commodity, (ii) determine its places of departure and destination and its location as it moves between those places, or (iii) provide information to persons who could be affected by it.

According to Consolidation on Food and Drugs Act (1985), no person shall sell an article of food that (a) has in or on it any poisonous or harmful substance; (b) is unfit for human consumption; (c) consists in whole or in part of any filthy, putrid, disgusting, rotten, decomposed or diseased animal or vegetable substance; (d) is adulterated; or (e) was manufactured, prepared, preserved, packaged or stored under unsanitary conditions. Where a person is prosecuted under this Part for having manufactured an adulterated food or drug for sale, and it is established that the person had in his possession or on his premises any substance the addition of which to that food or drug has been declared by regulation to cause the adulteration of the food or drug, the onus of proving that the food or drug was not adulterated by the addition of that substance lies on the accused.

In Food and Drug Regulations (2013), a food is adulterated if (a) a pest control product of the Pest Control Products Act or its components or derivatives, for which no maximum residue limit has been specified under sections 9 or 10 of that Act for that food, are present in or on the food, singly or in any combination, in an amount exceeding 0.1 part per million; or (b) an agricultural chemical or its components or derivatives, other than a pest control product as defined in the Pest Control Products Act or its components or derivatives, are present in or on the food, singly or in any combination, in an amount exceeding 0.1 part per million. Fish, except fish protein, and meat products or preparations thereof are adulterated if any of the following substances or any substance in one of the following classes is present therein or has been added thereto: (a) mucous membranes, any organ or portion of the genital system, or any organ or portion of a marine or fresh water animal that is not commonly sold as an article of food; (b) preservatives, other than those provided for in this Division, except (i) sorbic acid or its salts in dried fish that has been smoked or salted, and in cold processed smoked and salted fish paste, and (ii) benzoic acid or its salts, methyl-p-hydroxy benzoate and propyl-p-hydroxy benzoate in marinated or similar cold-processed, packaged fish and meat products; and (c) food colour except as provided for in this Division. Poultry meat, poultry meat by-products or preparations thereof are adulterated if any of the following substances or any substance in the following classes is present therein or has been added thereto: (a) any organ or portion of poultry that is not commonly sold as food; (b) preservatives, other than those provided for in this Division; and (c) colour, other than caramel.

Following Meat Inspection Regulations (1990) "adulterated" means, in respect of a meat product intended for sale, use or consumption as an edible meat product in Canada, (a) containing or having been treated with (i) a pesticide, heavy metal, industrial pollutant, drug, medicament or any other substance in an amount that exceeds the maximum level of use prescribed by the Food and Drug Regulations, (ii) an ingredient, a food additive or any source of ionizing radiation not permitted by or in an amount in excess of limits prescribed by these Regulations or by the Food and Drug Regulations, (iii) any poison, decomposed substance or visible contamination, or (iv) any pathogenic microorganism in excess of levels published in the Manual of Procedures, or (b) failing to meet the standards set out in Part I. No adulterated meat product shall be identified as edible. Where an adulterated meat product in a registered establishment can be made to conform to the standards prescribed by this Part for an edible meat product, the meat product shall be held by an inspector until it is made to conform to those standards by the operator.

8.5 Australia—New Zealand Legislation

According to Standard 3.2.2 regarding Food Safety Practices and General Requirements in chapter 3 of the Code covers the "one step back and one step forward" elements of traceability under Clause 5 (2) Food receipt and Clause 12 Food recall. In relation to food receipt, a food business must be able to provide information about what food it has on the premises and where it came from: A food business must provide, to the reasonable satisfaction of an authorised officer upon request, the following information relating to food on the food premises: the name and business address in Australia of the vendor, manufacturer or packer or, in the case of food imported into Australia, the name and business address in Australia of the importer; and the prescribed name or, if there is no prescribed name, an appropriate designation of the food. This means that a food business must not receive a food unless it is able to identify the name of the food and the name of the supplier. A food business engaged in the wholesale supply, manufacture or importation of food must have a system, set out in a written document, to ensure it can recall unsafe food. The system should include records covering: (a) production records; (b) what products are manufactured or supplied; (c) volume or quantity of products manufactured or supplied; (d) batch or lot identification (or other markings); and (e) where products are distributed any other relevant production records. This information should be readily accessible in order to know what, how much and from where product needs to be recalled.

Primary production and processing standards in Chapter 4 of the Food Standards Code also include traceability measures. There are specific traceability requirements for: (a) seafood businesses (Standard 4.2.1): A seafood business must maintain sufficient written records to identify the immediate supplier and immediate recipient of seafood for the purposes of ensuring the safety of the seafood; (b) dairy primary production, transport and processing businesses (Standard 4.2.4): As part of the documented food safety programme in clause 7, a dairy transport business must have a system to identify the immediate supplier and immediate recipient of the dairy product; (c) poultry processors (Standard 4.2.2): A poultry producer must be able to identify the immediate recipient of the poultry handled by the poultry producer. A poultry processor must ensure that it can identify the immediate supplier and immediate recipient of poultry product handled by the poultry processing business; and (d) egg producers and egg processors (Standard 4.2.5): An egg producer must not sell eggs unless each individual egg is marked with the producers' unique identification. An egg producer who supplies egg pulp must mark each package or container containing the pulp with the producers' unique identification. Both of them do not apply to eggs or egg pulp sold or supplied to an egg processor (the supplied product) if that egg processor complies with clause 20 in respect of the supplied product. In addition, an egg producer must have a system to identify to whom eggs or egg pulp is sold or supplied.

8.6 Conclusions

Food safety legislation makes it necessary for companies that manufacture food to have perceptibility throughout their processes and ensure safety throughout. Quick response to a non-conformance or recall is serious in protecting public health and safety. The ability to supervise a material or product through all stages of the production and distribution chain is important to consumer safety. For a food industry, it is vital to have traceability systems in place because it is practical, but also because it is a regulatory requirement for food industry to be able to trace product at least one step back and one step forward in many countries. They need to have systems in place to trace raw materials to the final food. Labelling is the key in this process because a food industry could be able to comprise and identify the food at any step in the process. Finally, technology (e.g., use of bar codes) allow companies to manage their processes and products to ensure food safety.

References

Abbott, H. 1991. Managing a Product Recall. Pitman: London.

Al-Jowder, O., M. Defernez and E.K. Kemsley. 1999. Mid-infrared spectroscopy and chemometrics for the authentication of meat products. Food Chemistry. 47: 3210–3218.

Arvanitoyannis, I.S., S. Choreftaki and P. Tserkezou. 2008. An update of EU legislation (Directives and Regulations) on food-related issues (Safety, Hygiene, Packaging, Technology, GMOs, Additives, Radiation, Labelling): presentation and comments. International Journal of Food Science and Technology. 40: 1021–1112.

Asioli, D., A. Boecker and M. Canavari. 2011. Perceived traceability costs and benefits in the Italian fisheries supply chain. International Journal on Food System Dynamics. 2(4): 357–375.

Borst, P. and H.M. Akkermans. 1997. Engineering Ontologies. International Journal of Human Computer Studies. 46(2-3): 365–406.

Canadian Federation of Agriculture (CFA). 2013. Standing Policy 2013 (http://www.cfa-fca.ca/sites/default/files/Policy%20Manual_E_2013_1.pdf).

Commission Regulation (EC) No 2065/2001 of 22 October 2001 laying down detailed rules for the application of Council Regulation (EC) No 104/2000 as regards informing consumers about fishery and aquaculture products (OJ L 278, 23.10.2001, p. 6).

Commission Regulation (EC) No 589/2008 of 23 June 2008 laying down detailed rules for implementing Council Regulation (EC) No 1234/2007 as regards marketing standards for eggs.

Consolidation on Safe Food for Canadians Act. 2013. 51. Governor in Council. Published by the Minister of Justice (http://laws-lois.justice.gc.ca/PDF/S-1.1.pdf).

Consolidation on Food and Drugs Act, R.S.C., 1985, c. F-27. Published by the Minister of Justice (http://laws-lois.justice.gc.ca/PDF/F-27.pdf).

Consolidation on Food and Drug Regulations, C.R.C., c. 870, 2013. Published by the Minister of Justice (http://laws-lois.justice.gc.ca/PDF/C.R.C.,_c._870.pdf).

Consolidation on Meat Inspection Regulations, 1990, SOR/90-288. Published by the Minister of Justice (http://laws-lois.justice.gc.ca/PDF/SOR-90-288.pdf).

Dabbenea, F., P. Gaya and C. Tortiab. 2014. Traceability issues in food supply chain management: A review. Biosystems Engineering. 120: 65–80.

Donnelly, A.-M. and M. Thakur. 2010. Food Traceability Perspectives from the United States of America and the European Union. Økonomisk fiskeriforskning Argang, 19-20: 1–8.

European Communities. 2007. Fact Sheet—Food traceability—Tracing food through the production and distribution chain to identify and address risks and protect public health. Health and Consumer Protection Directorate—General, June 2007.

European Parliament. 2002. Regulation (EC) No. 178/2002 of the European Parliament and of the Council. Official Journal of the European Communities. L31/1–L31/24.

Federal Meat Inspection Act—Title 21: Food and Drugs—Chapter 12: Meat inspection—Subchapter I: Inspection requirements, adulteration and misbranding. U.S. Food and Drug Administration.

Folinas, D., I. Manikas and B. Manos. 2006. Traceability data management for food chains. British Food Journal. 108(8): 622–633.

Food Safety Modernization Act Public Law 111–353—Jan. 4, 2011 124 STAT. 3885 (2011). US Food and Drug Administration (http://www.fda.gov/Food/GuidanceRegulation/FSMA/ucm247548.htm).

Food Standards Code, Chapter 4: Primary Production Standards, Part 4.2, 2003, Food Standards Australia—New Zealand (http://www.foodstandards.gov.au/code/Pages/default.aspx).

Gordijn, J. and H.M. Akkermans. 2001. Designing and evaluating e-business models. IEEE Intelligent Systems. 16(4): 11–17.

Marcos Lorenzo, I., J.L. Perez Pavon and M.E. Fernandez Laespada. 2002. Detection of adulterants in olive oil by headspace–mass spectrometry. Journal of Chromatography. 945: 221–230.

McEntire, J.C., S. Arens, M. Bernstein, B. Bugusu, F.F. Busta, M. Cole, A. Davis, W. Fisher, S. Geisert, H. Jensen, B. Kenah, B. Lloyd, C. Mejia, B. Miller, R. Mills, R. Newsome, K. Osho, G. Prince, S. Scholl, D. Sutton, B. Welt and S. Ohlhorst. 2010. Traceability (product tracing) in food systems: an IFT (Institute of Food Technology) report submitted to the FDA (Food and Drug Administration), volume 1: Technical aspects and recommendations. Comprehensive Reviews in Food Science and Food Safety. 9: 92–158.

Poultry Products Inspection Act—Title 21: Food and Drug—Chapter 10: Poultry and Poultry Products Inspection. US Food and Drug Administration, Section 1 of Pub. L. 85–172.

Pugh, N.R. 1973. Principles of product traceability, Product liability prevention conference, PLP(4) American Society Quality Control, Newark, NY (USA). 65–69.

Regattieri, A., M. Gamberi and R. Manzini. 2007. Traceability of food products: General framework and experimental evidence. Journal of Food Engineering. 81(2): 347–356.

Regulation (EC) No 178/2002 of the European Parliament and of the Council of 28 January 2002 laying down the general principles and requirements of food law, establishing the European Food Safety Authority and laying down procedures in matters of food safety (OJ L 31, 1.2.2002, p. 1).

Regulation (EC) No 1760/2000 of the European Parliament and of the Council of 17 July 2000 establishing a system for the identification and registration of bovine animals and regarding the labelling of beef and beef products and repealing Council Regulation (EC) No 820/97.

Regulation (EU) No 1169/2011 of the European Parliament and of the Council of 25 October 2011 on the provision of food information to consumers, amending Regulations (EC) No 1924/2006 and (EC) No 1925/2006 of the European Parliament and of the Council, and repealing Commission Directive 87/250/EEC, Council Directive 90/496/EEC, Commission Directive 1999/10/EC, Directive 2000/13/EC of the European Parliament and of the Council, Commission Directives 2002/67/EC and 2008/5/EC and Commission Regulation (EC) No 608/2004.

Smith, G., J. Tatum, K. Belk, J. Scanga, T. Grandin and J. Sofos. 2005. Traceability from a US perspective. Meat Science. 71(1): 174–193.

Standard 3.2.2—Food Safety Practices and General Requirements,Food Standards Australia—New Zealand (http://www.comlaw.gov.au/Series/F2008B00576).

Thakur, M. and K.M. Donnelly. 2010. Modeling traceability information in soybean value chains. Journal of Food Engineering. 99(1): 98–105.

United Nations. 1986. UN Guidelines for Consumer Protection. General Assembly resolution, 39/248.

United States Department of Agriculture (Economic Research Service). 2002. Traceability for Food Marketing & Food Safety: What's the Next Step? Agricultural Outlook/January–February 2002. 21–25.

United States. 2011. Public Law 111-353 Food Safety Modernization Act, 4/1/2011, 111th Congress, 124 Stat 3885.

United States. 2002. Public Law 107-188. Public health security and bioterrorism preparedness and response act of 2002, 12/6/2002.

US Code (21 USC 342)—Title 2: Food and Drugs—Chapter 9: Federal Food, Drug, and Cosmetic Act—Subchapter IV: Food—§ 342. Adulterated food.

Whitsitt, V., C. Beehner and C. Welch. 2013. The role of good manufacturing practices for preventing dietary supplement adulteration. Analytical and Bioanalytical Chemistry. 405(13): 4353–4358.

Zhang, J., X. Zhang, L. Dediu and C. Victor. 2011. Review of the current application of fingerprinting allowing detection of food adulteration and fraud in China. Food Control. 22(8): 1126–1135.

Conclusions, Future Trends and Suggestions

Conclusions, Future Trends and Suggestions for Improving Food Authenticity and Traceability

Ioannis S. Arvanitoyannis

9.1 Introduction

Over the last 20 years there has been an impressive progress in inventing novel analytical techniques and methods towards improving both food and feed authenticity and traceability effectiveness.

In this chapter an attempt is made to summarise the most important advances in food analysis and their related potential applications with regard to authenticity and traceability.

The establishment of novel regulations and directives within the frame of EU member states has triggered the efforts for both enhancing the efficiency of the currently applied analytical techniques in conjunction with the application of multivariate analysis in an attempt to group the samples according to their geographical origin, variety, processing or preservation techniques, and contaminant (i.e., pesticides, hormones) residues among others.

9.2 Novel Methods for Improving Food Authenticity

The methods applied for food authenticity have been considerably improved mainly due to employment of either novel techniques or combination of two or more techniques. The results obtained with these novel techniques are considerably much more accurate than those reported in the previous decades. Two Tables (9.1 and 9.2) reporting test methods, investigated foods, authenticity parameters, are given here. According to Sass-Kiss (2008), "one of the objectives in the fight against adulteration

School of Agricultural Sciences, Department of Agriculture, Ichthyology and Aquatic Environment, University of Thessaly, Fytoko St., 38446 Nea Ionia Magnesias, Volos, Hellas, Greece.

Table 9.1. Test methods, investigated foods, authenticity parameters and related sources.

Test method	Investigated food	Authenticity parameter	Reference
PCR	Cow, ewe, goat, buffalo	Breed identification	Plath et al., 1997
PCR	Fish	Detection of white species	Dooley et al., 2005
AFLP	Salmon	Distinguishing 10 different salmon-like species	Russell et al., 2000
AFLP	Fish and seafood	Identification of species of origin	Maldini et al., 2006
Qualitative PCR	Meat	Presence or absence of pork	Popping, 2002
Quantitative PCR	Meat	Beef meat quantification	Popping, 2002
Quantitative PCR	Meat	Meat quantification	Lopez-Andreo et al., 2006
RT-PCR	Fish	Quantification of fish nuclear DNA	Rodriguez et al., 2011
RT-PCR	Red Deer	Quantification of red deer content	Grandits et al., 2011
Analysis of co-enzyme Q10	Meat from different animal species	Animal species determination	Mancheno et al., 2011
RFLP	Fish	Characterisation of origin	Mueller et al., 2011
Physicochemical profile	Honey	Geographical Indications	Madas et al., 2011
Metabolomic fingerprinting	Honey	Authenticity	Cajka et al., 2011
Mineral content determination	Honey	PDO classification	De Alda et al., 2011
DNA test (Protein based BSE test)	Cattle	Geographical traceability	Popping, 2002
HPLC	Cow, ewe, goat, buffalo	Identification of species of origin	De la Fuente and Juarez, 2005
HPLC	Seafood	Detection of species	Arvanitoyannis et al., 2005
HPLC	Cheese	Detection of rennet whey	De la Fuente and Juarez, 2005
HPLC/GC	Milk	Triacylglycerol stereospecific composition	Lombardi et al., 2011
ELISA	Cow, ewe, goat, buffalo	Identification of species of origin	De la Fuente and Juarez, 2005
UPLC-MS/MS	Slimming food supplements	Determination of synthetic adulterants	Gadaj et al., 2011
FT-IR		Detection of contaminated or counterfeit ingredients	Perston and Goth, 2011

is to develop sensitive analytical methods to detect the presence or determine the lack of specific compound(s) which characterise the raw material of foods of plant or animal origin. The main advantage of HPLC regarding authenticity is its ability to separate and detect compound(s) or compound groups that are at low levels in food products of interest. Relatively low molecular mass compounds or those of higher molecular mass, such as proteins, are chosen as targets of analysis, and by using these compounds good results can be obtained in the determination of authenticity." In a

Table 9.2. Representative examples of food authenticity issues [Kvanicka (2005), Arvanitoyannis (2008)].

Food commodity	Authenticity issue
Fat	Butter adulterated with hydrogenated oil and animal fat
Milk and Dairy	i) Undeclared addition of water to milk ii) Cow's milk in sheep's, goats', or buffalo milk yogurt or cheese iii) Differentiation between cheese produced from raw or heat-treated milk
Meat and Fish	i) Erroneous declaration of species ii) Labelling previously frozen meat as fresh iii) Undeclared water addition to bacon iv) Undeclared addition of horse meat
Honey	i) Incorrect declaration of floral or geographical origin of honey ii) Undeclared sugar addition to honey iii) Undeclared use of genetically modified food

chapter I had written almost 10 years ago, titled "Trends in food authentication" edited by Da-Wen Sun (Publ. Academic Press), I referred to several emerging technologies such as PCR on cow, ewe, goat, buffalo, fish, meat and beef meat (Plath et al., 1997; Dooley et al., 2005; Popping, 2002; Lopez-Andreo et al., 2006), AFLP on salmon, fish and seafood (Maldini et al., 2006; Popping, 2002), DNA test (protein based) on BSE for geographical traceability of cattle (Popping, 2002), HPLC on cow, ewe, goat and buffalo (identification of species of origin, De la Fuente and Juarez, 2005), seafood (detection of species, Arvanitoyannis et al., 2005), cheese (detection of rennet whey, De la Fuente and Juarez, 2005), and ELISA on cow, ewe, goat and buffalo (identification of species of origin, De la Fuente and Juarez, 2005). However, in the meantime, the above mentioned techniques have evolved considerably and the new trend of food analysis is rather focused on the combination of two or more techniques in order to improve both the rapidity and the accuracy of the obtained results. Some representative techniques are the following: GC-HRMS, LC-MS/MS, GC-MS/MS, RT PCR, Metabolomic fingerprinting, UPLG-MS/MS, LC-GC, RFLP, HPLC-FLDA, GCxGC, SPE Clean-up and Solvent Vent PTV GC-MS Analysis, Fast Low Pressure-GC with tandem MS (FLP-GC-MS/MS).

9.3 Effect of Food Contaminants (due to environmental and human pollution) on Foods and Feeds

The continuously increasing industrialisation in conjunction with the augmenting human population resulted in gradually greater pollution of the environment. Some representative contaminants are: Polycyclic Aromatic Hydrocarbons (PAHs), Organohalogen Pollutants (OHPs), Polychlorinateddibenzodioxins (PCDDs), Polychlorinated dibenzofurans (PCDFs), Dioxin-Like Polychlorinated biphenyls (DL-PCBs), Polychlorinated biphenyls/Polybrominated diphenyl Ethers (PCBs/PBDEs), PolyBrominated diphenyl ethers (PBDEs), Hexabromocyclododecane (HBCD), Di(2-ethylhexyl) phthalate (DEHP) and heavy metal contents.

Tables 9.3 and 9.4 summarize the results reported both in oral and poster presentations (5th and 6th, in 2011 and 2013, respectively) on Recent Advances in Food Analysis.

Table 9.3. Determination of PAHs contained in foods with several analysis tools.

Food/ Feed	Contaminant/ ContamContent	Analysis tool	Method specifications	Reference
Honey	**PAHs**	GC-MS	A solid–liquid extraction at first, then the extract was cleaned up using a silica gel chromatography column.	Vavrova et al., 2011
Seafood	**PAHs**	GC-MS	NA*	Drabova et al., 2011
Foods (model solutions)	**HCs**	LC-GC	NA*	Tranchida et al., 2011
Model solutions	**PAHs**	HPLC columns GC, LC	Development of solutions for isolating 29 PAHs, two HPLC columns in series; a PAH specific phase and an aromatic bond selective phase were used. Problems with interfering compounds.	Kowalski et al., 2011
Cow milk, Human milk	**POPs**	GC × GC	Comprehensive 2-dimensional gas chromatography (GC × GC) with an electron capture detector (ECD) may offer a more cost-effective alternative.	de Zeeuw et al., 2011
Olive oil	**PAHs**	**GC-MS and HPLC-FLDA**	A new SPE cartridge containing two different sorbent layers was evaluated in the simultaneous extraction and cleanup of PAHs from olive oil. The layers consist of Florisil and a mix of Z-Sep/C18.	Stenerson et al., 2013
Edible fats	**PAHs**	**SPE Clean-up and Solvent Vent PTV GC–MS Analysis**	The GC separation of all 16 PAHs became feasible on a DB-17MS (20 m × 0.18 mm × 0.18 μm) capillary column in a 6890NAgilent Technologies GC coupled to a 5973 inert single Quadrupole MS.	Lestingi et al., 2013

Polychlorinated Biphenyls (**PCBs**), Polybrominated Diphenyl Ethers (**PBDEs**), Persistent Organic Polycyclics (**POPs**), Hydrocarbons (**HCs**), Not available (**NA***)

Table 9.4. Determination of PCDD, PCB/PBDE, HBCD, DEHP, PCDF and DL-PCBs contained in foods with several analysis tools.

Food/Feed	Contaminant/Contam. content	Analysis tool	Method specifications	Reference
Poultry	HBCD	**LC-MS/MS SPME/GC-MS**	Abdominal fat of animals slaughtered throughout the experiment was collected and analysed with LC-MS tandem MS (LC-MS/MS) for HBCD direct quantification and with solid phase micro-extraction SPME-GC-MS	Ratel et al., 2011
212 different food types were covered, [circa 88% of dietary consumption in the adult and child populations]	17 polychlorodibenzo-p-dioxins and polychlorinated dibenzofurans (PCDD/Fs), 12 "dioxin-like" and six "non-dioxin-like" polychlorinated biphenyls (dl-PCBs, ndl-PCBs), 16 perfluorinated alkyl acids (PFAAs), eight polybrominated diphenyl ethers (PBDEs), three poly-brominated biphenyls (PBBs), three hexabromocyclododecane enantiomer pairs (HBCDDs) and 20 PAHs	**GC-HRMS, LC-V.S/MS or GC-MS/MS**	A significant four-fold decrease in exposure to PCDD/Fs and PCBs in the French population compared to the previous 2005 and 2007 available assessments	Cariou et al., 2013
Multi-residue analysis of contaminants in Fish	POPs, PCBs, PAHs, PBDEs, and novel flame retardants	**Fast Low Pressure-GC with Tandem MS**	FLP-GC-MS-MS resulted in fast separation of more than 200 analytes	Sapozhnikova and Lehotay, 2013b
Fingerprint analysis of flavour compounds in a mountain cheese	Headspace SPME with low and high resolution GC/MS	**HPLC columns GC, LC**	37 compounds belonging to various chemical classes such as aldehydes, ketones, esters, fatty acids, alcohols, terpenes and hydrocarbons were identified	Kaplan et al., 2013
Dioxins in fish from Brazil	Application of an isotope dilution method for detecting 17 toxic dioxins (PCDD/F). The fish samples were freeze dried, extracted with hexane and cleaned up by passing through a SiO2 column	**GC-HRMS**	The method was applied within the frame of Brazilian National Residue and Contaminants Control Plan (PNCRC) on 95 samples of different fish species with satisfactory repeatability and within-laboratory reproducibility	Pissinatti et al., 2013

Table 9.4. contd....

Table 9.4. contd.

Food/Feed	Contaminant/Contam. content	Analysis tool	Method specifications	Reference
Foods	PBDE mixture	GC	Combination of a new column with GC column was shown to prolong by 50% the lifetime of GC column. The analysis of PBDEs is demanding in view of various structural isomers that have to be isolated and thermally labile	Sellers et al., 2013
Fish	POPs, PBDD/DFs, PXDD/DFs, PBDEs, PCBs	GC-HRMS	The GC separation of all POPs presupposes intensive clean-up and fractionation procedures in conjunction with optimized instrumental parameters	Zacs et al., 2013

HRMS: High Resolution Mass Spectroscopy

The US Department of Agriculture, Agricultural Research Service (USDA-ARS) and Food Safety and Inspection Service (USDA-FSIS) conducted statistical surveys for dioxins (PCDDs, PCDFs, and PCBs) and polybrominated diphenyl ethers every five years since the mid-1990s (Lupton et al., 2013). The contents of PBDEs in US beef/veal, pork and poultry were analysed. The Environmental Protection Agency (EPA) disclosed in 2012 the chronic oral reference dose (RfD) for 2,3,7,8–TCDD of 0.7 pg/kg-day for human exposure. Calculations were undertaken to find out the dioxin trigger levels (2 pg TEQ/g fat for beef and 4 pg TEQ/g fat for pork and poultry). The currently reported dioxin contents are 0.088–6.46 pg TEQ/g fat, 0.051–0.36 pg TEQ/g fat, 0.046–0.39 pg TEQ/g fat, and 0.052–1.32 pg TEQ/g fat, for beef, pork, chicken and turkey, respectively.

A facile and fast technique for carrying out analysis simultaneously for 143 pesticides and 50 environmental contaminants [POPs, PCBs, PAHs, PBDEs, and novel flame retardants (NFR)]. The fish samples were extracted with acetonitrile and SPE and cleaned up with a sorbent based on zirconium. Sapozhnikova and Lehotay (2013a) reported that application of fast low pressure vacuum outlet gas chromatography tandem mass spectrometry (LP-GC/MS-MS) on pre-treated fish samples resulted in rapid separation of more than 200 analytes in 10 minutes.

Kotz and his co-workers (2013) applied an Atmospheric Pressure Gas Chromatography in conjunction with tandem MS (APGC-MS/MS and GC-HRMS) in an attempt to determine the PCDD/Fs and PCBs contents in feed and food matrices. Employment of APGC-MS/MS system revealed that the system displayed adequate sensitivity to samples analysis. Any reported deviations of the results obtained with APGC-MS/MS from reference values fell below 20% for the WHO-PCDD/F-TEQ. Since the suggested amendments of EU regulations for usage of GC-MS/MS as confirmatory method can be addressed by APGC-MS/MS, the latter is expected to find more widespread applications taking into account the rapidity of this method which in combination with rapid clean-up and appropriate extraction methods will improve considerably the quality of the analytical results.

Application of fingerprint analysis of flavour compounds in Kars Kashkaval (mountain cheese) was carried out with GC-MSD and High Resolution-TOFMS by Kaplan et al. (2013). Headspace solid phase matrix extraction (SPME) in conjunction with high and low resolution GC/MS (HR-GC/MS, LR-GC/MS) were used to identify the volatile compounds. The detected and identified compounds amounted to thirty seven (37) belonging to various chemical classes (i.e., aldehydes, ketones, esters, fatty acids, alcohols, terpenes and hydrocarbons).

Recently, an isotope dilution method was validated to detect 17 toxic dioxins (PCCD/F) in fish in an attempt to improve the monitoring procedure (Pissinatti et al., 2013). Fish samples were initially freeze-dried, extracted with hexane [Pressurised Liquid Extraction (PLE)] and/or Soxhlet. Clean up was conducted by using acid SiO_2 column, and dichloromethane and hexane as elution solvents. Extracts were analysed with GC-HRMS. Limits of quantification (LOQ) ranged from 0.05 pg/g (TCDD/F) to 0.40 pg/g (OCDD/F) and the method was shown to be linear within the range 0.05 to 1.000 pg/g (TCDD/F). The method was employed within the frame of the Brazilian National Residue and Contaminants Control Plan (PNCRC). Ninety-five samples of various species were gathered from producers coming from different parts of Brazil and examined over 12 months period. It is noteworthy that 21% of the above-mentioned samples gave values greater than the LOQ.

Polybrominated diphenyl ethers (PBDEs) were determined to be recalcitrant and bio-accumulative in the environment thereby confirming the fact that they stand for a serious threat for food contamination. The interest in the analysis of PBDEs resides in the presence of several structural isomers that must be chromatographically separated. PBDEs contained in EPA Method 1614 were well-separated when a short thin film column was opted for by Sellers et al. (2013).

Zacs et al. (2013) established and validated a methodology for the analysis of five groups of persistent organic pollutants (POPs) such as polybrominated, polychlrominated, and mixed brominated-chlorinated dibenzo-p-dioxins and dibenzofurans, polychlorinated biphenyls and polybrominated diphenyl ethers in fish samples. The applied method complies with EU Regulation 252/2012. GC was coupled with high resolution mass spectroscopy (GC-HRMS) for detection of PCDD/DFs and PCBs. The limits of detection within the selected POP groups reached with this method varied within a range from 0.02 to 2.17 pg/g fat for fish.

The impact of cooking conditions and fat content on the bio accessibility of PCBs, PCDDs, and PCDFs in meat products was investigated by Planche and her co-workers (2013a). Examination of "food products such as consumed" that is after cooking for meat includes both the impact of cooking conditions and the presence of fat content on the bio accessibility of the 209 congeners of PCBs and the 17 toxic congeners of PCDD/Fs in meat products. An artificial gastrointestinal system was elaborated and coupled to a two dimensional GC coupled to Time of Flight Mass Spectrometry (GCxGC-TOF/MS). The developed methodology proved to be effective toward detecting and identifying 209 PCBs and 17 toxic PCDD/Fs in a very short analysis time, that is 74 minutes. This method was validated with regard to resolution, linearity and limit of detection.

According to Neugebauer et al. (2013), "a complete approach was developed using a specialized multistep-cleanup in combination with a modern HRGC column and HRMS detection, which enables a comprehensive analysis of all PCB congeners with around 180 peak separations. They presented results of this method as applied to biota and biota-related samples, e.g., fish oils from South American origin from different fish species. They were able to show patterns of all PCBs congeners as well as congener groups, opening access to examine preferences in transport and enrichment as well as metabolic behaviour."

The effect of the chicken hen breeding method on the presence of PCDD/F, PCB and PBDEs in eggs was studied by Roszko et al. (2013). Hens are rather susceptible to get contaminated by dioxins mainly because of the feed of plant origin they consume (De Vries, 2006). This trend gets intensified in the case of free range hens that consume more grass, worms and soil particles. The average concentrations of six indicator PCBs and 13 PBDEs in eggs from hens bred in cages were 475.2 ± 252.5 pg/g and 111.1 ± 75.8 pg/g of fat. The calculated average PCDD/F/PCBs TEQ value was 0.956 ± 0.952 pg/g of fat. The average concentrations calculated for free range hen eggs were as follows: 3952.2 ± 10204.8 pg/g, 164.5 ± 161.8 pg/g and 2.656 ± 3.908 pg/g TEQ of fat, respectively for six indicator PCBs, 13 PBDEs and PCDD/F/PCBsTEQ, for free range and cages, respectively. It was thereby confirmed that the free range eggs have higher dioxin levels than the ones coming from the cages. However, in some cases the contamination values for the eggs of the two categories converged toward similar values. It was found that most samples produced under free range and caged conditions have the congener profile and concentration values corresponding to the environmental background of the investigated POPs.

EU regulation 1259/2011 established for the 1st time maximum levels for non-dioxin-like PCBs (NDL-PCB) in foods. Six PCB congeners (28, 52, 101, 138, 153 and 180) were determined as indicators to evaluate the human exposure upon food consumption. According to Tavoloni and her co-workers (2013), "the applied method involved a pressurized ASE extraction of the fat and a clean-up on an Extrelut NT3 column acidified with H_2SO_4 and connected on top of a silica SPE column. The sample was injected on a GPC column to perform Size Exclusion Purification (SEP). The chromatographic separation took place in a capillary column with a GC-MS single quadrupole with internal standards. Employment of a new GC-MS/MS simplified considerably the pre-treatment procedure since no clean-up is required any more." The improved method was described by increased sensitivity and selectivity and the LOQ was brought to 0.5 ng/g fat. The obtained results were in good agreement with those of Proficiency Testing (PTs).

Although Atmospheric Pressure GC (APGC) was first developed in 1970s, it has made an impressive comeback providing an alternative technique to high resolution EI-GC/MS and EI-GC/MS/MS. The obtained results clearly showed that the sensitivity reached employing APGC-MS/MS led to similar results to those obtained with HR-GC/MS (dioxin analysis standard). Sensitivity, repeatability and relative standard deviations (RSD) were acceptable range-wise; comparable to HR-GC/MS, less than 10% RSD, and RSD varying from 2.5 up to 9.6%, respectively (Roberts et al., 2013).

The Second French Total Diet Study (TDS 2) was a large scale investigation targeting to provide exposure evaluation data of the French population to a wide range of chemical hazards due to their diet. The two major criteria considered were (a) the most often consumed foods and (b) foods less consumed but most heavily contaminated. A high number of various food types (212) were selected thus covering 88% of dietary consumption in adult and child populations. Several analytical methods such as GC-HRMS and LC-MS/MS, and GC-MS/MS with isotopic dilution approach were employed for detection and quantification purposes. The analytes consisted of 17 polychlorodibenzo-p-dioxins and polychlorinated dibenzofurans (PCDD/Fs), 12 "dioxin like" and six "non-dioxin like" polychlorinated bisphenyls (dl-PCBs, ndl-PCBs), 16 perfluorinated alkyl acids (PFAAs), eight polybrominated diphenyl ethers (PBDEs), three polybrominated biphenyls (PBBs), three hexabromocyclododecane enantiomer pairs (HBCDs), and 20 polycyclic aromatic hydrocarbons (PAHs). It is noteworthy that the obtained results revealed an increase of 400% in exposure to PCDD/F and PCBs in the French population in comparison to the previous assessments (2005 and 2007) (Cariou et al., 2013).

The extensive application of hexabromocyclododecane (HBCD) as an additive flame retardant in expanded (EPS) and extruded (XPS) polystyrene foams resulted in wide contamination and accumulation of this substance in both biotic and abiotic environments (Priority Existing Chemical Assessment Report No. 34, 2012 Australia). Lara et al. (2013) proposed a non-complicated and fast method which combines a single step of sample pre-treatment with chiral liquid chromatography coupled to tandem mass spectrometry. This method was elaborated for the determination of HBCD stereoisomers [(+)α, (−)α, (+)β, (−)β, (+)γ, (−)γHBCD] in fish. According to these authors "HBCD stereoisomers were extracted from fish samples with a supramolecular solvent (SUPRAS) made up of reverse aggregates of decanoic acid (DeA) and generated with adding water to a solution of the surfactant in tetrahydrofurane. The sample was vortex shaken with SUPRAS for five minutes. The SUPRAS was separated from sample extract containing the analytes with ultracentrifugation and then diluted 1:1 with CH_3OH prior to its analysis. Separation of the HBCD stereoisomers was conducted on a stationary phase of β-cyclodextrin and their quantification in a MS equipped with electron ionization source (EI-MS) and a triple quadrupole mass analyser. The quantitation limits for the determination of HBCD stereoisomers in hake, code, sole, panga, whiting and sea bass were within the intervals 0.5–5.6 ng/g and recoveries for fish samples ranged between 87 and 114% with RSD from 1 to 10%."

Kuc and Grochowalski (2013) used a sensitive ID-LC-MS/MS method for the detection and identification of the three HBCD isomers (α-, β-, γ-). The quantitative determination of HBCD isomers was conducted by means of the SRM transition of the chlorine adduct [M-H+Cl]-(676.6 *m/z*) of the HBCD isomers. Application of Cluster Analysis (CA) to obtained data facilitated the differentiation of groups of objects in three categories (pending on isomer content) according to the Euclidean distance.

The determination of polychlorinated dibenzo-p-dioxins, dibenzofurans and dioxin-like biphenyls in fish and canned fish and investigation of the congener profile and similarity between the samples was carried out by Grochowalski et al. (2013). According to the current WHO guidelines (EU Regulation 1259/2011) for the determination of dioxins and PCBs in foodstuffs is required to mark the 17 most toxic congeners of PCDDs/PCDFs and 12 congeners of DL-PCBs. The content of these compounds expressed their total value known as toxic equivalent (TEQ). The acceptable level of dioxin in fish is 3.5 pg WHO-TEQ/g fresh weight and the limit value of the sum of dioxin and DL-PCBs amounts to 6.5 pg WHO TEQ/g f.w. The analysed samples amounted to 103 consisting of marine fish (fresh, frozen and smoked) and canned fish. These samples were homogenized, dried, extracted in Soxhlet apparatus and clean-up by means of membranes and SiO_2 columns. A sensitive isotope dilution technique in GC was coupled with tandem MS (ID-GC-MS/MS) and employed for the detection and identification of congeners of PCDDs/PCDFs and DL-PCBs. It was found that the marine fish was, on several occasions, contaminated with dioxins and DL-PCBs at levels greater than the acceptable limits. Cluster analysis based on Euclidean distance was used effectively for grouping the objects. It was suggested that the involved contaminants (PCDDs, PCDFs and DL-PCBs) may originate from same sources and/or bioaccumulation behaviour in the marine environment.

Beser and Yusa (2013) made use of a design of experiment approach (DOE) for the optimisation of PBDEs analysis. For the optimization of Microwave Assisted Extraction (MAE) parameters, the three major attributes were solvent volume, exposure time and temperature and their optimal values were 50 mL, 2 min. and 75°C. Further investigated conditions were GC-MS/MS parameters, extraction solvent and matrix effect. The finally opted method composed of the following stages: extraction using MAE with n heptane:acetone (1:1 *v/v*), acid treatment, SPE clean-up, GC-MS/MS detection and separation with capillary column. The MS functioned on the electron impact mode (EI) at a voltage of 70 eV and a filament current of 50 μA. Validation of the method took place by investigating the linearity, accuracy, precision and limit of quantification (LOQ). Accuracy and precision were analysed with regard to recovery and RSD. The validated method was tested with 12 real samples.

The presence of dicyandiamide (DCD) in dairy products/ingredients alarmed the Ministry of Primary Industries (MPI) in New Zealand which asked for withdrawing the fertilizer containing DCD from distribution in NZ. DCD had been in use since 2004 for direct application to pasture and for reduction of nitrate leaching to waterways and greenhouse emissions. Dubois et al. (2013) put in place a method based on LC-MS/MS to provide a reliable tool towards assessing the presence of DCD in dairy ingredients. The analytical procedure includes the following steps: (i) reconstitution of dairy powder in water, (ii) dilution with CH_3CN, (iii) acidification, (iv) partitioning with sodium chloride upon centrifugation, (v) washing the supernatant with hexane, (vi) analysis with LC-MS/MS in SRM applying the positive electrospray ionisation mode and (vii) quantification with the isotopic dilution approach. An analysis of more than 220 milk-based ingredients from NZ showed that 77% were below the reporting limit at 0.05 mg/kg whereas 22% of the samples were in the range 0.05–1.0 mg/kg and the highest value recorded reached 1.39 mg/kg. Although the determined DCD values are not a threat issue to humans, DCD has just recently been added to the list of hazardous compounds like melamine and as such there is fear that it may be used as economic adulterant in foods.

Michel et al. (2013) suggested an improved approach to detect and identify the present persistent organic pollutants (POPs) such as dioxins, PCBs, PAHs and chlorinated pesticides in fatty foods and beverages by means of Quechers extraction/cleanup and GC/MS. The majority of these compounds are lipophilic and they accumulate in fatty tissues. Moving up the hierarchy of the food chain the phenomenon of bioaccumulation is enhanced since the compounds go from smaller organisms to larger ones (US EPA, 2013). The determined enhanced values of PAHs might be also due to grilling of meat and meat products. In this research work, the QuEChERS method developed by Anastassiades and Lehotay (2003) was applied for extraction and clean-up of PAHs from raw salmon, grilled burger as well as for PCBs from cow's milk. The GC analysis of the above-mentioned compounds was carried out in capillary columns designed to provide the optimum selectivity, temperature range and analysis time. It was shown that the employment of zirconia-coated silica increased the removal of matrix components with an enhanced recovery for the POPs.

Kobothekra et al. (2013) investigated the optimization of Quechers extraction method for the determination of Endocrine Disrupting Chemicals (EDCs) including alkyphenols and bisphenolA residues in dairy products via LC-Hybrid LTQ Orbitrap MS. Since the presence of residues in foods is of great importance for the well-being of the average human being, there has been a strong motive for introducing a new method based on U-HPLC-LTQ FT Orbitrap MS. Employment of the latter revealed excellent sensitivity and made feasible the high mass resolution and accuracy identification of all compounds involved in dairy products. In all investigated compounds, precision was lower than 6% (expressed as RSD). The method involved microscale extraction with CH_3CN and dispersive solid phase extraction with a major reagent. Although employment of different sorbents may lead to diverging results, the experimental design approach proved to be of great help to screening and estimation of parameters like extraction solvent, the adsorbent and the sample size on the method extraction yield. It is noteworthy that the sensitivity obtained falls below the maximum residue levels (MRLs) established by the relative EU regulation for food monitoring programs.

HBCD is an emerging toxic micropollutant due to human activities and occurring both in the environment and in animal tissues. Direct quantification of HBCD is very difficult because of its fast metabolism in biota. Berge et al. (2011) investigated the relationship between the volatile compound metabolic signatures in chicken liver for back-tracing a dietary exposure to rapidly metabolized xenobiotics. Planche et al. (2013b) studied three groups of laying hens (n = 56) which were administered the same feed regime [either non-contaminated (control sample) or contaminated over 71 days with either 0.1 µg/g or 10 µg/g HBCD]. Hens were slaughtered at regular intervals during the experiment and their liver was removed for analysis. Solid Phase Extraction-Gas Chromatography-Mass Spectroscopy (SPE-GC-MS) was employed to identify and quantify the volatile compounds in liver. The identified volatile compounds in poultry liver could be used for tracing back an exposure to HBCD in laying hens (Planche et al., 2013b).

9.4 Conclusions and New Trends

Both traceability and authenticity are issues that have been gaining ground over the last 15 years. This is due to globalization and continuously increasing number of adulterations in the chain of food products. Furthermore, the number of Directives and Regulations (EU), Acts (US, Canada, Australia-NZ) that have been voted increased considerably. The introduced automatization such as Entreprise Resource Planning (ERP) occurred in an attempt to limit the adulteration cases along the food chain. On the other hand, the appearance of novel rapid techniques which combine more than one method proved to be very effective toward increasing the effectiveness of authenticity (Ch. 1 and Ch. 9). Since nowadays foods can be transported from one country to the rest of the world, and quite often from one continent to others which means that the legislation might be quite different between countries. Therefore, an effort should be made to rectify the differences in legislation (Ch. 8) among the various countries so that the imports and exports are facilitated. In other words, authenticity has recently emerged as a major issue requiring the authorities' attention both in terms of legislation and detection and identification analysis so that the number of such incidents gradually decreases still it stops to exist.

References

Anastassiades, M. and S.J. Lehotay. 2003. Fast and easy multiresidue method employing acetonitrile extraction/partitioning and "dispersive solid-phase extraction" for the determination of pesticide residues in produce. J. AOAC International. 86(22): 412–431.

Arvanitoyannis, I.S. 2008. Trends in Food Authentication. pp. 616–643. *In*: Da-Wen Sun (ed.). Modern Techniques in Food Authentication. London, UK: Academic Press (An Imprint of Elsevier).

Arvanitoyannis, I.S. 2003a. Genetically engineered/modified organisms in foods. Applied Biotechnology, Food Science and Policy. 1(1): 3–12.

Arvanitoyannis, I.S., E.V. Tsitsika and P. Panagiotaki. 2005. Implementation of quality control methods (physicochemical, microbiological and sensory) in conjunction with multivariate analysis towards fish authenticity. International Journal of Food Science and Technology. 40: 237–263.

Arvanitoyannis, I.S. 2003b. Wine authenticity. pp. 426–450. *In*: M. Lees (ed.). Food Authenticity and Traceability. New York, NY: Woodhead Publishing Ltd.

Arvanitoyannis, I.S. and M. van Houwelingen-Koukaliaroglou. 2003. Implementation of chemometrics for quality control and authentication of meat and meat products. Critical Reviews in Food Science and Nutrition. 43: 173–218.

Arvanitoyannis, I.S. and N.E. Tzouros. 2005. Implementation of quality control methods in conjunction with chemometrics toward authentication of dairy products. Critical Reviews in Food Science and Nutrition. 45: 231–249.

Berge, P., J. Patel, A. Fournier, C. Jondreville, C. Feidt, B. Roudault, B. Le Bizec and E. Engel. 2011. Use of volatile compound metabolic signatures in poultry liver to back-trace dietary exposure to rapidly metabolized xenobiotics. Environmental Science & Technology 45: 6584–6591.

Beser, M. and V. Yusa. 2013. Design of experiment approach (DOE) for the optimization of polybrominateddiphenyl ethers (PBDEs) determination in fish samples by microwave-assisted extraction and GC-MS/MS (F-43). p. 264. *In*: Jana Pulkrabová, Monika Tomaniová, Michel Nielen and Jana Hajšlová (eds.). Book of Abstracts of 6th International Symposium on Recent Advances in Food Analysis, held in Prague (Czech Republic), Nov. 5–8, 2013.

Cajka, T., H. Danhelova, Riddellova K. Katerina, J. Hajslova, M. Bednar and D. Titera. 2011. Application of metabolomics fingerprinting/profiling for honey authenticity (B-44). p. 156. *In*: Jana Pukrabová and Monica Tomaniová (eds.). Book of Abstracts, 5th International Symposium on Recent Advances in Food Analysis, Nov. 1–4, 2011, held in Prague (Czech Republic).

Cariou, R., N. Bemrah, Marchand P. Philippe, V. Sirot, G. Dervilly-Pinel, J.-C. Jean-Charles Leblanc and B. Le Bizec. 2013. Human Dietary Exposure to Persistent Organic Pollutants: Results of the second French total diet study (F-28). p. 256. *In*: Jana Pulkrabová, Monika Tomaniová, Michel Nielen and Jana Hajšlová (eds.). Book of Abstracts of 6th International Symposium on Recent Advances in Food Analysis, held in Prague (Czech Republic), Nov. 5–8, 2013.

Cherta, L., J. Beltran, T. Portolés, J.G.J. Mol, J. Nácher, M. Ábalos and F. Hernández. 2013. Potential of Gas Chromatography – (Triple Quadrupole) Mass Spectrometry coupled to Atmospheric Pressure Chemical Ionization for POP analysis (L-22). p. 117. *In*: Jana Pulkrabová, Monika Tomaniová, Michel Nielen, and Jana Hajšlová (eds.). Book of Abstracts of 6th International Symposium on Recent Advances in Food Analysis, held in Prague (Czech Republic), Nov. 5–8, 2013.

De Alda, C., A. Galego, J.C. Bravo, P. Fernandez and J.S. Durand. 2011. Characterization of Spanish honeys with Protected Designation of Origin "Miel de Granada" (B-21). p. 151. *In*: Jana Pukrabová and Monica Tomaniová (eds.). Book of Abstracts, 5th International Symposium on Recent Advances in Food Analysis, Nov. 1–4, 2011, held in Prague, Czech Republic.

De la Fuente, M.A. and M. Juarez. 2005. Authenticity assessment of dairy products. Critical Reviews in Food Science and Nutrition. 45: 563–585.

De Vries, M., R.P. Kwakkel and A. Kijlstrat. 2006. Dioxinsin organic eggs: a review. NJAS. 54(2): 207–221.

De Zeeuw, J., J. Cochran, M. Misselwitz and J. Kowalski. 2011. The quechers extraction approach and comprehensive two – dimensional gas chromatography of halogenated persistent organic pollutants in cow milk and human breast milk. *In*: Jana Pukrabová and Monica Tomaniová (eds.). Book of Abstracts, 5th International Symposium on Recent Advances in Food Analysis, Nov. 1–4, 2011, held in Prague (Czech Republic).

Dooley, J.J., H.D. Sage, M.A.L. Clarke et al. 2005. Fish species identification using PCR-RFLP analysis and lab-on-a-chip capillary electrophoresis: application to detect white fish species in food products and an interlaboratory study. Journal of Agricultural and Food Chemistry. 53: 3348–3357.

Drabova, L., J. Pulkrabova, K. Kalachova, K. Mastovska, V. Kocourek and J. Hajslova. 2011. Rapid GC-MS method for analysis of polycyclic aromatic hydrocarbons (PAHs) in seafood: AOAC collaborative study. *In*: Jana Pukrabová and Monica Tomaniová (eds.). Book of Abstracts, 5th International Symposium on Recent Advances in Food Analysis, Nov. 1–4, 2011, held in Prague (Czech Republic).

Dubois, M., M. Frank, K. Gartenmann, A. Tarres, E. Gremaud, P. Mottier and T. Delatour. 2013. Determination of Dicyandiamide by LC-MS/MS in dairy ingredients: A response to a decision of the ministry of primary industries in New Zealand (F-44). p. 264. *In*: Jana Pulkrabová, Monika Tomaniová, Michel Nielen and Jana Hajšlová (eds.). Book of Abstracts of 6th International Symposium on Recent Advances in Food Analysis, held in Prague (Czech Republic), Nov. 5–8, 2013.

Focardi, C., E. Droghetti, M. Nocentini and G. Smulevich. 2011. Official Food Control in Italy during the years 2007–2011 to detect fraudulent treatment of fish with CO_2 using a spectrophotometric method (B-14). p. 147. *In*: Jana Pukrabová and Monica Tomaniová (eds.). Book of Abstracts, 5th International Symposium on Recent Advances in Food Analysis, Nov. 1–4, 2011, held in Prague (Czech Republic).

Gadaj, A., D. Rai, A. Furey and M. Danaher. 2011. Application of UPLC-MS/MS for determination of synthetic adulterants in slimming food supplements (B-36). p. 158. *In*: Jana Pukrabová and Monica Tomaniová (eds.). Book of Abstracts, 5th International Symposium on Recent Advances in Food Analysis, Nov. 1–4, 2011, held in Prague (Czech Republic).

Grandits, S., W. Mayer, R. Hochegger and M. Cichna-Markl. 2011. Quantification of the Red Deer content by real-time PCR to detect food adulteration (B-23). p. 152. *In*: Jana Pukrabová and Monica Tomaniová (eds.). Book of Abstracts, 5th International Symposium on Recent Advances in Food Analysis, Nov. 1–4, 2011, held in Prague (Czech Republic).

Grochowalski, A., R. Chrząszcz, A. Maślanka, M. Węgiel and J. Kuc. 2013. Determination of polychlorinated dibenzo-p-dioxins, dibenzofurans and dioxin-like biphenyls in fish and canned fish and investigation of the congener profile and similarity between the samples (F-36). p. 260. *In*: Jana Pulkrabová, Monika Tomaniová, Michel Nielen and Jana Hajšlová (eds.). Book of Abstracts of 6th International Symposium on Recent Advances in Food Analysis, held in Prague (Czech Republic), Nov. 5–8 (2013).

Kaplan, M., E.O. Olgun, W. Peters and J. Wendt. 2013. Fingerprint analysis of flavor compounds in a mountain cheese by GC-MSD and High Resolution GC-TOFMS (E-19). p. 240. *In*: Jana Pulkrabová, Monika Tomaniová, Michel Nielen and Jana Hajšlová (eds.). Book of Abstracts of 6th International Symposium on Recent Advances in Food Analysis, held in Prague (Czech Republic). Nov. 5–8, 2013.

Kobothekra, V., V. Boti and T. Albanis. 2013. Optimization of Quechers extraction for the determination of EDCS in dairy products via LC-hybrid LTQ Orbitrap MS (F-46). p. 265. *In*: Jana Pulkrabová, Monika Tomaniová, Michel Nielen and Jana Hajšlová (eds.). Book of Abstracts of 6th International Symposium on Recent Advances in Food Analysis, held in Prague (Czech Republic), Nov. 5–8, 2013.

Kotz, A., W. Traag, Winterhalter H. Helmut and R. Malisch. 2013. Application of APGC–MS/MS for the determination of PCDD/Fs and PCBs in Feed and Food Matrices (L-87). p. 150. *In*: Jana Pulkrabová, Monika Tomaniová, Michel Nielen and Jana Hajšlová (eds.). Book of Abstracts of 6th International Symposium on Recent Advances in Food Analysis, held in Prague (Czech Republic), Nov. 5–8, 2013.

Kowalski, J., S. Lupo, T. Kahler, S. Stevens and J. Cochran. 2011. Searching for the Holy Grail: Separation of all priority polycyclic aromatic hydrocarbons and their known interferences by serial combination of different HPLC columns. *In*: Jana Pukrabová and Monica Tomaniová (eds.). Book of Abstracts, 5th International Symposium on Recent Advances in Food Analysis, Nov. 1–4, 2011, held in Prague (Czech Republic).

Kuc, J. and A. Grochowalski. 2013. Investigation of content of Hexabromocyclododecane isomers in marine fish and clustering of samples (F-37). p. 261. *In*: Jana Pulkrabová, Monika Tomaniová, Michel Nielen and Jana Hajšlová (eds.). Book of Abstracts of 6th International Symposium on Recent Advances in Food Analysis, held in Prague (Czech Republic), Nov. 5–8, 2013.

Kvasnicka, F. 2005. Capillary electrophoresis in food authenticity. Journal of Separation Science. 28: 813–825.

Lara, A.B., M.D. Sicilia and S. Rubio. 2013. Determination of hexabromocyclododecane stereoisomers in fish by supramolecular solvent-based microextraction and LC/Tandem MS (F-35). p. 260. *In*: Jana Pulkrabová, Monika Tomaniová, Michel Nielen and Jana Hajšlová (eds.). Book of Abstracts of 6th International Symposium on Recent Advances in Food Analysis, held in Prague (Czech Republic), Nov. 5–8, 2013.

Lestingi, C., T. Tavoloni, E. Bastari and A. Piersanti. 2013. 16 EU PAH in edible fats by styrene divinylbenzene SPE Clean-Up and solvent vent PTV GC-MS analysis (F-25). p. 255. *In*: Jana Pulkrabová, Monika Tomaniová, Michel Nielen and Jana Hajšlová (eds.). Book of Abstracts of 6th International Symposium on Recent Advances in Food Analysis, held in Prague (Czech Republic), Nov. 5–8, 2013.

L'Homme, B., G. Scholl, G. Eppe and J.-F. Focant. 2013. Validation of GC-MS/MS Confirmatory Method for the EU Official Control of levels of PCDDS and DLPCBS in Feed Material of Plant Origin (L-21). p. 117. *In*: Jana Pulkrabová, Monika Tomaniová, Michel Nielen and Jana Hajšlová (eds.). Book of Abstracts of 6th International Symposium on Recent Advances in Food Analysis, held in Prague (Czech Republic), Nov. 5–8, 2013.

Lombardi, G., F. Blasi, P. Damiani, L. Giua and Cossignani L. Lina. 2011. Linear Discriminant Analysis (LDA) on triacylglycerol stereospecific composition for the detection of milk adulteration (B-15).

p. 148. *In*: Jana Pukrabová and Monica Tomaniová (eds.). Book of Abstracts, 5th International Symposium on Recent Advances in Food Analysis, Nov. 1–4, 2011, held in Prague (Czech Republic).

Lopez-Andreo, M., A. Garrido-Pertierra and A. Puyet. 2006. Evaluation of postpolymerase chain reaction melting temperature analysis for meat species identification in mixed DNA samples. Journal of Agricultural and Food Chemistry. 54: 7973–7978.

Lupton, S., M. O'Keefe and P. Basu. 2013. Dioxin, Furan, PCB, and PBDE Levels in U.S. Foods: Survey Trends and Consumer Exposure (L-83). p. 130. *In:* Jana Pulkrabová, Monika Tomaniová, Michel Nielen and Jana Hajšlová (eds.). Book of Abstracts of 6th International Symposium on Recent Advances in Food Analysis, held in Prague (Czech Republic), Nov. 5–8, 2013.

Madas, M.N., L.A. Marghitas, S.D. Dezmirean, O. Bobis, B. Kim-Nguyen and E. Haubruge. 2011. Georaphical indications for honey: A physico-chemical profile of acacia honey produced in Romania (B-41). p. 161. *In:* Jana Pukrabová and Monica Tomaniová (eds.). Book of Abstracts, 5th International Symposium on Recent Advances in Food Analysis, Nov. 1–4, 2011, held in Prague (Czech Republic).

Maldini, M., F.N. Marzano, G. González-Fortes et al. 2006. Fish and seafood traceability based on AFLP markers: elaboration of a species database. Aquaculture. 261: 487–494.

Mancheño, D., M.-C. Aristoy and F. Toldra. 2011. Analysis of coenzyme Q_{10} in meats from different animal species (C-24). p. 176. *In:* Jana Pukrabová and Monica Tomaniová (eds.). Book of Abstracts, 5th International Symposium on Recent Advances in Food Analysis, Nov. 1–4, 2011, held in Prague (Czech Republic).

Michel, F., K.K. Stenerson, M. Halpenny, M. Ye, O. Shimelis, E. Barrey and L.M. Sidisky. 2013. Improved approach for the determination of persistent organic pollutants (POPs) in fatty acids and beverages using Quechers extraction/Cleanup and GC/MS (F-45). p. 265. *In:* Jana Pulkrabová, Monika Tomaniová, Michel Nielen and Jana Hajšlová (eds.). Book of Abstracts of 6th International Symposium on Recent Advances in Food Analysis, held in Prague (Czech Republic), Nov. 5–8, 2013.

Mueller, S., J. Bahrs-Widsberger and P. Buss. 2011. Fish species identification by RFLP on the Agilent 2100 Bioanalyzer (B-35). p. 158. *In*: Jana Pukrabova and Monica Tomaniova (eds.). Book of Abstracts, 5th International Symposium on Recent Advances in Food Analysis, Nov. 1–4, 2011, held in Prague (Czech Republic).

Neugebauer, F., C. Ast, O. Paepke and N. Lohmann. 2013. The total PCB task: A comprehensive HRGG-HRMS Method for analysis of all 209 PCB congeners in fish matrices (F-19). p. 252. *In;* Jana Pulkrabová, Monika Tomaniová, Michel Nielen and Jana Hajšlová (eds.). Book of Abstracts of 6th International Symposium on Recent Advances in Food Analysis, held in Prague (Czech Republic), Nov. 5–8, 2013.

Papachristopoulou, C., K. Stamoulis, P. Tsakos, V. Vozikis, C. Papadopoulou and K. Ioannides. 2013. Measurement of trace metal concentrations in the liver of sheep and goats from northern Greece, using radioisotope-excited energy-dispersive XRF spectrometry (F-39). p. 262. *In:* Jana Pulkrabová, Monika Tomaniová, Michel Nielen and Jana Hajšlová (eds.). Book of Abstracts of 6th International Symposium on Recent Advances in Food Analysis, held in Prague (Czech Republic), Nov. 5–8, 2013.

Perston, B. and S. Goth. 2011. FT-IR Spectroscopy and Chemometrics for detection of contaminated or counterfeit ingredients (B-16). p. 148. *In:* Jana Pukrabová and Monica Tomaniová (eds.). Book of Abstracts, 5th International Symposium on Recent Advances in Food Analysis, Nov. 1–4, 2011, held in Prague, Czech Republic.

Pissinatti, R., C. Nunes, E. Santos, R. Prates, E. Magalhães and D. Augusti. 2013. Method Validation and Occurrence of PCDD/FS in Fish from Brazil (F-5). p. 245. *In:* Jana Pulkrabová, Monika Tomaniová, Michel Nielen and Jana Hajšlová (eds.). Book of Abstracts of 5th International Symposium on Recent Advances in Food Analysis, held in Prague (Czech Republic), Nov. 5–8, 2013.

Planche, C., J. Ratel, P. Blinet and E. Engel. 2013a. Impact of cooking conditions and fat content on the bioaccessibility of PCBs, PCDDs and PCDFs in Meat Products (F-16). p. 250. *In:* Jana Pulkrabová, Monika Tomaniová, Michel Nielen and Jana Hajšlová (eds.). Book of Abstracts of 6th International Symposium on Recent Advances in Food Analysis, held in Prague (Czech Republic), Nov. 5–8, 2013.

Planche, C., J. Ratel, A. Fournier, P. Blinet, C. Jondreville, C. Feidt, B. Le Bizec and E. Engel. 2013b. Back-tracing an emerging environmental toxicant (Hexabromocyclododecane, HBCD) in animal-derived food chain based on foodomics (G-1). p. 273. *In:* Jana Pulkrabová, Monika Tomaniová, Michel Nielen and Jana Hajšlová (eds.). Book of Abstracts of 6th International Symposium on Recent Advances in Food Analysis, held in Prague (Czech Republic), Nov. 5–8, 2013.

Plath, A., I. Krause and R. Einspanier. 1997. Species identification in dairy products by three different DNA-based techniques. Zeitschriftfür Lebensmittel-Untersuchung und -Forschung. 205: 437–441.

Popping, B. 2002. The application of biotechnological methods in authenticity testing. Journal of Biotechnology. 98: 107–112.

Ratel, J., A. Fournier, P. Berge, P. Blinet, C. Jondreville, C. Feidt, B. Le Bizec and E. Engel. 2011. Volatile Compound Metabolic Signatures in Poultry fat for back-tracing dietary exposure to Hexabromocyclododecane (HBCD). *In*: Jana Pukrabová and Monica Tomaniová (eds.). Book of Abstracts, 5th International Symposium on Recent Advances in Food Analysis, Nov. 1–4, 2011, held in Prague (Czech Republic).

Roberts, D., J. Dunstan, M. McCullagh, G. Kendon, I.E. Jogvesten and B. van Bavel. 2013. Selective and sensitive detection and quantification of Stockholm convention POPs, including dioxins, using atmospheric pressure GC-MS/MS (F-27). p. 256. *In*: Jana Pulkrabová, Monika Tomaniová, Michel Nielen and Jana Hajšlová (eds.). Book of Abstracts of 6th International Symposium on Recent Advances in Food Analysis, held in Prague (Czech Republic), Nov. 5–8, 2013.

Rodriguez, M.P., A. Boix and Ch. von Holst. 2011. Development of two complementary Real Time PCR Methods for the quantification of fish nuclear DNA (B-18). p. 149. *In*: Jana Pukrabová and Monica Tomaniová (eds.). Book of Abstracts, 5th International Symposium on Recent Advances in Food Analysis, Nov. 1–4, 2011, held in Prague (Czech Republic).

Romanotto, A., Urbansky R. Rene and R. Noak. 2013. MMM GC-OIL: The simultaneous clean-up and measurement of 200 pesticides, EPA PAHS, 18 Plasticizers, and congener PCBs in Fat and Oil samples (R-35). p. 428. *In*: Jana Pulkrabová, Monika Tomaniová, Michel Nielen and Jana Hajšlová (eds.). Book of Abstracts of 6th International Symposium on Recent Advances in Food Analysis, held in Prague (Czech Republic), Nov. 5–8, 2013.

Roszko, M., K. Szymczyk and R. Jędrzejczak. 2013. The influence of the chicken hen breeding method on the presence of PCDD/F, PCB and PBDES in eggs (F-20). p. 252. *In*: Jana Pulkrabová, Monika Tomaniová, Michel Nielen and Jana Hajšlová (eds.). Book of Abstracts of 6th International Symposium on Recent Advances in Food Analysis, held in Prague (Czech Republic), Nov. 5–8, 2013.

Russell, V.J., G.L. Hold and S.E. Pryde. 2000. Use of restriction fragment length polymorphism to distinguish between salmon species. Journal of Agricultural and Food Chemistry. 48: 2184–2188.

Sapozhnikova, Y. and S.J. Lehotay. 2013a. Multi-class, Multi-residue Analysis of environmental contaminants and pesticides in fish using fast low-pressure Gas Chromatography-Tandem Mass Spectroscopy (L-84). p. 130. *In*: Jana Pulkrabová, Monika Tomaniová, Michel Nielen and Jana Hajšlová (eds.). Book of Abstracts of 6th International Symposium on Recent Advances in Food Analysis, held in Prague (Czech Republic), Nov. 5–8, 2013.

Sapozhnikova, Y. and S.J. Lehotay. 2013b. Multi-class, multi-residue analysis of pesticides, polychlorinated biphenyls, polycyclic aromatic hydrocarbons, polybrominated diphenyl ethers and novel flame retardants in fish using fast, low-pressure gas chromatography–tandem mass spectrometry. Analytica Chimica Acta. 758: 80–92.

Sass-Kiss, A. 2008. Chromatographic Technique: High Performance Liquid Chromatography (HPLC). pp. 361–409. *In*: Da Wen-Sun (ed.). Modern Techniques for Food Authentication. Academic Press (An Imprint of Elsevier), London, UK.

Sellers, K., J. Cochran, M. Misselwitz and J. Thomas. 2013. Same separation with half the column: extending the lifetime of your GC Column with Column trimming maintenance and method translation (F-6). p. 245. *In*: Jana Pulkrabová, Monika Tomaniová, Michel Nielen and Jana Hajšlová (eds.). Book of Abstracts of 6th International Symposium on Recent Advances in Food Analysis, held in Prague (Czech Republic), Nov. 5–8, 2013.

Sreenivasa, R.J., K. Bhaskarachary and T. Longvah. 2013. Determination of heavy metal contents in Indian marine fish using ICP-MS Method after closed vessel microwave digestion (F-51). p. 268. *In*: Jana Pulkrabová, Monika Tomaniová, Michel Nielen and Jana Hajšlová (eds.). Book of Abstracts of 6th International Symposium on Recent Advances in Food Analysis, held in Prague (Czech Republic), Nov. 5–8, 2013.

Stenerson, K.K., O. Shimelis, K. Espenschied, M. Halpenny and C. Dumas. 2013. Efficient Extraction and Analysis of PAHs from Olive Oil using a new Dual-Layer SPE Cartridge. *In*: Jana Pulkrabová, Monika Tomaniová, Michel Nielen and Jana Hajšlová (eds.). Book of Abstracts of 6th International Symposium on Recent Advances in Food Analysis, held in Prague (Czech Republic), Nov. 5–8, 2013.

Stojisic, M., M. Tot, D. Klisara and B. Marosanovic. 2013. Analysis of As, Cd, Cu, Fe and Pb in chocolate and cocoa products by ICP-MS (F-13). p. 249. *In*: Jana Pulkrabová, Monika Tomaniová, Michel Nielen and Jana Hajšlová (eds.). Book of Abstracts of 6th International Symposium on Recent Advances in Food Analysis, held in Prague (Czech Republic), Nov. 5–8, 2013.

Tavoloni, T., C. Lestingi, E. Bastari and A. Piersanti. 2013. From GC–MS to GC–MS/MS Triple Quadrupole Analysis of NDL PCBS in food: Reduced Clean-up and increased sensitivity (F-26). p. 255. *In:* Jana Pulkrabová, Monika Tomaniová, Michel Nielen and Jana Hajšlová (eds.). Book of Abstracts of 6th International Symposium on Recent Advances in Food Analysis, held in Prague (Czech Republic), Nov. 5–8 (2013).

Tranchida, P., M. Zoccali, G. Purcaro, S. Moret, L. Conte, P. Dugo and L. Mondello. 2011. A rapid and sensitive multidimensional liquid-gas chromatography (LC-GC) method for the determination of hydrocarbon contamination in foods. *In:* Jana Pukrabová and Monica Tomaniová (eds.). Book of Abstracts, 5th International Symposium on Recent Advances in Food Analysis, Nov. 1–4, 2011, held in Prague (Czech Republic).

U.S. Environmental Protection Agency International Programs/Resistant Organic Pollutants in http://era. gov/international/toxics/pop.html (accessed July 2014).

Vávrová, M., S. Navrátil, M. Stoupalová and L. Wanecká. 2011. Determination of PAHs in Honey. *In:* Jana Pukrabová and Monica Tomaniová (eds.). Book of Abstracts, 5th International Symposium on Recent Advances in Food Analysis, Nov. 1–4, 2011, held in Prague (Czech Republic).

Zacs, D., J.R. Jabova, A. Viksna and V. Bartkevics. 2013. Elaboration of the GC-HRMS Method for simultaneous determination of polybrominated, polychlorinated, mixed polybrominated/chlorinated dibenzo-p-dioxins and dibenzofurans, polychlorinated biphenyls and polybrominated diphenyl ethers in fish samples (F-14). p. 249. *In:* Jana Pulkrabová, Monika Tomaniová, Michel Nielen and Jana Hajšlová (eds.). Book of Abstracts of 6th International Symposium on Recent Advances in Food Analysis, held in Prague (Czech Republic), Nov. 5–8, 2013.

Index